Institutional Partnerships in Multi-Hazard Early Warning Systems

Maryam Golnaraghi (Ed.)

Institutional Partnerships in Multi-Hazard Early Warning Systems

A Compilation of Seven National Good Practices and Guiding Principles

Springer

Editor
Maryam Golnaraghi
World Meteorological Organization
Geneva
Switzerland

ISBN 978-3-662-52138-0 ISBN 978-3-642-25373-7 (eBook)
DOI 10.1007/978-3-642-25373-7
Springer Heidelberg Dordrecht London New York

Die Deutsche Nationalbibliothek verzeichnet diese Publikation in der Deutschen Nationalbibliografie; detaillierte bibliografische Daten sind im Internet über http://dnb.d-nb.de abrufbar.

Softcover re-print of the Hardcover 1st edition 2012

Printed on acid-free paper

Springer is part of Springer Science+Business Media (www.springer.com)

Foreword

Michel Jarraud

Michel Jarraud,
Secretary-General World Meteorological Organization

In terms of disasters directly or in-directly caused by natural hazards, 2010 was one of the deadliest and most costly years on record. Among the most noteworthy events included, the deadly January earthquake in Haiti, with an estimated 220,000 lives lost, the unprecedented summer heat wave in Russia, with estimated 11,000 lives lost in Moscow alone, the extensive flooding in Pakistan, during which at one point nearly one-fifth of the country's total land area was underwater, with a death toll of close to 2,000, about 20 million people affected, and with massive destruction of property, infrastructure, and livelihoods. In 2011, a number of disasters such as floods in Australia, the complex earthquake, tsunami and nuclear disaster in Japan have showed that even the wealthiest and most prepared countries are not immune to disaster risks.

Every year natural hazards cause significant loss of life and set back economic and social development by years, if not decades. As expressed in the Fourth Assessment Report of the Intergovernmental Panel on Climate Change, which WMO co-sponsors, there is increasing evidence of greater vulnerability to the risks associated with hydrometeorological hazards, due to climate variability and change. On a global scale, as revealed by statistics for the past five decades, the economic losses attributable to hydrometeorological hazards have increased significantly over the last decade; however, the reported loss of life has decreased dramatically during the same period as a consequence of the development of early warning systems (EWS) in a number of high-risk countries, thanks in particular to advances in meteorological hazard monitoring and forecasting, as well as more effective and coordinated emergency preparedness and planning at national to local levels.

Disaster risk reduction has always been at the core of the mission and among the highest priorities of the World Meteorological Organization (WMO) and the National Meteorological and Hydrological Services (NMHSs) of its 189 Members. Moreover, in referring to the Hyogo Framework for Action, in 2011, the sixteen World Meteorological Congress highlighted the importance of Multi-Hazard Early Warning Systems (MHEWS) as a proven tool for saving lives while also stressing the value of appropriate documentation and the sharing of good practices and lessons learned.

Accordingly, this book includes a key collection of good practices developed over a period of four years through the collective and systematic efforts of WMO and its Members and the International Strategy for Disaster Reduction (ISDR) System partners. In particular, the book documents seven such practices developed in Bangladesh, mega city of Shanghai in China, Cuba, France, Germany, Japan, and the United States of America (USA), while also providing a set of common principles for MHEWS institutional coordination, based on a synthesis of lessons learnt, including the role of NMHS in the planning, coordination and operational cooperation with other agencies. Indeed, much of the material contained in this volume is already being used globally in supporting a number of MHEWS capacity development initiatives.

I am therefore indeed confident that this book will serve as a key source of information for all concerned WMO Members and partners to support their efforts in strengthening their multi-hazard early warning systems.

Michel Jarraud, Secretary-General of the World Meteorological Organization

Foreword

Margareta Wahlström

Margareta Wahlström,
Assistant Secretary-General for Disaster Risk Reduction

In 2007, cyclone Favio caused great damage in central Mozambique, but only ten were killed. A few years earlier, cyclones of similar intensity killed over 800 people. Experts attributed this to training conducted on evacuation procedures and a simple yet effective warning system which involved the communities in the measurement of the precipitation levels and river gauges to the spreading of warnings using megaphones or radio announcements.

Early warning systems, which ensure people are prepared and ready to act, are one of the most effective measures that Governments can undertake to achieve the goals they set themselves in the Hyogo Framework for Action, namely "the substantial reduction of disaster losses, in lives and in the social, economic and environmental assets of communities and countries".

We have collectively come a long way from the time when an effective early warning system consisted of a timely forecast. Today, it is good practice for an early warning system to be part of a broader disaster risk management approach, using available information on the location of vulnerable groups and the different types of hazards that can affect them, and actively preparing communities to react and respond when a warning is issued and potential disaster strikes.

The World Meteorological Organization (WMO), one of the most active partners of the International Strategy for Disaster Reduction (ISDR), and the National Hydrological and Meteorological Services (NHMS) have been at the forefront of this change and have championed the early warning cause on many occasions and in international fora.

I am therefore particularly pleased with this interesting publication, as it presents seven cases of development of national early warning systems, documented in a systematic manner. It is my hope that these success stories provide incentives and inspiration for Governments the world over to aim for similar standards in the development of their own early warning systems.

Margareta Wahlström, Assistant Secretary General for Disaster Risk Reduction, United Nations International Strategy for Disaster Reduction

Preface

Over the past decade, the international community has paid significant attention to the topic of early warning systems (EWS). Ten years after the adoption of the Yokohama Strategy,[1] in January 2005, just a few weeks after the tragic December-2004 Indian Ocean Tsunami, the United Nations International Strategy for Disaster Reduction (UN-ISDR) convened the Second World Conference on Disaster Reduction in Kobe, Japan. During this conference, 168 countries negotiated and adopted the "Hyogo Framework for Action 2005–2015" (HFA), shifting the paradigm for disaster risk reduction from post disaster response to a more comprehensive approach, also including prevention and preparedness measures.[2]

The second high-priority area of the HFA stresses the need for, "identifying, assessing and monitoring disaster risks and enhancing early warning." The HFA further stresses that EWS must be an integral component of any nation's disaster risk management strategy, enabling governments at national to local levels and the communities to take appropriate measures toward building resilience in anticipation of disasters.

Many good practices around the world have demonstrated that EWS should be developed with a multi-hazard, multi-sectoral and multi-level (national to local) approach. Effective EWS are comprised of four operational components, to ensure that,

- Hazards are detected, monitored, forecasted, and hazard warnings are developed;
- Risks are analyzed and this information is incorporated in the warning messages;
- Warnings are issued (by a designated authoritative source) and disseminated in a timely fashion to authorities and public at-risk;
- Community-based emergency plans are activated in response to warnings, to reduce potential impacts on lives and livelihoods.

These four components need to be coordinated across many agencies at national to local levels for the system to work. Failure in one component or lack of coordination across them could lead to the failure of the whole system. The issuance of warnings is a national responsibility; thus, roles and responsibilities of various public and private sector stakeholders for implementation of the EWS should be clarified and reflected in the national to local regulatory frameworks, planning, budgetary, coordination, and operational mechanisms.

In 2006, the Global Survey of Early Warning Systems[3] and the outcomes of the Third International Early Warning Conference (EWC-III)[4] concluded that though progress has been made, many gaps remained to be addressed to ensure that EWS are implemented in all countries, particularly those with least resources. The 2006 Global Early Warning Survey Report cited challenges on legislative, financial, organizational, technical, operational, training and capacity building fronts.

Throughout these international events and assessments, it has become clear that governments and various agencies could benefit from experiences of other governments, with good practices in EWS that had been demonstrated to reduce loss of lives and livelihoods. It also has been voiced in many international and regional forums that there is a need for systematic documentation of such good practices, lessons learned, and synthesizing the factors that have contributed to their successes.

To this end, the 15th World Meteorological Congress in 2007 requested that such an initiative be undertaken by the WMO in partnership with its Member States and UN partners, particularly, related to institutional and operational aspects of EWS for weather, water and climate-extremes.

[1] Yokohama Strategy for a Safer World: Guidelines for Natural Disaster Prevention, Preparedness and Mitigation (1994).

[2] Hyogo Framework for Action 2005–2015: Building the Resilience of Nations and Communities to Disasters (2005).

[3] Following the tragic 2004 Indian Ocean Tsunami, former Un Secretary General, Kofi Annan requested The Global Early Warning Survey. The Survey was implemented by an interagency task Team, Chaired by the World Meteorological Organization and the United Nations Office for the Coordination of Humanitarian Affairs (UN-OCHA), coordinated by the United Nations International Strategy for Disaster Reduction (UN-ISDR) (2006)
(www.reliefweb.int/rw/lib.nsf/db900sid/AMMF6VKH6Z/$file/UNISDR-Sep2006.pdf?openelement)

[4] Third International Early Warning Conference was sponsored by the Government of Germany and was held in Bonn, Germany, 27–29 March 2006.

The compilation of this book is the result of nearly four years of consultations and efforts, including, (1) two international Multi-Hazard Early Warning Systems Symposia to establish the criteria for good practice identification, documentation and synthesis processes, (2) a documentation process engaging many national agencies and ministries from each of the seven countries presented in this book; and, (3) a detailed synthesis process involving consultations with a large cadre of experts and practitioners in EWS, from around the world.

This book features cases from Bangladesh, megacity of Shanghai in China, Cuba, France, Germany, Japan and the United States of America, building on the expertise of WMO Members and relevant national ministries, UN and international partners. The book makes the case for greater integration of EWS in development, preparedness and planning at all levels of society. It provides the basis for a holistic and systematic approach to the mapping and evaluation of early warning systems including improvement and sustainability. It offers government officials, heads of agencies and their operational staff as well as other stakeholders in EWS with detailed information on policy and legal frameworks, institutional coordination and collaboration and operational aspects of EWS.

This work has clearly revealed that even though the specific design and implementation of EWS in each of the seven cases vary according to their governance mechanisms, specific history, culture, socio-economic conditions, institutional structure, capacities and resources for sustainability of their respective systems, there are principles common to all them that have led to the reduction of the impacts of hazards, particularly through saving of lives. This synthesis is provided in Chapter 10. It is noteworthy to highlight that while these cases are treated as good practices, no system is ever perfect. A critical commonality to the cases presented in this book is that there are feedback mechanisms and investments for continuous evaluation and improvement of as various aspects of these systems, over time.

As the editor of this book, it is my hope that this book will provide the motivation for the continuation of a systematic and holistic approach to the study and strengthening of EWS, with engagement of the public and private sectors, NGOs and the academic community.

I would like to thank the WMO Members, and in particular their National Meteorological and Hydrological Services and Disaster Risk Management Agencies for their commitment and contributions to this work. I also would like to thank the team within the WMO Secretariat that have supported this work. I am also grateful to the Secretary-General of WMO for his continuous support, encouragement and deep commitment.

Geneva, Switzerland, 20 May 2011 Maryam Golnaraghi

Table of Content

Acknowledgements

World Meteorological Organization (WMO) and the editor would like to extend their deep appreciation to the team of experts from national agencies of a number of countries and territories as well as international agencies who contributed to the drafting and/or peer review of the chapters in this book, including, Dilder Ahmed, Curtis B. Barrett, Catherine Borretti, Daniel Bovitz, Steve Buan, Philippe Cova, Michael C. Curry, Jean-Noel Degrace, Laura Diamond, Lei Feng, Atsushi Goto, Arjumand Habib, Satoshi Harada, Naoyuki Hasegawa, Cyrille Honoré, Michael Hudson, Chris Maier, Harold 'Jim' Keeney Jr., Lynn Maximuk, Frank Musonda, Haizhen Mu, Satoshi Ogawa, Yuichi Ono, Sergio Pasquini, Miguel A. Puig, Yutaka Sasagawa, Md. Shahidullah, Gerhard Steinhorst, Shotaro Tanaka, Axel Thomalla, Bhupindar Tomar, José Maria Rubiera Torres, Firoz Verjee, Rainer Vogelgesang, Norihisa Washitake, Joseph Mark Widner, Caroline Witter, Tang Xu and Yongjie Zou.

This work was motivated by the outcomes of three International Early Warning Conferences hosted by the Government of Germany,[5] and the 2006 "Global Survey of Early Warning Systems Report".[6] The methodology for the systematic identification and documentation of the national good practices in Multi-Hazard EWS (MHEWS) presented in this book, was established through two International Experts' Symposia on Multi-Hazard Early Warning Systems (MHEWS Symposia), organized by the WMO in cooperation with a number of United Nations and International agencies and Member States.[7] Specifically, the following national and international experts provided advice and feedback through their participation in the working groups and review teams associated with the two MHEWS Symposia: Shadekul Alam, Rosario Alfaro O, André Bachoc, P.M. Jayatilaka Banda, Curtis B. Barrett, Francoise Bénichou, Ariane de Billy, Michael Bittner, Catherine Borretti, Fredrick Branski, David S. Burridge, Ezio Bussoletti, Jean-Marie Carrière, Tanja Cegnar, Carlos Costa-Posada, Hilda Aguilar Cressent, Philippe Dandin, Keiji Doi, Alan Douglas, Charles Dupuy, Philippe Estiez, Victor M. Fallas Chinchilla, René Feunteun, Kazuhiko Fukami, Hussein Gadain, Luis Garcia, Pierre-Alexandre Genillon, Francois Gérard, David S. Green, Xavier Junior Gulele, Mu Haizhen, Sri Woro B. Harijono, Paulus Joseph Prih Harjadi, Chris Hill, Cyrille Honoré, Branka Ivancan-Picek, François Jacq, Michel Legrand Jean, Yves Le Bars, Magda Little, Lynn Maximuk, Mireille Mayoka-Massengo, Olivier Moch, Eunice Mucache, Mussa Mustafa, Sergio Pasquini, Andreas Pfaffernoschke, Eugene Poolman, Miguel A. Puig González, Sophie Purdey, Paolo Reggiani, David Robinson, Daniel Roux, José Rubiera, Michel Legrand Saah, Michael D. Selves, Pierluigi Soddu, Alain Soulan, Vesna Stajcic, Bruce Stewart, Bernard Strauss, Jean-Michel Tanguy, Akira Terakawa, Elda Vásquez de Godoy, Fazlul Wahab, Caroline Wittwer, Zou Yongjie, Steve Zebiak, Karl-Otto Zentel, Shameza Abdulla, Keith Alverson, Gabriel Arduino, Liliana Balbi, Reid Basher, Patricio Bernal, Joanne Burke, Stefanie Dannenmann, Melissa Arditto Delgado, Maxx Dilley, Glenn Dolcemascolo, William Effah, Norberto Fernandez, Simão Ferraz de Campos Neto, Brad Garanganga, Menghestab Haile, John Harding, Terry Jeggle, Esteban Leon, Horst Letz, Alexia Massacand, Paolo Mattei, Ricardo Mena, Michèle Meunier, Sujit Mohanty, Ian O'Donnell, Paul Paredes-Portella, Pascal Peduzzi, Heidi Peugeot, Francesco Pisano, Everett Ressler, Alain Retière, Magali Salazar, Johan Schaar, Rahul Sengupta, Gurpreet Singhota, Michael Staudinger, Yvette Stevens, Tang Xu, Akihiro Teranishi, Giacomo Teruggi, Katharina Thywissen, Bhupinder Tomar, Zoe Trohanis, Olivier Van Damme, Juan Carlos Villagran de Leon, Masahiro Yamamoto, Hong Yan, Yuichi Ono, Cosmas L. Zavazava.

[5] Three International Early Warning Conferences are as follows: i) First International Conference on Early Warning in 1998 (http://www.geomuseum.com/ewc98/); ii) Second International Conference on Early Warning in 2003 (http://www.ewc2.org/pg000001.htm); (iii)Third International Conference on Early Warning in 2006

[6] Global Early Warning Survey, commissioned by the Former UN Secretary General, Kofi Annan in 2005, following the Indian Ocean tsunami, coordinated by ISDR Secretariat with support by a multi-agency task team, co-chaired by WMO and OCHA. The report of this survey can be accessed at: http://www.reliefweb.int/rw/lib.nsf/db900sid/AMMF-6VKH6Z/$file/UNISDR-Sep2006.pdf? openelement.

[7] The two MHEWS experts symposia included: i) Symposium on Multi-Hazard Early Warning Systems for Integrated Disaster Risk Management Geneva, Switzerland, 23–24 May, 2006: http://www.wmo.int/pages/prog/drr/events/ews_symposium_2006/index_en.html ii) Second Experts' Symposium on Multi-Hazard Early Warning Systems with focus on the Role of National Meteorological and Hydrological Services, Toulouse, France, 5–7 May 2009:
http://www.wmo.int/pages/prog/drr/events/MHEWS-II/index_en.html

The seven good practices documented in this book were presented in a series of training workshops on "Multi-Hazard Early Warning Systems with Focus on Institutional Coordination and Cooperation,"[8] in Southeast Europe, Central America, the Caribbean, and South America. These Workshops engaged directors and senior staff of Disaster Risk Management Agencies, National Meteorological and Hydrological Services, Red Cross and Red Crescent Societies and other EWS stakeholders at national and regional levels. Through this process, we solicited valuable inputs, ideas and feedback on the themes covered by this book. We are thankful to all the participants.

In addition, we are grateful for cooperation and technical support of a number of regional and international agencies, including, Food and Agriculture Organisation (FAO), International Federation of Red Cross and Red Crescent (IFRC), International Telecommunication Union (ITU), United Nations Office for the Coordination of Humanitarian Affairs (UN-OCHA), Provention Consortium, UNESCO and its Intergovernmental Oceanographic Commission (UNESCO-IOC), United Nations Children's Fund (UNICEF), United Nations Development Programme (UNDP), United Nations Environment Programme (UNEP), United Nations Habitat, United Nations Institute for Training and Research (UNITAR), United Nations International Strategy for Disaster Reduction (UN-ISDR), United Nations University (UNU), The World Bank and its Global Facility for Disaster Reduction and Recovery (GFDRR), World Food Programme (WFP), and World Health organization (WHO).

This book was compiled by Dr. Maryam Golnaraghi, with support from a team of WMO staff and consultants, including, Charles Baubion, Jean-Baptiste Migraine, James Douris, Desmond O'Neal, and Rainer Dombrowsky.

[8] These training workshops included: (i) Training Workshop on Multi-Hazard Early Warning Systems with focus on Institutional Coordination and Cooperation, Pula, Croatia, 1–3 October, 2009 (ii) Training Workshop on Multi-Hazard Early Warning Systems with focus on Institutional Partnerships and Coordination, San Jose, Costa Rica, 22–26 March, 2010, (iii) Technical Conference on Multi-hazard Early Warning Systems during the XVth Session of WMO Regional Association III, Bogota, Colombia, 20–21 Sept 2010, (iv) Technical Cooperation Workshop for the Development of the Caribbean Regional Cooperation Programme in Multi-Hazard Early Warning System Christ Church, Barbados, 2–5 Nov, 2010, (v) National Disaster Coordinators and Meteorologists Dialogue Advancing Multi-Hazards Early Warning Systems in the Caribbean, Montego Bay, Jamaica, 6 Dec. 2010. For more information on these workshops please go to: http://www.wmo.int/pages/prog/drr/events_en.html.

List of Authors

Dilder Ahmed	Disaster Management Bureau (DMB), Bangladesh
Catherine Borretti	Météo-France, France
Steve Buan	National Weather Service (NWS), United States of America
Jean-Noel Degrace	Météo-France, Martinique, France
Laura Diamond	National Weather Service (NWS), United States of America
Lei Feng	Shanghai Meteorological Bureau, China Meteorological Administration, China
Maryam Golnaraghi	World Meteorological Organization
Atsushi Goto	Office of International Affairs, Japan Meteorological Agency, Japan
Arjumand Habib	Bangladesh Meteorological Department (BMD), Bangladesh
Satoshi Harada	Office of International Affairs, Japan Meteorological Agency, Japan
Naoyuki Hasegawa	Office of International Affairs, Japan Meteorological Agency, Japan
Harold 'Jim' Keeney Jr.	National Weather Service (NWS), United States of America
Haizhen Mu	Shanghai Climate Center, China Meteorological Administration, China
Satoshi Ogawa	Office of International Affairs, Japan Meteorological Agency, Japan
Miguel A. Puig	Department of Civil Defense, Cuba
José Maria Rubiera Torres	National Forecast Center, Institute of Meteorology, Cuba
Md. Shahidullah	Cyclone Preparedness Programme (CPP), Bangladesh
Gerhard Steinhorst	Deutscher Wetterdienst, Germany
Shotaro Tanaka	Office of International Affairs, Japan Meteorological Agency, Japan
Yutaka Sasagawa	Office of International Affairs, Japan Meteorological Agency, Japan
Xu Tang	Shanghai Regional Meteorological Center, China Meteorological Administration, China
Rainer Vogelgesang	Deutscher Wetterdienst, Germany
Norihisa Washitake	Office of International Affairs, Japan Meteorological Agency, Japan
Yongjie Zou	Shanghai Emergency Response Management Office, China.

Chapter 1
An Overview: Building a Global Knowledge Base of Lessons Learned from Good Practices in Multi-Hazard Early Warning Systems

Maryam Golnaraghi

Every year natural hazards cause significant loss of life, and set back economic and social development by years if not decades in many countries, particularly those with least resources. From 1980 to 2008, weather-, water- and climate-related hazards and conditions accounted for 90% of total number of disasters, 70% of the two million casualties, and 75% of total economic losses.[9]

According to the Fourth Assessment Report of the Intergovernmental Panel on Climate Change (IPCC), new scientific evidence has revealed that potential risks associated with weather-related hazards are on the rise, owing to the increasing frequency and severity of droughts, extreme temperatures, severe precipitation and severe storms, linked to climate variability and change. In addition, over the last five years the world has witnessed a number of "unusual" extreme events such as:

i. The severe cyclone storm GONU that affected Oman, Iran and the United Arab Emirates in 2007, noting that it was the second cyclonic storm making landfall in Iran in more than 100 years, since 1898,
ii. Tropical cyclone NARGIS the first such storm to hit Myanmar in 40 years resulted in death of over 140,000 people and estimated losses at nearly 3% of the national GDP for 2008, and;
iii. Tropical cyclone Catarina that affected Brazil in March 2004, was the first hurricane-intensity tropical cyclone ever recorded in the Southern Atlantic Ocean.

Could these become more of the rule than the exception? Are new patterns of risks emerging? If countries are challenged with managing the current level of risk associated with these hazards; how can they address the increasing and changing patterns of risks, in the future?

However, as illustrated in Fig. 1.1, over the last five decades, while economic losses associated with hydro-meteorological hazards have increased, there has been a significant decrease in loss of life. This has been attributed to the development of effective early warning systems, based on advancements in monitoring and forecasting of weather-related hazards (Box 1.1, Box 1.2), combined with effective communication and emergency preparedness at national to local levels in several countries with a history of high-impact weather-related hazards, such as Cuba, Bangladesh, France, and the United States.

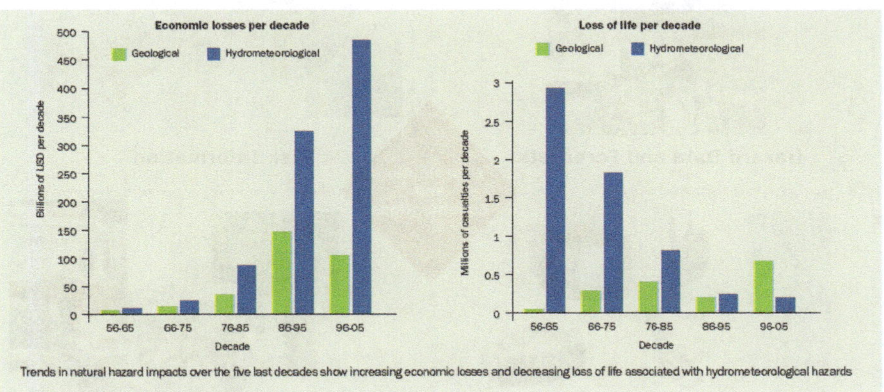

Source: EM-DAT: The OFDA/CRED International Disaster Database

Fig. 1.1 Decadal loss of life and economic losses related to geological versus hydro-meteorological hazards based on data from EM-DAT – The OFDA/CRED International Disaster Database

Maryam Golnaraghi, Chief of Disaster Risk Reduction Programme, World Meteorological Organization

[9] EM-DAT: The OFDA/CRED International Disaster Data-base.

M. Golnaraghi ed., *Institutional Partnerships in Multi-Hazard Early Warning Systems*,
DOI 10.1007/978-3-642-25373-7_1, © Springer-Verlag Berlin Heidelberg 2012

1.1 International Attention to Early Warning Systems

Over the past decade, the international community has paid significant attention to the topic of early warning systems, including, three international conferences (1998, 2003, 2006, hosted by the government of Germany), two international experts' symposia on Multi-Hazard EWS (2006 and 2009, World Meteorological Organization), the 2006 United nations Global EWS Survey Report, and the World Disaster Report (2009, International Federation of red Cross and Red Crescent Societies) (see references).

Ten years after the adoption of the Yokohama Strategy,[10] in January 2005, during the Second World Conference on Disaster Reduction (Kobe, Japan), 168 countries adopted "Hyogo Framework for Action 2005–2015" (HFA) shifting the paradigm for disaster risk reduction from post disaster response to a more comprehensive approach that would also include prevention and preparedness measures.[11] The HFA outlines five priority areas for action, including:

i. Ensure that disaster risk reduction is a national and a local priority with a strong institutional basis for implementation;
ii. Identify, assess and monitor disaster risks and enhance early warning;
iii. Use knowledge, innovation and education to build a culture of safety and resilience at all levels;
iv. Reduce the underlying risk factors;
v. Strengthen disaster preparedness for effective response at all levels.

The second high-priority area of the HFA stresses the need for, "identifying, assessing and monitoring disaster risks and enhancing early warning." The HFA further stresses that EWS must be an integral component of any nation's disaster risk management strategy, enabling governments at national to local levels and the communities to take appropriate measures toward building resilience in anticipation of disasters.

The Second International Conference on Early Warnings (2003) concluded that effective EWS are comprised of (Fig. 1.2):

- **Monitoring and warning service:** Hazards are detected, monitored, forecasted, and hazard warnings are developed;
- **Risks knowledge:** Risks are analyzed and this information is incorporated in the warning messages;
- **Dissemination:** Warnings are issued (by a designated authoritative source) and disseminated in a timely fashion to authorities and public at-risk;
- **Response capacity:** Community-based emergency plans are activated in response to warnings, to reduce potential impacts on lives and livelihoods.

Indeed, many good practices around the world have demonstrated that EWS should be developed with a multi-hazard, multi-sectoral and multi-level (national to local) approach.

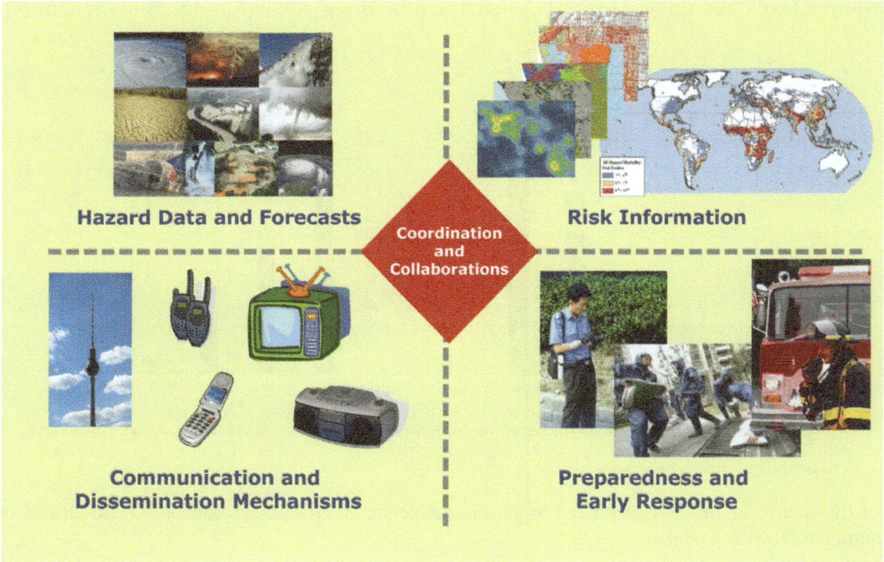

Fig. 1.2 WMO Schematic of the four operational components of effective EWS developed based on the UN-ISDR definition of EWS)

[10] Yokohama Strategy for a Safer World: Guidelines for Natural Disaster Prevention, Preparedness and Mitigation (1994).
[11] Hyogo Framework for Action 2005–2015: Building the Resilience of Nations and Communities to Disasters (2005).

Implementation of these components requires coordination across many agencies at national to local levels for the system to work. Failure in one component or lack of coordination across them could lead to the failure of the whole system. The issuance of warnings is a national responsibility; thus, roles and responsibilities of various public and private sector stakeholders for implementation of the EWS should be clarified and reflected in the national to local regulatory frameworks, planning, budgetary, coordination, and operational mechanism.

Various assessments and the outcomes of the mid-term HFA review in 2010 have revealed that many nations around the globe operate EWS for various natural and man-made hazards. However, the governmental priority, stage of development and overall effectiveness of the EWS at national to local levels, vary widely. Many countries, especially those with the highest risks and least resources, remain highly challenged in building and sustaining their EWS at national to community levels.

Throughout these international events and assessments, it has become clear that governments and various agencies could benefit from experiences of other governments, with good practices in EWS that had been demonstrated to reduce loss of lives and livelihoods. It also has been voiced in many international and regional forums that there is a need for systematically documentation of such good practices, lessons learned, and synthesizing the factors that had contributed to their successes (Golnaraghi et al. 2007, 2008).

Box 1.1 Sixty Years of International and Regional Cooperation in Meteorology to support National Early Warning Systems

Over the last few decades, meteorological, hydrological and climate forecasts have become increasingly accurate as a result of remarkable international co-operation, facilitated by the World Meteorological Organization (WMO). This involves coordinated research and an operational network, comprised of the WMO Global Observing System, Global Telecommunication System and Global Data Processing and Forecasting System that enable monitoring, detecting, forecasting and exchange of weather, water and climate related information, engaging the National Meteorological and Hydrological Services (NMHS) of 189 Members.

Through this coordinated operational network, a wide range of global and regional forecast products and services based on latest technologies and forecasting tools are provided to the support the National Meteorological and Hydrological Services in supporting them with the development of national products and services such as hazard analysis, early warnings and other products and services to support sectoral risk management decision-making. Examples where this coordinated network is supporting the WMO Member States include:

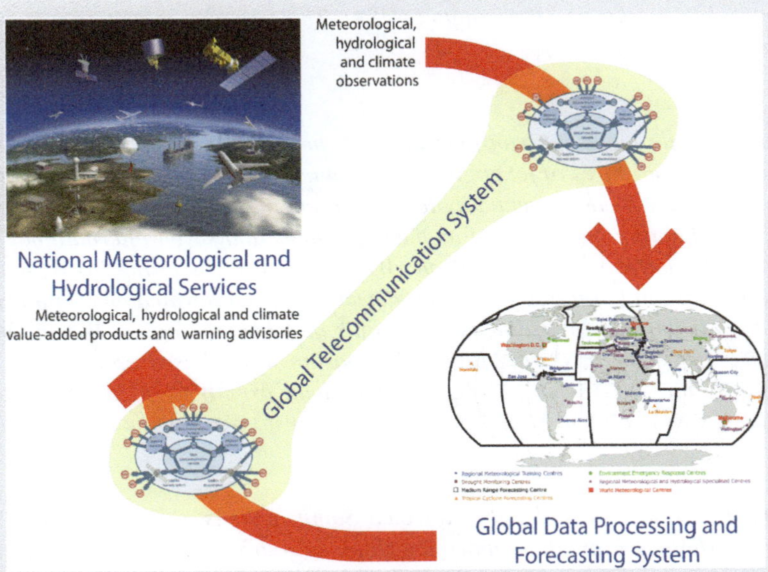

Internationally coordinated network of WMO involving Global Observing System, Global Telecommunication System and Global Data Processing and Forecasting System facilitating sharing of data, analysis and forecasts across 189 WMO Members through their National Meteorological and Hydrological Services.

Box 1.2 Applications of WMO Global Operational Network

Tropical Cyclone Programme

The Tropical Cyclone Programme is an example of cooperation using regionalcapacities to support national warning systems to promote disaster risk reduction strategies. Through the Programme, six Regional Specialized Meteorological Centres (RSMCs) are dedicated to providing tropical cyclone analysis, forecasts and alerts in support of National Meteorological Services' operational warnings. These include: RSMC Nadi-wTropical Cyclone Centre, RSMC La Reunion-Tropical Cyclone Centre, RSMC New Delhi – Tropical Cyclone Centre, RSMC Tokyo – Typhoon Center, RSMC Honolulu, and RSMC Miami – Hurricane Center. The Programme is supported by five regional committees, involving forecasters from the NMHSs, which ensure ongoing improvements in the tropical cyclone forecasting and warning systems. This has enabled availability of tropical cyclone warning capacities to all countries at risk.

Emergency Response Activities

The WMO programme of Emergency Response Activities (ERA) established in 1986 to assist NMHSs, governments and international organizations to respond effectively to environmental emergencies with large-scale dispersion of airborne hazardous substances is another example of regional cooperation. The programme is focussed on nuclear facility accidents, but also provides for meteorological support in emergency response to the dispersion of smoke from large fires, volcanic ash, dust and sand storms and chemical releases from industrial accidents. The WMO operational network of global, regional and national meteorological centres provides the infrastructure for specialized atmospheric dispersion-modelling that play a crucial role in assessing and predicting the spread of air- and water-borne hazardous substances. Some applications include:

Nuclear Accidents

The Chernobyl nuclear accident (April 1986) led to strengthened international cooperation in the event of a nuclear emergency through the Joint Radiation Emergency Management Plan of the International Organizations. The plan is coordinated by the International Atomic Energy Agency in cooperation with international organizations including WMO, the World Health Organization, and the Food and Agriculture Organization. WMO maintains a system of eight Regional Specialized Meteorological Centres which provide highly specialized computer-based simulations of the atmosphere that predict the long-range movement of airborne radioactivity to support environmental emergency response, when needed. These centres, which provide complete global coverage 24 hours a day, every day, are located in Beijing (China), Obninsk (Russian Federation) Tokyo (Japan), Exeter (United Kingdom), Toulouse (France), Melbourne (Australia), Montreal (Canada) and Washington (USA). This response system was activated on 12 March 2011 in the aftermath of the earthquake in Japan.

Volcanic ash

Volcanic ash is a direct safety threat to jet transport aircraft, primarily because the melting point of ash is around 1100°C, while the operating temperatures of jet engines are around 1400°C. The ash melts in the hot section of the engines and then fuses on the turbine blades, eventually leading to engine stall. The International Civil Aviation Organization is responsible for coordinating the efforts of its member states and seven international organizations, including WMO, which comprise the International Airways Volcano Watch (IAVW). Under the IAVW, international ground-based networks, global satellite systems and in-flight air reports detect and observe volcanic eruptions and ash cloud and pass the information quickly to appropriate air traffic services units and Meteorological Watch Offices, which provide the necessary warnings to aircraft before or during flight. The warnings are based on advisory information supplied by nine Volcanic Ash Advisory Centres (VAACs) designated upon advice from WMO. The designated VAACs are located in Anchorage, Buenos Aires, Darwin, London, Montreal, Tokyo, Toulouse, Washington and Wellington.

Wildfires

Following the worst smoke and haze episodes that affected South-East Asia in autumn 1997, which impacted many socio-economic sectors including civil aviation, maritime shipping, agricultural production, tourism and the health of populations, WMO joined with the Association of South-East Asian Nations (ASEAN) to set up the ASEAN Regional Specialized Meteorological Centre in Singapore. This Centre provides smoke/haze information and forecasts to NMHSs to assist in environmental emergency situations. It also displays weather and hot spots using satellite images on its website. Satellite imagery can provide information on the dryness of vegetation, location and size of major fires and smoke plumes, energy released by fires, and air pollutants in the smoke plumes.

1.2 Methodology for Identification and Documentation of Good Practices in Multi-Hazard EWS

The process for systematic identification of good practices might be said to have commenced during the Third International Early Warning Conference (EWC-III). At the EWC-III, a Checklist for Developing Early Warning Systems was distributed, providing a simple list of the main elements and actions to consider when developing or evaluating the EWS.[12] However, while the participants expressed appreciation for this guiding tool, they noted that many countries require more detailed guidance and assistance to implement holistic and effective EWS within a multi-hazard framework. Following the EWS-III, WMO, in cooperation with other UN and international partner organizations held two international experts symposia on multi-hazard EWS to develop a clear methodology for identification and documentation of good practices.

1.2.1 The First International Experts' Symposium on Multi-Hazard EWS

WMO hosted the First International Expert's Symposium on Multi-Hazard EWS (MHEWS-I) in May 2006, which brought together experts from various disciplines and organizations, working at national and international levels. The Symposium had the following main goals:

i. To provide recommendations for an integrated approach to warning systems, building on and linking existing capacities of different stakeholders, including identification of actions at national level to strengthen early warning capabilities and regional and international actions to support these national efforts.
ii. To explore further the concept of a "multi-hazard approach" to EWS, including potential economies and synergies from such approach, and to provide recommendations on additional studies and/or demonstrations that might be required to fully assess all aspects of this approach.
iii. To develop criteria for what constitutes a good practice and identify such cases in the world.

When considering the role of technical agencies, in particular those responsible for monitoring, detecting and forecasting of the hydro-meteorological and climate-related hazards, the symposium stressed that in many countries the National Meteorological Services (NMSs) and the National Hydrological Services (NHSs) are separate agencies, under different line ministries. The operational roles of such technical agencies could be divided into three categories based on their institutional structures and mandates for specific hazards;

- **Type I hazards:** NMS or NHS have sole mandate to monitor, detect and develop warnings for the hazard.
- **Type II hazards:** NMS and NHS have a joint mandate between them and/or with another specialized technical agency to monitor, detect and develop warnings for the hazard.
- **Type III hazards:** NMS and NHS are required to provide data and forecast products or lend their infrastructure (e.g., communication systems) to other agencies that have the official mandate to monitor, detect and develop warnings for the hazard.

Technical and operational cooperation of these agencies would be fundamental to the development of hazard warnings. Furthermore, many other agencies are responsible for managing the risks and ensuring coordination for emergency preparedness within and across many sectors. These agencies would also monitor and document the impacts of hazards on various sectors such as health, agriculture and food security, infrastructure, water resource management, transportation. Development of effective partnership among technical and sectoral agencies is critical for the development of warnings that incorporate risk sectoral information.

The Symposium also provided significant input for the development of a standard template that could be utilized by the national ministries and agencies to document systematically their respective national EWS. A "simplified" outline of this template is provided in Box 1.3.

During MHEWS-I, four examples of good practices were identified using the above criteria, including Bangladesh, Cuba, France and City of Shanghai in China. It was recommended that these cases be documented for further analysis of commonalities and lessons learned.

[12] "Developing Early Warning Systems: A Checklist," UN/ISDR. 2006. Developed as an outcome of the Third International Conference on Early Warning (EWC III), 27–29 March 2006, Bonn, Germany. It was created in order to help governments and communities implement people-centred early warning systems. The checklist was translated into 19 Indian Ocean country languages. Link: http://www.unisdr.org/2006/ppew/info-resources/ewc3/checklist/English.pdf.

Box 1.3 A "Simplified" Outline of the Template Used for Systematic Documentation of Good Practices in EWS

1. ***Overview of the Early Warning Systems (EWS)***
2. ***Background in the establishment of EWS***
3. ***Governance and Institutional Arrangements (national to local levels)***
   a) *Policy, intuitional and legal frameworks to support emergency planning and response*
   b) *National to local emergency planning and related linkages to EWS*
   c) *Organizational structure for implementing the plans*
   d) *Institutional capacities and concept of operations (coordination and operational collaboration)*
   e) *Financial and budgetary aspects*
4. ***Utilization of risk information in emergency contingency planning and warnings***
   a) *Organizational responsibilities and arrangements for the development of risk information*
   b) *Hazard assessment, quantification and mapping (national to local)*
   c) *Assessment of vulnerabilities and exposure (national to local)*
   d) *Storage and accessibility of disaster and national hazard risk information*
   e) *Development and utilization of hazard/risk information to support emergency planning and warnings*
5. ***Hazard Monitoring, forecasting, and mandates for warning development***
   a) *Organizational responsibilities for monitoring, forecasting and development of hazard warnings*
   b) *Organizational collaboration and coordination for development of hazard warnings*
6. ***Development of understandable, authoritative, recognizable and timely warnings***
   a) *Warning message development cycle*
   b) *Warning message improvement cycle*
7. ***Warning dissemination mechanisms (national to local)***
8. ***Emergency preparedness and response activities (national to local)***
   a) *Disaster preparedness and response planning and emergency response activation*
   b) *Community response capacities*
   c) *Public awareness and education*
9. ***Sustainability, resources and budgetary commitments***
10. ***Improvement of overall operational framework of EWS through on-going drills and feedback and evaluations during and after an event***
11. ***Examples of previous events where the operational EWS has led to improvements in emergency preparedness and prevention***
12. ***Overall lessons learned and future steps for improving Meteorological, Hydrological and Climate services contribution in EWS particularly focusing on institutional coordination and cooperation with the disaster risk management agencies and EWS stakeholders (public and private)***

Following the MHEWS-I, in each country, through a formal process, teams of policy and operational experts from the NMHS, Disaster Risk Management Agencies, other key ministries such as health, environment and water, and the Red Cross and Red Crescent Societies (RCRC)[13] were engaged to document their respective EWS.

1.2.2 The Second International Experts' Symposium on Multi-Hazard EWS: with focus on the Role of NMHS

In 2009, WMO convened the Second International Experts' Symposium on Multi-Hazard EWS, (MHEWS-II) in 2009.[14]

MHEWS-II reviewed the outcomes of the first four documented good practices, discussed lessons learned, and identified additional good practices in MNEWS including Japan, United States of America, Italy and Germany, for further expansion of this initiative.

[13] RCRC was specifically involved in the documentation of the Bangladesh Cyclone Preparedness Programme.

[14] MHEWS-II was organized by WMO with a number of partners and was hosted by Météo-France, in Toulouse, France on May 5–7, 2009. MHEWS-II brought together nearly 100 experts from the NMHS, disaster risk management agencies, various ministries and the international and regional partners.

1.3 Documentation of Good Practices

This book provides detailed analysis of seven of these good practices with particular focus on multi-hazard EWS for meteorological and hydrological hazards. These include, (i) Bangladesh Cyclone Preparedness Programme; (ii) Cuba Tropical Cyclone Early Warning System; (iii) The French "Vigilance" System; (iv) The Warning Management of The Deutscher Wetterdienst in Germany, (v) Multi-Hazard Early Warning System in Japan, (vi) Multi-Hazard Early Warning System of The United States' National Weather Service; and, (vii) Shanghai Multi-Hazard Emergency Preparedness Programme.

This work has clearly revealed that the specific design and implementation of EWS in each of the seven cases vary according to their governance mechanisms, specific history, culture, socio-economic conditions, institutional structure, capacities and resources for sustainability of their respective systems. However, there are principles common to all them that have led to the reduction of the impacts of hazards, particularly through saving of lives. These 10 common principles are provided with more details in Chap. 9 of this book. They include:

1. There is a strong political recognition of the benefits of EWS reflected in harmonized national to local disaster risk management policies, planning, legislation and budgeting;
2. Effective EWS are built upon four components: (i) hazard detection, monitoring and forecasting; (ii) analyzing risks and incorporation of risk information in emergency planning and warnings: (iii) disseminating timely and "authoritative" warnings, and, (iv) community planning and preparedness and the ability to activate emergency plans to prepare and respond, with coordination across agencies involved in EWS, at national to local levels;
3. EWS stakeholders are identified and their roles and responsibilities and coordination mechanisms clearly defined and documented within national to local plans, legislation, directives, MOUs, etc, including those of the technical agencies such as the NMHS;
4. EWS capacities are supported by adequate resources (e.g., human, financial, equipment, etc.) across national to local levels and the system is designed and implemented accounting for long-term sustainability factors;
5. Hazard, exposure and vulnerability information are used to carry-out risk assessments at different levels, as critical input into emergency planning and development of warning messages;
6. Warning messages are; (i) clear, consistent and include risk information, (ii) designed with consideration for linking threat levels to emergency preparedness and response actions (e.g., using colour, flags, etc) and understood by authorities and the population, (iii) issued from a single (or unified), recognized and "authoritative" source;
7. Warning dissemination mechanisms are able to reach the authorities, other EWS stakeholders and the population at risk in a timely and reliable fashion;
8. Emergency response plans are developed with consideration for hazard/risk levels, characteristics of the exposed communities (e.g., urban, rural, ethnic populations, tourists, and particularly vulnerable groups such as women, children, the elderly and the hospitalized), coordination mechanisms and various EWS stakeholders;
9. Training on risk awareness, hazard recognition and related emergency response actions is integrated in various formal and informal educational programmes and linked to regularly conducted drills and tests across the system to ensure operational readiness at any time;
10. Effective feedback and improvement mechanisms are in place at all levels of EWS to provide systematic evaluation and ensure system improvement over time.

It is noteworthy to highlight that while these cases are treated as good practices, no system is ever perfect. A critical commonality to the cases presented in this book is that there are feedback mechanisms and investments for continuous evaluation and improvement of as various aspects of these systems, over time.

References

EM-DAT is the database of Université Catholique de Louvain – Brussels – Belgium – The OFDA/CRED International Disaster Database – www.em-dat.net.

First International Conference on Early Warning (Potsdam, 1998) www.geomuseum.com/ewc98/

Second International Conference on Early Warning (Bonn, 2003) www.ewc2.org/pg000001.htm

Third International Conference on Early Warning (Bonn, 2006) (www.ewc3.org)

First Experts' Symposium on Multi-Hazard Early Warning Systems (Geneva, 2006) (www.wmo.int/pages/prog/drr/events/ews_symposium_2006)

Second Experts' Symposium on Multi-Hazard Early Warning Systems (Toulouse, 2009) (www.wmo.int/pages/prog/drr/events/MHEWS-II/index_en.html)

Global Early Warning Survey (2006)
(www.reliefweb.int/rw/lib.nsf/db900sid/AMMF6VKH6Z/$file/UNISDR-Sep2006.pdf?openelement)

Global Assessment Report 2009,
(http://www.preventionweb.net/english/hyogo/gar/?pid:34&pil:1)

World Disaster report 2009, Focus on early warning, Early Action, Published by The International Federation of Red Cross and Red Crescent Societies

Golnaraghi, M, J. B. Migraine (2007) "Disaster Risk Management: An Investment in Development," The Full Picture, Published by Tutor Rose, Page 160–163.

Golnaraghi, M., J. Douris, J. B. Migraine (2008) "Saving Lives Through Early Warning Systems and Emergency preparedness," Risk Wise, Published by Tutor Rose, Page 137–141.

Golnaraghi, M., J. Douris, C. Baubion (2011) "Good Practices in Multi-Hazard Early Warning Systems," Risk Returns, Published by Tutor Rose, Page 95–97.

List of Acronyms

DRR	Disaster Risk Reduction
EWC-III	Third International Early Warning Conference
EWS	Early Warning Systems
HFA	Hyogo Framework for Action 2005–2015: Building the Resilience of Nations and Communities to Disasters
IPCC	Intergovernmental Panel on Climate Change
MHEWS	Multi hazard Early Warning Systems
MHEWS-I	First International Experts' Symposium on Multi-Hazard EWS
MHEWS-II	Second International Experts' Symposium on Multi-Hazard EWS
MOU	Memorandum of Understanding
NMHS	National Meteorological and Hydrological Organization
RCRC	Red Cross and Red Crescent Societies
UN	United Nations
UN-ISDR	UN – International Strategy for Disaster Reduction
WMO	World Meteorological Organization

Chapter 2
The Tropical Cyclone Early Warning System of Cuba

José Maria Rubiera Torres and Miguel A. Puig

Abstract For more than 40 years, the Cuban Government has been investing material resources and human capital in the creation and improvement of its early warning system for tropical cyclones and other alert systems. As a result, the resolution of meteorological and hydrological monitoring systems has been strengthened through the acquisition of new equipment, the training of specialists and the development of working tools that improve understanding of the hazards. Plans have been refined and structures have been perfected to ensure their implementation along with the population's preparation, at national to local levels. An extensive network of radio and television stations, newspapers and other facilities has been established to ensure the rapid dissemination of alert messages. Recently, local access and distribution points for early warnings have also been set up in communities at greater risk. Associated with the local administration centers, these outlets contribute to ensuring the timely transmission of information for the reduction of municipal risk. In summary, Cuba's system has the following key components:

- The technological capabilities of the meteorological service to predict the occurrence of hazardous events and issue specialized warnings for the Civil Defense and a clear explanation of the current and future situation to the population at large.
- Ensuring the application of plans elaborated from modeled risk scenarios to advance preparations by institutions trained for their implementation and reinforced by a prepared population with an appropriate perception of the hazard. All of this facilitates rapid mobilization.
- A network that uses all available means, from advanced technology to people themselves, to ensure the rapid transmission of alert messages.

2.1 Overview

Cuba's island condition and geographic location in the path of most tropical cyclones that develop in the Atlantic basin and the Caribbean Sea make these hydrometeorological events the greatest hazard faced by the nation. At the same time, official United Nations reports indicate that Cuba is one of the less vulnerable countries in relation to these weather phenomena. This has resulted, to a large extent, from the fact that, for several years, Cuba has had an Early Warning System (EWS) in place, supported by numerous institutions including an efficient Meteorological Service with an extensive monitoring network of meteorological and hydrological stations and weather radars. The Early Warning System ensures continuous monitoring, timely warning and effective communication links between the monitoring systems and Civil Defense organs, from the national down to the local level. The dissemination of forecasts and warnings is underpinned by a secure communications system and by the involvement of all mass media, alternative resources, and the people. Plans have been designed for different situations on the basis of the most likely disaster scenarios. All existing resources are made available during these contingencies, ensuring that people are effectively protected against the different hazards that they may face where they live.

Dr. José Maria Rubiera Torres, Director National Forecast Center, Institute of Meteorology, National Meteorological Service
Miguel A. Puig, Chief of Operations, National Staff of the Civil Defense, Cuba

M. Golnaraghi ed., *Institutional Partnerships in Multi-Hazard Early Warning Systems*,
DOI 10.1007/978-3-642-25373-7_2,

2.1.1 Background to the Establishment of EWS in Cuba

In Cuba, a set of institutions is entrusted with the monitoring of all events threatening the country. They ensure surveillance against hydrometeorological events, droughts, forest fires, floods, earthquakes, epidemics, and animal and plant diseases. They also, systematically, keep the relevant state authorities informed, with the reporting frequency ranging from daily to monthly depending on the variables and phenomena involved. These institutions have branches in all provinces, with some having municipal representation, and supply data on the variables they measure to both national and territorial authorities. This surveillance network provides the basis for Cuba's Early Warning System and is part of its Civil Defense System.

Prior to 1995, tropical cyclone activity in the Atlantic, Gulf of Mexico and Caribbean Basin was relatively low. During the decade of the 80's, for example, only hurricane "Kate" hit Cuba, in 1985, and it was a weak storm. An active period began in 1995 and that year registered the second most active hurricane season of the 20th Century, with 1933 being in first place. Although Cuba itself was not affected by any hurricanes in 1995, neighboring countries were badly hit by some of them. In consequence, the Cuban National Meteorological Service identified a need to provide early alerts on tropical cyclones, to increase awareness and preparation time, rather than simply issuing information when a hurricane directly threatened the country. Early Warning Messages were, therefore, initiated in 1996, being added to the regular warnings system. The first Early Warning Message was issued on 14 October 1996, several days before hurricane "Lili" crossed over the central provinces of Cuba. Success was immediate and since that time Early Warnings have been developed and institutionalized so that, today, not only text information is provided but also graphs depicting probabilities. Moreover, continuous interaction takes place with the Civil Defense, the Central Government of the country and other interested organizations.

2.2 Governance and institutional Arrangements (National to Local Levels)

2.2.1 Policy, institutional and legal Frameworks to Support Emergency Planning and Response

Cuba has a broad legal framework that regulates the functioning of Early Warning Systems at all levels. Those laws and directives which provide guidance in this area are mentioned below:

- Law No. 75/94 (December 21, 1994) of National Defense establishes the main missions and measures of the Civil Defense System and the principles for its territorial and institutional organization.
- As a complementary document to Law 75, Decree-law No. 170/97 (May 8, 1997), on the Civil Defense system of measures, regulates the role and position of state agencies and organizations, economic entities and social institutions in the disaster reduction process; the organization and implementation of these measures to protect the people and the economy; the establishment of phases during the response and the funding for disaster reduction plans.
- Guideline No. 1/05 (June 20, 1995) of the Vice President of the National Defense Council[15] on the planning, organizing and preparing by the country for disaster situations, establishes the regulations for the disaster reduction process and the guidelines to organize response and recovery at all levels. This document provides for the supply of updated information by the surveillance and early warning systems and its contribution to the actions implemented during the response, as one of the most important elements within disaster reduction plans. It also stipulates that in case of tropical storms, "Early Warning notices will be issued prior to the establishment of the response phases, so that the necessary measures are taken in advance". According to this guideline, the National Staff of the Civil Defense is the agency in charge of ensuring the implementation of civil defense measures and the observance of international standards and agreements regarding the civilian population signed by Cuba. It is also responsible for coordinating with the Ministry of Foreign Investment and Economic Cooperation concerning international cooperation and aid programs in cases of disasters caused by natural hazards or other types of catastrophe. In addition, its powers and functions include organizing, coordinating and controlling the work of state agencies and organizations, economic entities and social institutions with a view to protecting the people and the economy, as well as acting as a national organizing platform for the system.
- Resolution No. 43/06 (August 8, 2006) of the Ministry of Science, Technology and the Environment establishes that the Environment Agency is the body that has the mandate to organize, lead and conduct hazard, vulnerability and disaster risk studies. There are other legal texts that complement this guiding document at all levels. These include: Law No. 81/97 (July 11, 1997) on the Environment, Law No. 41/83 (July 13, 1993) on Public Healthcare, Law No. 77/95

[15] National Defense Council is the highest level in the response structure of the country.

(September 5, 1995) on Foreign Investment and Law No.85/98 (July 21, 1998) on Forestry, as well as ministerial and local resolutions which address specific aspects relating to the functioning of the early warning system for tropical cyclones and other hydro-meteorological events.

- Resolution 106 /99 (December 6, 1999) of the Ministry of Science, Technology and Environment establishes the General Norms of Direction, Organization and Operation of the Institute of Meteorology (INSMET), subordinated to the Ministry of Science, Technology and the Environment (CITMA). This document describes the general structure of the Institute of Meteorology of Cuba as the National Meteorological Service with the main mission of "giving authorized, reliable and opportune meteorological and climatic information on the state and future behavior of the atmosphere. This information is aimed at protecting the security of human life and reducing losses of material goods from disasters of meteorological origin, contributing directly to the well-being of the community and sustainable socioeconomic development". One of the functions of the Cuban Meteorological Service laid down in the Eleventh section of the document is to improve meteorological and climatic prediction, especially of phenomena that constitute hazards to human life, and to material goods and the national economy. Furthermore, this Resolution assigns to the Institute of Meteorology nation-wide responsibilities to issue through the Media, as the only authorized institution, meteorological and climatic information that may be required, especially warnings and forecasts of variables, processes and meteorological phenomena that constitute a hazard for human life, material goods, and the economy and for the development of the country. A somewhat similar role is defined for Provincial Meteorological Centers, as the only institutions authorized to disseminate and issue through the provincial Media meteorological and climatic information that may be required, particularly the warnings and forecasts of processes and meteorological phenomena that could constitute a hazard for human life, the loss of material goods, the economy and for the development of the county.

On a broader, overarching level, the Executive Committee of the Council of Ministers issued Ordinance Law No. 279/07 (March 19, 2007) "On General Principles, Organization, Preparation and Provisions of the Hydrometeorological System of Cuba for Exceptional Situations". This Law lays down that the Hydrometeorological System for Exceptional Situations is the group of hydrological and meteorological entities deployed in the territory of the country that has as its main mission to obtain, analyze, evaluate, process and issue the necessary hydrological and meteorological information for the execution of protective measures directed at the mitigation of the effects of disasters caused by natural hazards.

2.2.2 National to Local Emergency Planning and Related Linkages to EWS

The success achieved by Cuba with warnings on the impact of tropical cyclones is based, to a great extent, on the response planning processes. These involve the following:

- Plans for disaster reduction are elaborated down to the lowest level of the administrative political division of the country and are updated every year between January and April.
- At each level, the necessary resources are earmarked to ensure the protection of people, their goods and the economic resources (transportation, foods, medical care, emergency electric generators for vital services, etc.), based on the degree of risk for each community. Shelters for evacuees are identified and certified along with houses that receive people for their protection during an emergency.
- The state and readiness of the system of communications is refined in each territory, through the employment of all local means.
- The resources existing in the reserves established in each territory are checked to ensure efficient and fast recovery action based on a rapid re-establishment of affected vital services. The supplies in the national reserve are also inventoried.
- With the participation of journalists and the national and local media, radio and television, and the social organizations, the communications networks required to keep the people informed are upgraded to ensure delivery of early warnings. These facilities include the existing television rooms in rural and mountainous communities that work with solar panels and dish satellite antennas.
- As part of this process, the authorities that lead the response actions upgrade their plans and inform the population about the changes made to those plans for their protection. These preparations culminate with "Exercise Meteor" that is carried out during one weekend in the month of May, just prior to the beginning of the hurricane season.
- In April of each year, journalists and leaders of the media (i.e. newspapers, television and radio stations) at all levels meet together in a Seminar to upgrade knowledge of their roles and to clarify the national and local strategy for the hurricane season. The National Forecast Center also participates in these Seminars.
- Specialists from the meteorological and hydrological services report on the state of their technical resources to address the hurricane season and hold meetings with the Civil Defense organs at all levels.

The Cuban Meteorological Service is an active participant in the preparation and planning procedures of the Early Warning System (EWS). Its fundamental function is to raise awareness among the people and the institutions of the country by providing information on what is a hurricane; the various hazardous phenomena associated with it; the various risks to be faced and how to avoid them; how the warnings system is organized; and how to interpret warning messages. The Meteorological Service also participates in the preparation phase for the nationwide exercise "Meteor" that is organized annually by the Civil Defense prior to the hurricane season. Meteorologists also participate through frequent delivery of talks to provincial and local Governments, Civil Defense bodies and journalists on hurricanes and the predicted behavior of the next hurricane season. These activities are covered by radio and the television and serve to prepare the population for the upcoming hurricane season. Coordination plans and communications links between the Meteorological Service and the Civil Defense are also updated, in order to place the system in a state of complete readiness prior to the start of the hurricane season.

The Cuban Civil Defense is organized throughout the national territory on the basis of the country's political-administrative divisions and State structure. Its activity is supported by the use of the human and material resources available in government agencies and organizations, economic entities and social institutions, i.e. the organized forces of society. The President of the Council of State heads the Civil Defense System, through the Minister of the Revolutionary Armed Forces. In this capacity, the Minister is supported by the National Staff of the Civil Defense which is the lead agency in this System. The presidents of local governments are, correspondingly, the heads of the Civil Defense in their territories. In carrying out this responsibility, they have the support of the local professional Civil Defense entities which, along with state organization at all levels, coordinate, organize and plan the periodic assessment of risks associated with each event, the disaster reduction measures, the people's preparedness, the dissemination of information regarding actions to take and the behavior to be observed during different situations. They also control the implementation of protective measures for the different segments of the population, their property and the economy. Similarly, the top authorities of the central state administration and those of social institutions and entities are the heads of the Civil Defense for these bodies and are responsible for the implementation of the measures contained in the approved disaster reduction plans in their respective areas. Those individuals at the heads of ministries, industries, companies, education centers, hospitals, banks, cooperatives, stores, workshops and other production, service or research centers are also the heads of the Civil Defense in their respective institutions. They are in charge of planning, organizing and implementing Civil Defense measures which are mandatory for all institutions.

Political and mass organizations at all levels play an important role in the implementation of Civil Defense measures because of their autonomy and characteristics. These organizations have always actively participated in evacuation, rescue operations, and citizen's orientation and information in the event of disasters. At the People's Councils[16], the lowest government representation, measures are implemented to respond to and recover from disasters in accordance with the disaster reduction plans for each municipality. Disaster reduction plans are drafted for every territory and for all economic entities. These plans include an assessment of the risks at every location and are updated yearly using the data provided by the entities regarding the vulnerability parameters established by the methodology. The plans include measures to mitigate vulnerability and preparedness, response and recovery actions. They are prepared according to the instructions provided by the entities' higher government level and the decisions adopted by the local governments.

Disaster Reduction Plans that are prepared at each territorial level as well as within each organization and ministry, start from an updated assessment of the difference in risk between each location. This is defined on the basis of the magnitude of the hazard associated with each meteorological event and the vulnerabilities identified at that level. Starting from the update and appropriate zoning of the risk, actions for mitigation of vulnerabilities are established for that year, taking into account the available material and financial resources and prioritizing the most risky areas. Taking into account the level of risk at each location, measures for the protection of the population and economic resources are refined and appropriate response actions are planned to address each event, including actions needed to ensure a rapid and efficient recovery. The national organisms and ministries update the guidance for their entities, which also refine their plans on the basis of the risks associated with the territory where they are located, taking into account the specific guidance from their ministry or organism.

The Cuban Meteorological Service is engaged from the very beginning in the process of planning an Early Warning. The Meteorological Service issues the first early warning signal by means of a document entitled "Early Warning Message" that is sent to the Civil Defense and the Central Government when a meteorological situation is seen by the Service as being potentially dangerous for the country during the upcoming 120 hours. This "Early Warning Message" is issued in clear language that is understandable to non-meteorological personnel. It enables them to appreciate the uncertainties associated with a meteorological process that may or could happen and affect the country within a relatively long time frame of 120 hours. This type of information is transmitted through the media to the population in order to ensure an increasing level of concern while, at the same time, not creating immediate alarm. The precise purpose of these "Early Warning Messages" is to inform the Civil Defense, the High Authorities of the Country, and the population in general, in a timely manner, that it is necessary to monitor the meteorological situation over the coming days so that if a hazardous meteorological system develops they will be ready to take the required preventive measures.

[16] People's Councils are constituted in cities, towns, neighbourhoods and rural areas.

2.2.3 *Organizational Structure for Implementing the Plans*

The Early Warning System for tropical cyclones includes the following elements:

- An effective meteorological and hydrological surveillance system with appropriate human and material resources to ensure permanent monitoring and timely warning, both at the national and local levels.
- Effective communication between meteorological and hydrological surveillance systems and Civil Defense institutions, both at the national and local levels.
- An effective network for transmitting information, supported by secure communications systems.
- The use of all of the mass media to disseminate warning notices, including radio, television, newspapers, along with alternative means and people, both at the national and local levels.
- Plans designed for different situations, on the basis of likely disaster scenarios, and supported by all resources available in each territory, in order to ensure the effective protection of people under different levels of risk.

Figure 2.1 maps the EWS organizational structure at national and provincial to local levels, identifying ministries, disaster risk management agencies and authorities, other technical agencies, media and non-governmental organizations (NGOs).

In the context of risk management and the handling of disaster hazards, the Early Warning System (EWS) plays an important role in reducing human and material losses in Cuba. The EWS is considered a major Civil Defense asset and is systematically used and strengthened. The Cuban EWS takes advantage of the existing socio-economic structure, the strength of the institutions, and the levels of organization and education of both the authorities and the general population, among other aspects that assist it's functioning. In general terms, the main participants in the Cuban EWS are:

- The central surveillance entities responsible for monitoring the hazards along with their territorial branches responsible for this work at regional and local levels. The central surveillance entity for meteorological systems is the National Forecast Center of the Institute of Meteorology (National Meteorological Service).
- The authorities entrusted with the provision and dissemination of disaster related information at the different levels and with implementing the relevant protective measures, advised by officials and experts of the Civil Defense. These authorities are the top officials at provincial and municipal levels (Governors or Presidents of the Provincial Government; Mayors or Presidents of the Municipal Government).
- The media and mass and social organizations at the local level which help to disseminate information. These include local media, newspapers, radio and television stations that exist in all the provinces and most of the municipalities in the country. The mass and social organizations that participate in the dissemination of the information in urban and rural areas, respectively, are the associations of neighbors known as the Committees for the Defense of the Revolution (CDR) that bring together residents and exists in all neighborhoods of the country, and the National Association of Small Farmers (ANAP).
- The people who are well organized and prepared. These are, fundamentally, people of the mass and social organizations that have been trained to disseminate alert messages in communities far from the main cities and towns. They also include the people in charge of the operational sites for early warning located in key places with difficult access, who are prepared to measure rainfall and river levels and flows. These latter individuals possess means of communication enabling them to rapidly inform their respective risk management centre located at the municipal government site.

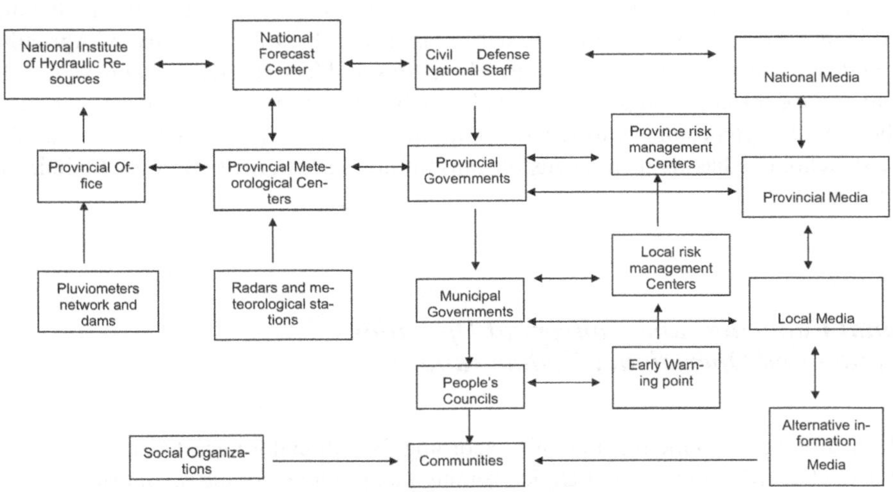

Fig. 2.1. The functional structure of the early warning system in Cuba

The EWS for tropical cyclones is a well-structured, coherent, nation-wide system that works closely with the Institute of Meteorology's National Forecast Center and the National Staff of the Civil Defense. These two organizations exchange and analyze information, allowing authorities to take the necessary measures and establish the relevant phases. At the national level, the exchange of information between the National Forecast Center of the Institute of Meteorology and the National Staff of the Civil Defense facilitates decisions on the phasing of the response for threatened territories, which presupposes a certain time frame for the adoption of measures in those places. The more effectively the Early Warning System works, the greater the opportunity it provides for provinces and municipalities to react, enabling them to protect the lives of people and economic resources exposed to different levels of risk. Whenever a province receives an early warning message, or notice of the establishment of one of the phases of the response, the authorities begin immediately to undertake the measures for that phase, outlined in the plan for disaster reduction, taking into account the characteristics of the threatening event, the level of risk to people and the exposed economic resources. In this decision-making process, a similar exchange to that at the national level, also takes place within the province. This engages the authorities and the meteorological service of that province in determining the magnitude of the impact of winds, rains and storm surge in the territory, building on the guidance that is received from the National Meteorological Service. Having completed this assessment, the planned protection measures, developed during the preparations for each possible scenario, begin to be implemented, making this a fast and efficient process.

Starting from the information provided by the Forecast Center of the Institute of Meteorology, the National Staff of the Civil Defense analyzes the initial situation. A note from the Civil Defense can be transmitted to provide an early warning to the governments of the threatened territories. The exchange of information between the Forecast Center and the National Staff of the Civil Defense continues while the hazard threat remains over any portion of the country. Response action in Cuba is organized into three phases or stages – the "Informative", "Alert" and "Alarm" phases. The protection measures to be undertaken in each phase are planned for each event. The early warning system functions according to the phase that has been established, taking into account the fact that as the hurricane comes closer to Cuban territory, the ways of disseminating the messages also change. The establishment of each phase is approved by the Central Government, having been proposed by the National Staff of the Civil Defense following consideration of the recommendations provided by the National Meteorological Service, along with other factors. Similarly, the meteorological and hydrological services in each province interact with the authorities and are responsible for keeping both them and the provincial population informed through all available means of communication.

There are three levels of local government in Cuba, namely, province, municipality and people's council. These three levels of government each have a President and a structure that facilitates the organization of the government's administration. The National Forecast Center (Meteorological Service) interacts with the Government at National level. The Provincial Meteorological Centers interact with the government at provincial, municipal and people's council levels, offering information and advice during the different Civil Defense phases. The Institute of Meteorology (Meteorological Service of Cuba) interfaces directly with the Central Government, the National Civil Defense, the National Media, the Provincial Meteorological Centers and the National Institute of Hydraulic Resources, through the National Forecast Center. The provincial meteorological centers provide information on their territory to the provincial and municipal governments, the Civil Defense at that level, the provincial and municipal media and the provincial delegations of the National Institute of Hydraulic Resources. The system is, moreover, a two-way system as the National Forecast Center also receives information from those entities.

The key stakeholders of the National Meteorological Service (NMS) include not only the Civil Defense, the Government, and the various media, at national, provincial and municipal levels, but also the National Institute of Hydraulic Resources, the Civil Aviation Weather Office and different ministries such as the Ministry of Agriculture, the Ministry of the Sugar Industry and the Ministry of Construction. These stakeholders receive meteorological Early Warnings and Warnings as a special service through computer links and e-mail. Once a message has been received by them, the stakeholders speak by phone with meteorologists to receive more detailed information that may be useful in their respective fields. Each of these stakeholders has its own means of transmitting this information down to their lowest level of its activity and receiving feedback.

2.2.4 *Institutional Capacities and Concept of Operations (Coordination and Operational Collaboration)*

The exchange of information between the Institute of Meteorology's National Forecast Center and the National Staff of the Civil Defense is regulated by a Memorandum of Understanding that provides for the monitoring of tropical cyclones as described above. The Operational Plan of the National Forecast Center outlines all the technical procedures required to ensure the implementation of this agreement, including the frequency of reports and their content. The different stakeholders mentioned earlier receive the information and then contact meteorologists directly to obtain additional details that can help in their decision making process (e.g. forecast wind speeds associated with the hurricane could exceed thresholds for an agriculture cropping activity). Such consultations with meteorologists are made by phone, though written advice may sometimes also be required depending on the information needed and its implications for the decision making process. This is a standard procedure of the Meteorological Service in dealing with these special users. Following the occurrence of a storm, substantial information is also requested to assess the damage and document claims to insurance.

Provincial Meteorological Centers fulfill a role similar to the National Forecast Center except that they relate to provincial and municipal governments, Civil Defense and local users including people's councils. In practice, the Provincial Meteorological Centers the represent the Institute of Meteorology at the provincial level, though only the National Forecast Center can issue Early Warnings and Warnings in the event of large scale weather systems such as hurricanes. In such circumstances, the Provincial Centers serve as advisors in assessing the local impact that the hurricane may have in a specific province. Where severe local storms or other rapidly developing local weather features are concerned, however, a Provincial Meteorological Center can issue its own local warning to local authorities, while merely informing and receiving advice from the National Forecast Center.

The roles of the National Staff of the Civil Defense in relation to drafting the early warning reports, the operation of the mass media, and activities undertaken by local authorities to ensure that the warning messages reach the entire population and that necessary measures be taken, are defined in additional documents issued at each level and included in the disaster reduction plan. Acting in accordance with Guideline No. 1 of the Vice-president of the National Defense Council, the heads of all state organisms and local authorities issue resolutions and instructions to regulate the functions assigned to their respective agencies and entities, within the general framework of the Early Warning System and the execution of other activities within the disaster reduction process. The Chief of the National Staff of the Civil Defense, the Minister of the Institute of Hydraulic Resources and the Minister of Radio and Television have issued corresponding instructions in this regard. Local authorities also issue guidance and directions to subordinate levels, specifying the functions to be carried out by these different structures in relation, for example, to the transmission of alert messages down to the level of people's councils; to reports that should be provided by the early warning access and distribution points; and to the maintenance of the flow of information. A large number of additional documents on this topic also exist but will not be summarized here.

Following the response to an event, an analysis is undertaken into how effectively the Early Warning System worked. Measures are taken to strengthen the system based on the experience gained in dealing with the event, thereby ensuring the refinement of procedures. The operation of the EWS is also assessed during the yearly "METEOR" drill that lasts for two days and involves all structures from the national to the local levels. In addition, the system is reviewed prior to the beginning of each hurricane season to ensure that everything is in place. "Exercise Meteor" is held annually during a weekend in May, prior to the beginning of the upcoming Hurricane Season. During the first day of this drill (Saturday), general preparations and all planned response and recovery measures are reviewed and exercised, at all levels of the country (nation, province, municipality, people's council, community and entities). On the second day (Sunday), practical activities such as evacuation and protective measures are carried out with the population and the forces that participate in the response. The Meteorological Service prepares a set of simulated warnings to be disseminated by all communications links that also serve to test the system. In addition, the Meteorological Service plays a role in increasing public awareness by reviewing the main characteristics of hurricanes, the warning service, and the correct interpretation of warnings, and by providing an overview of the coming season. These talks are presented at the commencement of the exercise (on Saturday) at provincial and municipal levels and this activity receives coverage by national, provincial and municipal media.

The Cuban Meteorological Service plays an active lead role in all aspects of the Early Warning process, as well as in the development and implementation of coordination and cooperation with stakeholders. Of top priority are the relationships with the Government and the Civil Defense at national, provincial and municipal levels that have been established in national legislation (see Sect. 2.2.1) and a Memorandum of Understanding (MOU) with the media and the National Institute of Hydraulic Resources (NIHR). Cooperative activities also exist with other entities, however, mainly concerning the enhancement of knowledge on hazardous weather phenomena and increasing awareness prior to a new hurricane season. The MOUs with the media and NIHR and the MOU with the Civil Aviation Weather Office are the only specific agreements of this type. However, the National Forecast Center's Operational Manual details the services to be provided to ministries and other important organizations as well as how the information for them should be transmitted and feedback obtained. The National Forecast Center produces a wide range of detailed information for these special interests, such as forecast winds, storm surge, rainfall, etc for specific locations.

The Cuban Committee of the Red Cross Society is another important stakeholder. The Society has several detachments that participate, at the community level, in the preparation of the population and operational groups that take part in rescue and salvage during the response stage, along with other territorial forces. Although no formal MOU exists between the National Meteorological Service (NMS) and the Cuban Committee of the Red Cross Society, both organizations participate jointly in workshops, seminars and training sessions. The Cuban Red Cross also receives operationally detailed information from the National Forecast Center so that they can update their contingency plans for specific areas in the event of threat. Their contingency plans do not mention these informal agreements because the information to be received from the National Forecast Center is normally delivered as part of the Operational Plan of the Meteorological Service and is included in this Plan. Since Cuba has chosen to use the MOU process as its basis for documentation of operations and coordination, the Meteorological Service would, however, like to see a statement as to the role the NMS plays included in the development of any overarching MOU regarding EWS operations. The NMS takes part in the planning and implementation process, participating in seminars, meetings and training sessions with the Cuban Society of the Red Cross and other stakeholders, such as the Latin American Center for Disaster Medicine (CLAMED) which has implemented a Masters Degree course on Disasters in which some of the professors are specialists from the NMC.

2.2.5 Key Factors considered for Sustainability of different Components of Early Warning Systems with a Multi-Hazard Approach

The development, maintenance, sustainability and strengthening of early warning systems is ensured by means of the allocation of the necessary resources at two levels – institutional and territorial. In Cuba, there are regulations issued by the Ministry of Economy and Planning and the Ministry of Finance and Prices that allow national organizations and local governments to plan the allocation of resources for disaster reduction, as well as those needed for the operation of the early warning systems. In the specific case of the EWS for tropical cyclones, the Ministry of Science, Technology and the Environment plans and receives resources for the maintenance and development of the National Weather Service. Correspondingly, the Ministry of Informatics and Communications and the Cuban Institute of Radio and Television are responsible for the investments that ensure the proper functioning of TV and radio stations at the national, provincial and municipal levels. This is also the case for newspapers. By law, local governments have at their disposal all the resources needed to undertake the protection measures during the response phase that begins the moment they receive the Early Warning Notice. The Cuban Meteorological Service receives funding for all its activities from the State, through the Central Government and the National Budget that is approved by Law every year by the National People's Power Assembly (National Assembly or Parliament). All required funds for the operation of Provincial Meteorological Centers and for the maintenance of radars and weather stations are also included in the National budget allotted to the Institute of Meteorology.

Development of EWS has a number of financial costs associated with it. These include long-term, on-going, investments in the four components of the early warning systems, including:

- Observing, monitoring and forecasting of hazards;
- The development of risk information and inclusion of risk information in the warning messages;
- Communications systems for dissemination of warnings to the authorities and the population at risk, and
- Emergency preparedness and response planning at the community level (training, emergency planning and drills), with support from provincial and national governments.

Furthermore, operations of the EWS during a potential event have many costs including the coordination, cost of preparedness prior to an event, evacuations, etc. In addition, there are other human resource and training requirements for the development and maintenance of these systems over time. As in any other Cuban institution, the budget for the coming year begins to be discussed at the Meteorological Service from April to June of the previous year. A proposed budget is presented in July each year. The proposed NMS budget is included in the overall budget of the Ministry of Science, Technology and Environment and forms part of the National budget proposal. The National State Budget is approved by the People's Power National Assembly (Parliament) during its December session. Budget allocations are then made at the Ministry level with the National Meteorological Service subsequently receiving its own budget along with responsibility for its implementation. The NMS obtains all funding requirements for the fiscal year in its State Budget through this process. The human resources needed for the maintenance of the Service are obtained each year from the University, taking into account assessments of requirements from the preceding three years and the progress that students have made in practical work at the NMS during three years prior to graduation. There is currently no methodology developed for cost-benefit assessments although an overall assessment of damage is made by local, provincial and national governments, with the aid of the Civil Defense and the insurance sector.

2.3 Utilization of Risk Information in Emergency Planning and Warnings

2.3.1 Organizational Responsibilities and Arrangements for the Development of Risk Information

Fulfilling a mandate under Guideline No. 1 of the Vice-president of the National Defense Council, the Environment Agency of the Ministry of Science, Technology and the Environment has been assigned the responsibility, under Resolution No. 43 of August 8, 2006 of that ministry, to organize, lead and conduct hazard, vulnerability and disaster risk studies. In carrying out these tasks, the Environment Agency can draw upon the nation's scientific potential and is assisted by other organizations of the central government administration. This agency has a group of specialists that have elaborated the methodology to assess the risk using vulnerability indicators which are determined in each municipality. These facilitate monitoring the reduction of major vulnerabilities. Guideline No. 1 and the Resolution No. 43 constitute the legal mandate for this agency to create the group of specialists that elaborated the methodology for risk assessment in each municipality and to prepare the specialists in each territory. At the local level, groups of specialists are being created from municipal organizations such as Physical Planning, Statistics, Water Resources, Public Health, Agriculture, Construction, Housing, Project Enterprises and others, under the supervision of the local office of the Ministry of Science, Technology and the Environment. This risk assessment program is undertaken in coordination with the National Staff of the Civil Defense.

Using a guide drafted by the National Staff of the Civil Defense for carrying out the risk studies, the Environment Agency, assisted by a group of experts from different scientific institutions, has compiled a methodology for risk assessment that considers the impact of the rain, the wind and the sea. This methodology has already been applied in the municipalities exposed to greatest hazard. The local office of the Ministry of Science, Technology and the Environment in each province coordinates with the other organizations in the territory to obtain the necessary data to determine the risk. Among the organizations involved are the territorial divisions of Housing, Physical Planning, Statistics, Water Resources, Public Health, Education and others. The findings are all stored in a database, supported by a Geographic Information System (GIS), which is updated as actions are implemented to reduce vulnerabilities. The methodology allows the local governments to determine the risk periodically so that its reduction can be monitored. This necessitates that the specialists from each organization update the information related to each indicator of vulnerability.

As regulated by law, the Meteorological Service of Cuba is the only entity in the country that issues meteorological information on dangerous phenomena. Its researchers and technicians participate in hazard, vulnerability and risk studies that are undertaken by multidisciplinary technical groups in the Agency of the Environment, a component of the Ministry of Science, Technology and Environment (CITMA) that has the national mandate for such studies. The participation of the NMS in these studies relates, mainly, to investigations of hazards and to numerical simulation of storm surge and winds performed for vulnerability and risk studies.

2.3.2 Hazard Assessment, Quantification and Mapping (National to Local)

Hydrometeorological events, specifically tropical cyclones, pose the greatest hazard for Cuba. During the past 158 years, Cuba has been affected by 205 cyclones for an average of 1.2924 annually. There is a 60 percent probability that a tropical cyclone will strike Cuba in any one year. The corresponding probability is 71.3 percent that a tropical cyclone will affect Cuba without the eye making landfall on the island. The probability of a major hurricane striking Cuba is only 13.3 percent annually while the corresponding figure for a hurricane coming close to Cuba increases to 17.3 percent. On average, therefore, a major hurricane passes over Cuba every 7.5 seasons and one comes close to Cuban territory every 5.8 seasons.

In Cuba, October is the most hazardous month in terms of the frequency and the intensity of tropical cyclones, followed by August and September with hazard probabilities that are quite similar. Overall, 97.3 percent of the tropical cyclones that have hit Cuba have done so within this season. All tropical cyclones that have affected the country during October have made landfall from the south coast and most of these have been formed in the central and western zones of the Caribbean Sea. In October 1963, Hurricane Flora affected the eastern region of the country for 5 days, with heavy rains unleashing great floods in the basin of Cauto River. Over one thousand people died and the economy and environment were seriously damaged. Addressing the causes for the disaster, then Prime Minister Fidel Castro outlined steps that should be taken to begin building reservoirs, canals and other water works, regulate flows and to prevent floods from occurring in future. This policy was rapidly implemented all over the country that today boasts 239 water dams, hundreds of kilometers of canals and other works. All of these facilities are monitored daily in relation to their operating standards and also receive appropriate technical maintenance. In October 2007, the combination of a long spell of rain plus intense precipitation associated with the passage of tropical storm Noel resulted in the accumulation of a volume of water similar to that of 1963 in the same areas. This time, however, the water works built during previous years and the prompt and organized action of the population and the authorities prevented the loss of human life.

The Cuban Meteorological Service (NMS) participates in studies of hazards, contributing data, processing and analysis and obtaining, as final output, hazard maps for different hazardous weather systems or weather elements. Several weather systems and elements have been addressed in this way, including hurricane storm surge maps and maps of strong winds, coastal flooding and floods due to intense rains. There are also return period maps and maps that try to take into account the effect of climate change, for future planning purposes. These studies are already in a very advanced stage but have not yet been completed for the whole country. More specifically, the maps for coastal flooding and hurricane storm surge have been completed for all Cuban coastal areas and an electronic Atlas has been created for rapid consultation. The NMS also plays an important role in national post-storm impact analysis, jointly with the Government and the Civil Defense. Here, the main contribution of the NMS is in the assessment of impacts in the areas affected by winds, rains, rain flooding, storm surge, coastal flooding and other aspects of the storm, as well as in the evaluation of the forecasts that were made. This impact assessment of the meteorological factors and the damage that they caused are very important inputs for updating hazard maps and related information.

2.3.3 Assessment of Vulnerabilities and Exposure (National to Local)

Starting from the studies carried out on the hazards of rains, the state of the sea and winds in hydrometeorological events that threaten the country, a group of national level specialists developed a methodology to determine the associated risk of each of these events. This methodology establishes indicators of vulnerability such as the technical condition of buildings, the drainage situation, the level of people's exposure in flood areas, etc. This enables multidisciplinary groups that have been created in each municipality with the participation of all the involved organisms to estimate the reduction of risk that constitutes the basis for annual updating of disaster reduction plans. All information related to the update of the databases for each vulnerability indicator and the determination of the risk for each area is carefully stored at the risk management centers in each municipality and is supported by a Geographic Information System (GIS). The results of the update of the risk levels in each people's council of the municipality are relayed to the provincial and national levels. This annual analysis enables the weakest aspects of the risk assessment to be determined and these are then corrected for next period. The communities with greater exposure to the hazards generally coincide with those at greater risk because they also concentrate the main vulnerabilities. Consequently, these places require more detailed planning of response actions and are, therefore, prioritized, generally being the first ones to be protected with the available resources in the territory.

The Environment Agency creates the groups of specialists that elaborate the methodology for risk assessment in each municipality and trains specialists in each territory. At the local level, groups of specialists from municipal organizations such as Physical Planning, Statistics, Water Resources, Public Health, Agriculture, Construction, Housing, Project Enterprises and others are created under the supervision of the local office of the Ministry of Science, Technology and the Environment. The risk assessment program is undertaken in coordination with the National Staff of the Civil Defense. Using a guide on the conduct of the risk studies drafted by the General Staff of the Civil Defense, and assisted by a group of experts from different scientific institutions, the Environment Agency put together a methodology for risk assessment that considers the impact of the rain, the wind and the sea. This methodology has already been applied in the municipalities exposed to greater hazard. The methodology allows for the local governments to periodically determine the risk so that its reduction can be monitored. This requires that the specialists from each organization update the information related to each indicator of vulnerability. The vulnerabilities associated with the destructive effects of hurricanes and other hydrometeorological events are assessed (winds, storm surge, coastal flooding, rains, etc.) While these are, according to historical data, the most frequently recurrent events, vulnerability indicators have also been determined for drought, forest fires and earthquakes.

The Environment Agency has the mandate to produce hazard, vulnerability and risk maps for each municipality. The designed vulnerability indicators consider the special needs and the economic and social development of each municipality and their communities to assess the risk to each area. The social achievements of the Cuban Revolution – which wiped out illiteracy and promoted education, provided free health care, built roads connecting the most remote places, guaranteed electricity to more than the 95 percent of the homes, and facilitated women's involvement and participation – have progressively led to a reduction in the population's vulnerability to the impacts of hurricanes and other weather events. Today, the main elements of vulnerability to hurricanes relate to the technical state and the construction methodology of houses, mainly in the eastern region where more than 240 coastal settlements still exist and a large number of houses are adjacent to rivers or in the path of spillways. The Cuban Meteorological Service also participates in vulnerability and risk assessment studies with data, processing, analysis and the prediction of local hazards and by providing advice for these studies as well as to local authorities.

2.3.4 Storage and Accessibility of Disaster and National Hazard Risk Information

The National Meteorological Service has the responsibility of gathering all the relevant basic information on meteorological systems as well as on each of the hazardous weather elements such as hurricanes, intense rains, storm surge, coastal flooding, severe local storms and their associated phenomena (lightning, tornados, strong winds), cold fronts, and strong winds not associated to the meteorological systems already mentioned. The assessment of risks, the conscious and progressive reduction of associated vulnerabilities and the monitoring of their reduction is a mandate of the Cuban state that the government at its different levels is required to implement. With financial assistance from agencies of the United Nations and some non-governmental organizations, Cuba has undertaken a new initiative aimed at facilitating this effort by establishing Risk Management Centers in the most vulnerable municipalities in the country. These Centers are located within the head offices of municipal governments and allow local authorities to keep a record of all hazard and vulnerability studies that have been completed, the main indicators of vulnerability in the territory and all information needed to monitor risk reduction. Additional centers exist at the provincial capitals where the data provided by the municipalities is assembled. The General Staff of the Civil Defense is developing an Intranet that will allow access to all of this information, thus creating a large database with controlled access.

There are three or more Early Warning Points associated with each Early Warning Center from which people can be warned of a hazard or the strike of a new meteorological event. These are, generally, located in remote places with difficult access. Local Inhabitants are trained to operate and maintain these facilities and they are given means of communication and power generators that allow for uninterrupted monitoring of local meteorological situations. Presently, there are 40 Early Warning Centers in operation, seven in the provincial capitals and 33 in the municipalities, along with 129 Early Warning Points. An additional 25 Early Warning Centers and 195 Early Warning Points are being planned. The Risk Management Center also takes part in the response, providing advice to local authorities and making available the data on risks that they store and those data that they receive from the early warning points, thus assisting in the decision-making process. In addition, these data are used to produce hazard maps and underpin the drafting of norms and regulations regarding the use of the land for construction.

The Cuban Meteorological Service has the responsibility of archiving, processing and quality controlling all of the primary meteorological data. It is also responsible for conducting post storm studies, identifying the preferred tracks of tropical cyclones, as well as their impacts, updating and preserving records of hazardous weather, such as tropical cyclones, storm surge, coastal flooding, cold fronts, strong winds, severe local storms and tornados, publishing the results and providing all the available information to the user community.

2.3.5 Development and Utilization of Hazard/Risk Information to support Emergency Planning and Warnings

The value of the risk associated with each event is distributed within each people's council. This permits planning the resources for the reduction of vulnerabilities, prioritizing those places at higher risk. In areas with higher probability that the impact of an event will become a disaster, the response and recovery measures are given more detailed attention and greater priority, including the transmission of the warnings on the evolution of any danger. As a general rule, Early Warning Points are situated in these locations to ensure greater effectiveness in the dissemination of the information. The Cuban Meteorological Service has an advisory role during the development and use of the risk information in the planning and warnings stages. Information on risk is incorporated in the warnings, mainly in direct radio and TV broadcasts by meteorologists (e.g. the risk that a certain coastal area may experience a hurricane storm surge and that area should be evacuated). The inclusion of information on risk in meteorological warnings is not mandated by the government, however, though it assists Civil Defense in the mobilization of the people to be evacuated in the threatened region.

2.4 Hazard Monitoring, Forecasting, and Mandates for Warning Development

2.4.1 Organizational Responsibilities for Monitoring, Forecasting and Development of Hazard Warnings

The National Meteorological Service has the sole mandate to issue meteorological warnings for hazards 1–3, 5–10, and 11 in Table 2.1. More specifically, this includes providing all scientific information on the hazard, including its forecast behavior, communicating the information on the hazard to the public and promoting awareness of the fact that it is a threatening meteorological situation for a certain region of the country. The National Forecast Center supplies the basic information to the National Staff of the Civil Defense, following previous coordination with Civil Defense including a discussion on the different aspects of the hazardous situation. The National Staff of the Civil Defense then issue a warning note for the threatened area stating that, in view of the information from the NMS, there is the need to take all protective measures. The hazards listed above are considered Type II hazards. Note that hazards numbers 2–9, and 10 are usually associated with hurricanes although some of them may often occur individually in other weather systems. Where hazards 4 and 9 listed above are concerned, the National Forecast Center provides information that can be used by the National Institute of Hydraulic Resources to assess the probability of river flooding and by the Forest Keeper Body to assess the probability of forest fires. These hazards are classified as Type III from the NFC point of view. Hazard 12 is of primary concern to the Cuban National Center for Seismologic Research, which plays the same role in relation to seismologic hazards as does the National Forecast Center in the case of hydrometeorological hazards.

While the Cuban Meteorological Service has, from the scientific and operational perspectives, the organizational responsibility for monitoring, forecasting and developing the hazard warning and communicating it to the public, the National Civil Defense is responsible for the development of the warning in terms of the mobilization of all national and local resources, including all logistics for protective measures and evacuations. There is, however, a genuine partnership in the sense that the Cuban National Meteorological Service and the National Civil Defense act as part of a single national system in which all efforts are combined for the protection of life and material resources. Consequently, we consider that what we really have is a combination of Type I and Type II hazards.

Table 2.1 Major hazards that affect Cuba

Hazard	National Agency for Mandate	Type of the Hazard
1. Thunderstorm or lightning	NFC, EMNDC	II
2. Tropical cyclone	NFC, EMNDC	II
3. Flash flood	NFC, EMNDC	II
4. River flooding	EMNDC, NIHR, NFC	III
5. Strong winds	NFC, EMNDC	II
6. Landslide or mudslide	NFC, EMNDC	II
7. Tornado (rotational high winds)	NFC, EMNDC	II
8. Coastal flooding	NFC, EMNDC	II
9. Forest or wild land fire	EMNDC, GB, NFC	III
10. Storm surge	NFC, EMNDC	II
11. Drought	NFC, EMNDC	II
12. Earthquakes	EMNDC, CNIS	I

(NFC = National Forecast Center (National Meteorological Service); EMNDC = National Staff Civil Defense; FKB = Forest Keeper Body; CNIS = National Center for Seismologic Research (National Seismologic Service); NIHR = National Institute of Hydraulic Resources)

2.4.2 Organizational Collaboration and Coordination for Monitoring, Forecasting and Development of Hazard Warnings

The Cuban Meteorological Service has several coordination mechanisms with the Government and the National Civil Defense. These include very frequent operational meetings while the threat of any hazardous meteorological system such as a hurricane exists, coordination by direct voice line and by computer networks. The operational meetings take place at either the National Forecast Center or at the headquarters of the National Staff of the Civil Defense, since both institutions are located about 50 meters apart. The monitoring of meteorological observations is carried out by the National Meteorological Service using a network of 68 first order meteorological stations, 30 automatic stations, 8 meteorological radars (providing coverage of the whole country) and a meteorological satellite receiving station. All meteorological information is in digital format and is relayed via computer networks that link all Provincial Meteorological Centers and meteorological stations with the National Forecast Center. Forecasts and warnings are distributed to the end users by electronic means. In addition, these products are posted on a web site (http://www.insmet.cu) and are made available to the public through a toll-free meteorological telephone service. Live broadcasts by meteorologists are also transmitted directly from the National Forecast Center on radio and national TV stations.

2.5 Development of timely authoritative, recognizable, and understandable Warnings

2.5.1 Warning Message Development Cycle

The Early Warning System for tropical hurricanes is organized and functions in the following sequential manner.

1. The Institute of Meteorology's National Forecast Center continuously monitors the formation and development of tropical cyclones and tropical waves from their genesis off the West African coast through their travel across the Atlantic towards the Caribbean. Within the framework of the WMO RA IV Hurricane Committee, international cooperation is maintained with several countries in the area, including with the United States National Hurricane Center, which is also a WMO Regional Specialized Center for Tropical Cyclones in WMO's Regional Association IV (RA IV). It has been established that no tropical cyclone that has passed outside the area South of latitude 24° N and limited by meridians 60° W and 89° W has affected Cuba. Consequently, any tropical cyclone that enters or is formed in this region (known as the "Area of Reinforced Surveillance" (Fig. 2.2) is closely monitored, even if it is merely a tropical wave with some potential to develop. An agreement exists with the National Forecast Center that, in such cases, the Center produces Early Warning Notices for use by the General Staff of the Civil Defense, sometimes as early as 120 hours before the system may strike Cuban territory. Such a Notice includes an assessment of the tropical system's future development. It is not intended for public broadcast but simply for hazard assessment. At the same time, however, public information will include some indication that the situation should be carefully monitored by everyone.
2. The National Staff of the Civil Defense evaluates the warning and issues a Notice to the governments of threatened provinces and to state organizations whose resources might be affected (e.g. ministries of Agriculture, Tourism, Information and Communications, and others). This Notice is not relayed to the national press due to the uncertainty that still exists regarding the developing system's track and organization. In their weather reports, however, TV channels do begin to refer to the fact that a weather system is likely to become a cyclone and is being watched.
3. The governments of threatened provinces which have received the Early Warning Notice begin to take the measures outlined in their disaster reduction plans to address the anticipated hazard, based on the risk level of each community. Working with the local meteorological and hydrological services, they assess the consequences of the possible impact of rains, waves and winds associated with the tropical event. The necessary logistics for these protection measures are identified but no resources are mobilized at this stage.
4. As the tropical cyclone continues to approach Cuba, the Meteorology Institute's Forecast Center increases the number of warnings, describing in detail the system's future trajectory and organization and the expected impact of the rains, waves and winds.
5. When it is anticipated that the tropical cyclone could affect the island territory within the following 72 hours, the phases foreseen in the response stage (Informative Phase (72 hours), Alert Phase (48 hours), and Alarm Phase (24 hours)) are announced by means of Bulletins issued by the Civil Defense National Staff and broadcast over national and local radio and television.
6. Building on the Forecast Center's assessment of the likely behavior of the rains associated with the tropical cyclone, the provinces' meteorological centers evaluate the probable local impact of the precipitation and forward this information to the hydrological service in the territory. In turn, the hydrological service estimates the potential for floods, based on the water resources situation (i.e. the water levels in reservoirs, canalization and drainage conditions, the degree of soil saturation, and the hydrological condition of rivers). When the rains begin to affect the territory, the local network of rain gauges records precipitation each hour and likely flood areas are mapped and reported to the local governments to confirm that the protective measures being taken are the correct ones under the circumstances.
7. A similar assessment of the likely impact of winds and waves is conducted, taking into consideration the structural vulnerabilities of housing, economic facilities and coastal settlements, each of which receive protection in accordance with their level of exposure and risk.
8. When the tropical cyclone is no longer a hazard for the country, the Recovery Stage is declared. At this stage, the restoration of damaged infrastructure and services commences, in accordance with territorial and national plans.

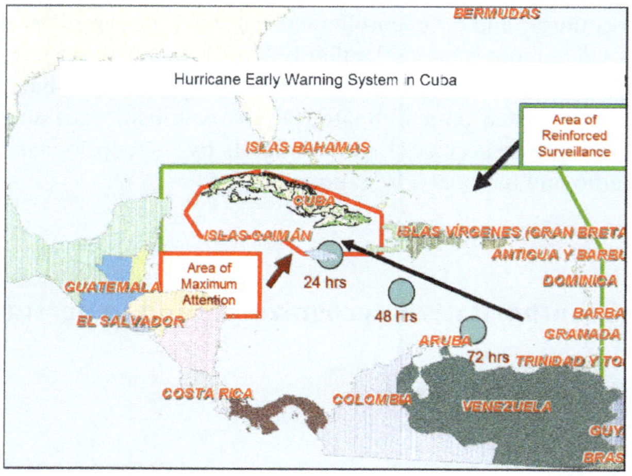

Fig. 2.2 Areas of Reinforced Surveillance and of Maximum Attention in the Hurricane Early Warning System in Cuba

The Program of Social Communication for disaster reduction provides the main guidance for the operation of the mass media during all stages of the disaster reduction cycle, including the conduct of surveys among the population regarding the reception and interpretation of the Early Warning message, immediately following the impact of each event. The methodical nature of this process has facilitated perfecting the content of EW messages and the mechanisms for their delivery to ensure that the whole population, including people who live in remote places, receive the EW messages in a timely and appropriate manner. During the process of updating plans previously outlined, checking is carried out to confirm whether corrective actions were included.

In Cuba, two approaches are taken to defining the threshold for warning readiness. One depends on space and the other on time:

The Space Approach

The space variable defines two areas around the country that determine the actions of Early Warning. Their boundaries were established on basis of a study of historical tracks dating back to 1851 that was undertaken by the Institute of Meteorology to determine the probabilities of hurricanes affecting Cuba. Two areas have been agreed upon – one for "reinforced surveillance" and one for "maximum attention" (Fig. 2.2). The first area is defined by meridians 60° W and 89° W, and for the parallel 24° N. The National Forecast Center of the Institute of Meteorology carefully monitors any tropical cyclone (including tropical waves with some possibilities for development) that is formed in or enters this area, whenever it represents a potential hazard for Cuba. This special monitoring lasts until the cyclone either leaves the area or the conditions for its development have disappeared. This makes possible the issue of an Early Warning, with an assessment of the future development of the system, although a reasonable amount of uncertainty may exist in the analysis. This Early Warning can be issued by the National Forecast Center from 120 to 72 hours before the tropical cyclone begins affecting any part of Cuban national territory. Response actions are modulated by several factors in addition to the storm strength threshold. In addition to the strength thresholds of the meteorological system, the degree of risk of a particular location is taken into account with other factors including the degree of isolation of the given location or difficulty in access to it and whether the beginning of the evacuation operation, as calculated from the already mentioned time variable, is going to be at night. In the latter circumstance, evacuation is scheduled to commence and be completed during daylight and never during the night. A modulation of the response is made in each these situations which are the subject of further studies. The other area, that of a "maximum awareness", is limited by an imaginary line located 200 kilometers off the Cuban coast line. All tropical cyclones that arrive at any point on this line can begin to affect the country in a 24-hour time frame. Consequently, the moment that the tropical cyclone being monitored penetrates into or makes a tangential approach to the limits of this area, protective measures should be rushed to completion in threatened territories within the 24-hour time frame corresponding to the alarm phase.

The Time Approach

The time approach specifies time-based stages of Early Warning, taking into account the space analysis outlined above. It is characterized by a step response with the levels of response being established on the basis of the time until the first effects of the tropical cyclone are expected to be felt in the threatened territory. Under this approach, the "Informative

Phase" commences 72 hours prior the onset of any destructive effects of the hurricane in Cuban territory; the "Alert Phase" commences 48 hours in advance of such impacts; and the "Alarm Phase" commences 24 hours prior to such effects being felt. Each Phase determines a set of actions to be executed to protect the population. These actions are defined in relation to the level of risk for each location and the strength and main features of the tropical cyclone (tropical storm, minor hurricane or major hurricane) as forecast by the National Forecast Center of the Institute of Meteorology. The specific Phases for each territory are established by the Civil Defense taking into account the future track and intensity of the tropical cyclone and the risk level of each area.

As explained previously, the tropical cyclone warnings issued by the National Forecast Center and the informative notes issued by the National Staff of the Civil Defense are relayed by the communication links to local authorities who disseminate the information through the mechanisms already mentioned. The meteorological information is broadcast live by meteorologists through TV channels and radio stations to the whole country. After that, the Civil Defense informative notes are delivered carrying the information and recommendations for the protection of lives and material goods. Both types of information have previously been coordinated between the National Forecast Center and the National Staff of the Civil Defense. The National Forecast Center interacts frequently with the National Staff of the Civil Defense. Both organizations are physically located some 50 meters apart, have high speed computer links, dedicated telephone lines and a special folder in a dedicated local network for all information to be shared. The two organizations discuss the performance of the watch, forecast and warning process immediately after a hurricane has affected the country to search for weaknesses and improve the mechanism for the next threat.

The National Forecast Center has an operational cycle for issuing warnings. This 6-hour cycle begins with the reception of all relevant data from the WMO Global Telecommunications System (GTS) and Cuban weather station network, meteorological radars, satellite imagery and reconnaissance aircraft, as well as numerical model guidance on hurricane track and intensity from Cuban and foreign forecast models. Meteorologists meet two hours prior to the issuance time to discuss all data and forecasts models, to identify the hurricane's position, strength, direction and speed of movement for the advisory issuance time. A draft of the advisory is written one hour before the issuance of the warning message. A half an hour prior to being issued, the text is discussed among meteorologists, including the Director of the Forecast Center who takes the final decision and makes any necessary changes. The Director then meets with the National Staff of the Civil Defense to discuss the results and coordinate the dissemination of the warning, which is delivered to users exactly at the issue time. Meteorologists then go "on air" on national television and radio with latest updates on the hurricane. This procedure is repeated every 6 hours when a hurricane is in the area of reinforced surveillance.

In addition to hurricane forecasts and warnings, the National Forecast Center issues different products such as specialized forecasts for ships and coastal interests, special services to agriculture, tourism, insurance companies, telecommunication companies, etc. All of these products are regarded as special information to address individual requirements of a special group of users. Meteorologists at the National Forecast Center have a plan for training each year that includes seminars, conferences, discussion of past cases, post graduate education, etc. The study of past cases is very important because the group chosen for that task has to discuss what happened and to assess whether the decisions that were made in an actual situation could have been better. The experience and knowledge of forecasters is being increased every year. Meteorologists also take part in meetings with stakeholders to increase their knowledge and perception of meteorological limitations so that they are able to interpret correctly any future warning information.

2.5.2 Warning Message Improvement Cycle

The operation of the Early Warning System is reviewed following each meteorological event, including its effectiveness during the response. The efficacy of the information transmitted is assessed by means of surveys conducted among the affected population. The Cuban Meteorological Service participates in this post-storm process. A review is undertaken of the whole forecast process and the hurricane "best track" is assessed after every storm. A critical review takes place jointly with the Civil Defense so that experience can be documented and taken into account during the next hurricane threat. This assists in improving the effectiveness of the forecast and warning processes as well as the procedures for Early Warnings and Warnings.

2.6 Warning Dissemination Mechanisms

The issue of Early Warning messages by the National Forecast Center commences 120 hours in advance of a possible impact and these are repeated every 24 hours. When a hurricane penetrates inside the area of surveillance in the Caribbean

Sea, updated warnings are then issued every 12 hours. In situations where the hurricane presents a potential threat to Cuban territory within 72 hours or less, warnings are issued every 6 hours. Finally, when the hurricane is very near to the territory of the country, warnings are issued every 3 hours or more frequently.

Radio and television are very important tools for warnings. The country has more than a television set per household and the TV signal reaches 98% of the national territory and almost all of the population. Widespread awareness and interest is generated through frequent live broadcasts by meteorologists, direct from the National Forecast Center that explain, in simple and plain language, the overall meteorological situation including an analysis of the uncertainties and different probabilities that are seen at that time.

Several entities participate in the issuance and distribution process for the Early Warning messages:

- The National Forecast Center of the Institute of Meteorology issues Early Warnings and Warnings on tropical cyclones for the National Staff of the Civil Defense, the Central Government and the public as well, with meteorologists presenting the meteorological information through national and local radio and TV.
- The National Staff of the Civil Defense issues Informative Notes with guidance and recommendations for actions to ensure the protection of lives and material goods. These are distributed to all levels of government as well as to both national and local radio and TV.
- Both national and local television channels and radio stations transmit special broadcasts on a 24-hourly basis. These include reports, interviews with specialists and authorities, information on the evolution of the hurricane and on the protective measures being adopted in each location, along with guidance on measures that should be completed. Commencing 48 hours before the forecast strike of any hurricane, National radio and TV install a remote unit facility at the National Forecast Center and a TV set is activated at the National Staff of the Civil Defense for the same purpose.
- The International Press Center disseminates information for the foreign press agencies and coordinates interviews with forecasters and specialists.
- All information is public and is delivered to end-users by the Meteorological Service, with the exception of some very technical and detailed, model-based, information on rainfall intensity, winds or storm surge that is delivered to the National Civil Defense.

2.7 Emergency Preparedness and Response Activities (National to Local)

2.7.1 Disaster Preparedness and Response Planning and Emergency Response Activation

In Cuba, Disaster Reduction Plans are drafted at all levels from the very basic people's council to the provincial governments and from local to national economic entities and organizations, based on an assessment of the risk at each level. The disaster reduction plans are updated every year based on risk assessments that take into account the vulnerabilities that have been reduced during the period. The results of this updating process are made public in each community as part of preparedness measures prior to the hurricane season. The National Forecast Center participates in the preparation of the preparatory work for the next Hurricane Season. This includes the preparation of the nationwide "Exercise Meteor" carried out every year by the Civil Defense during the month of May, just prior to the hurricane season. In preparation for the beginning of the upcoming hurricane season, joint training and combined work sessions are undertaken on a nation-wide scale with the involvement of the personnel of the Meteorological Service, the Civil Defense National Staff and journalists from different national and provincial media.

2.7.2 Community Response Capacities

Community preparedness is planned and organized by the Civil Defense in every municipality with the assistance of the local social and mass organizations. The Cuban Red Cross and other non-governmental entities also participate. The population's awareness and skills are put to the test every year during the Civil Defense exercise "Meteor". Experiences derived from the strike of a tropical hurricane are constantly analyzed in each community, locality, and province and at the national level, which facilitates improving the existing plans. Personnel from the Provincial Meteorological Centers meet with authorities and people in their communities, providing advice on the warnings issued by the National Forecast Center and tailoring this information to take account of local conditions and vulnerabilities.

2.7.3 Public Awareness and Education

In Cuba, citizen preparedness for disaster situations extends from the highest authorities to the people in their work places, schools and communities. It is aimed at making each participant capable of organizing or carrying out the planned actions according to their responsibility. It is also aimed at making them aware of the risk to which they may be exposed and the measures they must take to protect their lives and their property. Preparedness at the level of the different sectors of society is supported by the legal framework in place for the work of the Civil Defense. Every year, the chief of the National Staff of the Civil Defense issues a new set of Methodological and Organizational Instructions for Preparedness by means of a Resolution. This outlines the ways in which preparedness is to be undertaken by each sector, which rules and procedures to follow, what control to exercise, and other relevant information. The Civil Defense circulates this Resolution to the heads of national institutions and to Civil Defense in the provinces (i.e. provincial government presidents). These individuals use the information to draft their own specific directives and guidance, adapted to their assessment of risks, preparedness priorities and local characteristics.

Civil Defense preparedness is pursued through exercises, drills, training, practical activities and demonstrations, workshops, courses, lectures, and events. Special mention should be made of the yearly, two-day long METEORO (Meteor) National Exercise for Disaster Case Actions, referred to earlier. Usually conducted during a weekend in May, this Exercise assists authorities and the population at large to prepare for disaster situations. It is also used to test the warning, communication and information systems and check the logistics needed for the different protective measures such as the evacuation of people, goods, and economic resources, the vulnerability mitigation actions. The Exercise receives extensive coverage by all communication media (television, radio, newspapers, etc) regarding aspects that the population ought to know for every territory. In addition, authorities and officials at the different levels receive training and the readiness of operation centers and specialized forces is also reviewed (i.e. fire fighters, Red Cross groups, rescue teams, doctors and paramedics, military detachments and construction brigades, etc.). A subject dealing with Civil Defense topics is taught in the national school system during 3rd and 5th grades of elementary education and 8th grade in junior high, complemented by extracurricular activities. Theoretical-practical classes are taught in senior high schools both as part of the curriculum and in extracurricular lessons. During higher education, all career programs include a subject on Civil Defense, in addition to other subjects dealing with related issues. In short, the preparation system for disaster reduction in Cuba is comprehensive, systematic and planned. Furthermore, it is constantly perfected through systematic analysis of its advances and deficiencies with a view to creating a disaster reduction culture that results in fewer and fewer human lives being lost and in reduced damage to the national economy.

A Communication Program for Disaster Situations has been designed for use by the media that contains general guidelines aimed at making the messages and reports more preparedness effective. Every year, mass media staff members attend updating training courses on disaster reduction that contribute to improving their work. Media managers and journalists from different television and radio stations and the newspapers attend the national workshop. Later, the same course is taught at the provincial level with the participation of provincial and municipal media. During the response and recovery phases, all the resources in the territory are used to keep the people informed and oriented. TV and radio units are set up and connected to power plants or solar panels; mobile and fixed public address systems are installed; the radio ham network is activated and radios with multiple sources of power supply are distributed. Furthermore, people to people information transfer also assists through the participation or grassroots organizations, social workers and local leaders. All of this is possible because there are television and radio stations at the municipal and community levels. There is one TV set in each school classroom and there are also VCR players and computers in all schools in the country. In addition, there are public television halls in remote communities. These facilities and initiatives contribute to the effectiveness of Cuba's Early Warning System by facilitating the dissemination of information to the people in a rapid, understandable and reliable way.

The Cuban Meteorological Service plays an important role in public awareness and educational activities. Hundred of conferences and talks are offered each year in work places, factories and in social organizations of different types. Several times a day, meteorologists present weather information through radio and national, provincial and municipal television stations across the whole national territory. In these presentations, they frequently explain different meteorological phenomena that occur in Cuba or other parts of the world, so that the population acquires more and more knowledge and awareness about weather. Through the educational channels of Cuban TV, short courses of 30 lessons each have also been presented by meteorologists on themes such as general meteorology and hurricanes. During hurricane season, meteorologists have also participated in brief spot broadcasts (only some seconds long) transmitted between regular program changes on television and radio, in order to raise awareness of hazardous weather elements, particularly hurricanes.

2.8 Development of Warnings and related Products and Services for Disaster Risk Management

The key users in Cuba's Tropical Cyclone Warning System are the Government (National, province, local), the Civil Defense System and the media (as an interface with the population). The main process which the NMS uses to understand users' needs and requirements for information, products and services is the input provided by major users in the form of reports and suggestions. These act as reliable feedback mechanisms that the NMS utilizes to develop requirements and specifications for NMS products and services jointly with users. Several assessment meetings are held between the NMS and the Civil Defense after every hurricane strike to review all operational work, forecasts and interactions between the two institutions, as well as the actions taken by the Civil Defense. They also include the joint work with the media. New requirements and specifications normally arise from these meetings and are taken into consideration in planning for future events. The main challenges in working with users relate to lack of knowledge by middle and low level Civil Defense officials and general journalists who are not directly linked with the NMS. The NMS attempts to overcome these challenges through the provision of out-of-season and pre-hurricane season training. Users' satisfaction with NMS products and services is assessed through inputs provided by users on their perceptions of positive and negative aspects and their degree of satisfaction with the services. These inputs are normally obtained through opinion polls and letters from users and are always given appropriate consideration.

2.9 Improvement of overall operational Framework of the EWS

The feedback mechanism that the NMS has been utilizing is direct dialogue with users, particularly special users such as the government and Civil Defense and others such as Ministries, the media, etc. Sometimes written suggestions are also received. All suggestions are taken into consideration and they help to improve the forecast and warning service. Congratulatory messages are received from many people and organizations after each hurricane impact as forecasts and warnings are generally successful.

2.10 Examples of previous Events where the operational EWS Led to Improvements in Emergency Preparedness and Prevention

An active period for tropical cyclones in the Atlantic, Caribbean and Gulf of Mexico area began in 1995. As illustrated in Table 2.2, Cuba experienced strikes from 17 tropical cyclones of different intensities during the subsequent 14 years, namely 6 tropical storms and 11 hurricanes. Five of the latter were minor hurricanes and 6 were major hurricanes. Over these 14 years, Cuba had 41 (direct and indirect) casualties from tropical cyclones or an average of 3 casualties per year.

The 2008 hurricane season was one of great activity for Cuba with two tropical storms and three hurricanes striking the country. Two major hurricanes during that year, "Gustav" and "Ike", resulted in extensive damage and, in the case of "Ike", some loss of life while tropical storm "Noel" also caused extensive flooding.

Early warning and preparedness saved lives during Hurricane "Gustav".

Category 4 hurricane "Gustav" struck Cuba in 2008. Warnings were issued several days before the hurricane's arrival and people were well prepared. Though great damage was caused by the hurricane, not a single life was lost.

- The strongest hurricane to hit Cuba during 2008, and also the strongest in the period since 1995, was Category 4 major hurricane "Gustav" that struck the Isle of Youth and the province of Pinar del Rio with winds of near Category 5 strength. The highest recorded wind speed was a gust of 340 km/h measured at the weather station of Paso Real de San Diego (WMO 78317). The National Forecast Center issued Early Warnings several days before the hurricane's arrival and Watches and Warnings were issued 72 hours prior to the hurricane's landfall. The increase in hurricane intensity, due to the high heat content in the Caribbean Sea, was well forecast. Meteorologists explained the main features of this deadly hurricane along with the track and intensity forecasts. People were well prepared and the Civil Defense performed superbly. Material losses were great but not a single life was lost.

Table 2.2 Tropical Cyclones that hit Cuba during the period 1995–2008

Name	Year	Saffir-Simpson Category	Casualties
Lili	1996	H2	0
Georges	1998	H1	6
Irene	1999	TT	2
Michelle	2001	H4	5
Isidore	2002	H1	0
Lili	2002	H2	1
Charley	2004	H3	4
Ivan	2004	H5	0
Dennis	2005	H4	16
Alberto	2006	TT	0
Ernesto	2006	TT	0
Noel	2007	TT	0
Fay	2008	TT	0
Hanna	2008	TT	0
Gustav	2008	H4	0
Ike	2008	H4	7
Paloma	2008	H2	0
TOTAL 1995 – 2008 (14 years)		TT = 6 H1, H2= 5 H3,H4,H5= 6	42/14 = 3

TT = Tropical Storm; H1, H2 = Minor Hurricane (Hurricane Category 1, 2; H3, H4, H5 = Major Hurricane (Hurricane Category. 3, 4, 5))

Hurricane "Ike" underlined the importance of training people how to react in the face of hazards.

Cuba was struck by Category 4 hurricane "Ike" a few days after "Gustav". While timely early warnings were issued there were 7 deaths associated with "Ike". The fact that most of these fatalities resulted from inappropriate actions by the victims reinforced the need for even greater efforts to advise people how to react in the face of warnings of hazards.

- A few days later, Category 4 major hurricane "Ike" arrived. This hurricane differed in that it affected the whole country. The Early Warning and the Watch and Warning process was a very good one. Everything was done to warn people to be prepared for another deadly hurricane. This was the first Category 4 hurricane to make landfall in the provinces of Holguin and Las Tunas. It subsequently continued to the west to affect the whole country and make landfall again in the area previously hit by hurricane "Gustav" a few days before. There were 7 deaths associated with "Ike", the majority of them provoked by negligence on the part of the victims and, in some instances, not directly related to the hurricane. The lesson here is that even greater efforts must be made to warn people not to do things that individuals in affected areas or areas under some type of warning are not supposed to do.

Evacuation in response to warnings of heavy rainfall prevented fatalities during floods caused by Tropical Storm "Noel".

Accurate warnings of intense rains accompanying Tropical Storm "Noel" enabled evacuations of people to be carried out before flooding occurred. No one was killed during the event though there were substantial losses to agriculture and economy.

- Tropical storm "Noel" during the same year resulted in substantial flooding and related damage in Cuba, highlighting that even tropical depressions that do not reach hurricane strength can present a major threat to life and property. Tropical Storm "Noel" had only light winds in its center. Some 300 kilometers to the East, however, huge bands of rain were forming. The warning issued in this case was for intense rains that were forecast to hit Eastern Cuba. Meteorologists explained in great detail that rain was the danger and that the location of the center of the tropical storm was not important. They stressed that the critical feature was the intense heavy rain that was going to arrive well after the center of the tropical storm had crossed Eastern Cuba. Rainfall totals in 24 hours were between 200 and 300 mm over that area. Since the soil was already saturated from heavy rainfall during the previous month, large areas were completely flooded. A massive evacuation was undertaken well before flooding occurred, however, and no one was killed during the event though there were large losses to agriculture and economy.

2.11 Overall Lessons learned and Future Steps for Improving NMHS Contribution in EWS Particularly Focusing on institutional Coordination and Cooperation

Hurricane "Lili" threatened Central and Western Cuba in 1996 when an increase in hurricane activity was taking place in the Atlantic, Caribbean and Gulf of Mexico. The Cuban NMS issued an Early Warning Message for the first time ever and it was successful. At that time, however, the NMS was not prepared to face a hurricane such as "Lili" as the computer and telecommunications infrastructure was still inadequate. Consequently, short wave radio contact was lost with all stations after their huge antennae collapsed one by one. Only by chance, was the NMS able to say that Hurricane "Lili" was making landfall at 11 am on October 18, 1996. Fortunately, everything went well with the forecast after that.

Following Hurricane "Lili", the Cuban NMS reviewed all procedures for transmitting weather data and warnings and the NMS received funding to upgrade and modernize the communications system. This facilitated a modernization of the Meteorological Service that still is underway, though the main actions have been completed several years ago. Radars have been automated, a computer network has been established and many technological improvements have been made, resulting in the modern National Forecast Center that we have today.

List of Acronyms

ANAP	National Association of Small Farmers
CDR	Committees for the Defence of the Revolution
CITMA	Ministry of Science, Technology and the Environment
CLAMED	Latin American Center for Disaster Medicine
CNIS	National Center for Seismologic Research (National Seismologic Service)
EMNDC	National Staff Civil Defense
FKB	Forest Keeper Body
GIS	Geographic Information System
GTS	(WMO) Global Telecommunications System
INSMET	Institute of Meteorology
MOU	Memorandum of Understanding
NFC	National Forecast Center
NIHR	National Institute of Hydraulic Resources
NMC	National Meteorological Center
NMS	National Meteorological Service
NGO	Non-Governmental Organization
RA IV	WMO Regional Association IV
VCR	Video Cassette Recorder
WMO	World Meteorological Organization

Chapter 3
The Bangladesh Cyclone Preparedness Program. A Vital Component of the Nation's Multi-Hazard Early Warning System

Arjumand Habib, Md. Shahidullah and Dilder Ahmed

Abstract Since the 1970 cyclone that killed an estimated 300,000 people and the 1991 cyclone which took over 138,000 lives, significant strides have been made in planning and preparedness that have resulted in greatly reduced losses from subsequent storms in Bangladesh. Central to the Bangladesh Early Warning System is the Cyclone Preparedness Programme (CPP), developed and improved through the efforts of the Government of Bangladesh (GoB), United Nations, International Red Cross and the Bangladesh Red Crescent Society. The CPP utilizes over 42,000 volunteers along with a transceiver telecommunications network to ensure rapid delivery of warnings to the population at risk. The success of the CPP arises from its simplicity, attention to socio-cultural aspects, and its extensive community-based volunteer and communications network that expedite the delivery of cyclone warning messages produced by the Bangladesh Meteorological Department (BMD). Vital warning information from the BMD can be disseminated to responding agencies and the public at-risk in less than 15 minutes from the time of warning issuance to the beginning of evacuation orders. Evacuation of the population at risk to more than 2,000 storm shelters is expedited through a highly organized volunteer network at the village level.

3.1 Overview

Bangladesh is a deltaic land of about 144,000 km^2 in area with the massive Himalayas to the north and the vast Bay of Bengal on the south. It is a South Asian country extending from 20° 45' N to 26° 40' N and from 88° 05' E to 92° 40' E and belongs to the South Asian Association for Regional Cooperation (SAARC). Most parts of the country are characterized by low-lying plains comprised of alluvial soil, with hills in the south-eastern and north-eastern parts. It is a land criss-crossed by 230 rivers, fifty-seven of which originate from outside its borders (India and Myanmar), that eventually flow into the Bay of Bengal (Fig. 3.1). The major rivers are the Ganges (the Padma), the Brahmaputra and the Meghna.

The complex coastline of Bangladesh extends for about 710 km along the continental shelf and its coastal waters are characterized by shallow bathymetry. The Bay of Bengal forms a funnel shape towards the Meghna estuary that significantly enhances storm surges which are known as being some of the highest in the world (Fig. 3.2).

The Government of Bangladesh (GoB) has a strong commitment to the reduction of the human, economic and environmental losses of disasters through enhancing overall disaster management capacity. Efforts have been continuing for optimum coordination and utilization of resources in addition to ensuring community-level involvement so that people are aware of the actions that they can employ in protecting their lives and properties against disasters. The objectives of the GoB in addressing disaster response are to coordinate agency efforts at different stages in the disaster management cycle, including disaster management practices, disaster mitigation, emergency preparedness, emergency response, disaster management mechanism, early recovery and immediate rehabilitation.

Bangladesh is vulnerable to a variety of hydrometeorological hazards.

Most disasters in Bangladesh are caused by hydrometeorological hazards such as cyclones, storm surges, thunderstorms/tornadoes/hailstorms, floods, drought, heat waves, cold waves, land slides, earthquakes and tsunami.

Arjumand Habib, Bangladesh Meteorological Department (BMD)
Md. Shahidullah, Cyclone Preparedness Programme (CPP)
Dilder Ahmed, Disaster Management Bureau (DMB)

M. Golnaraghi ed., *Institutional Partnerships in Multi-Hazard Early Warning Systems*,
DOI 10.1007/978-3-642-25373-7_3,

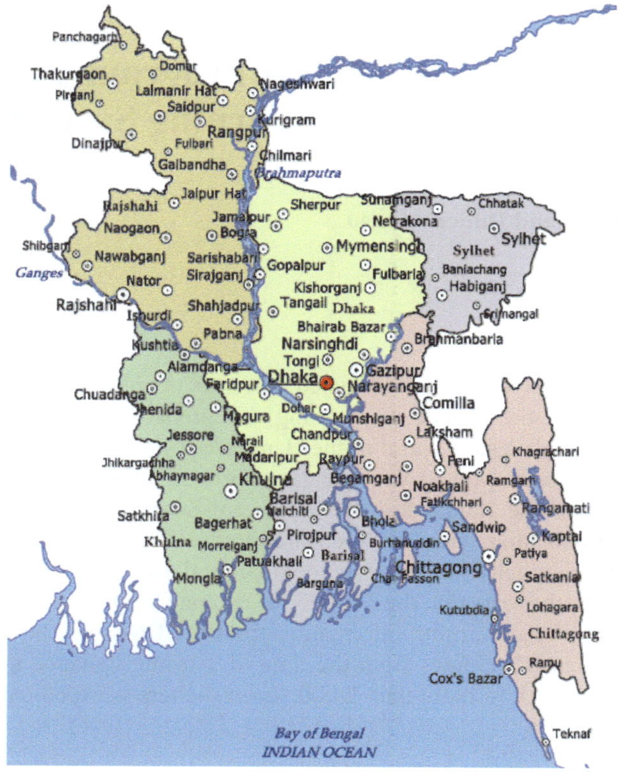

Fig. 3.1 Map of Bangladesh

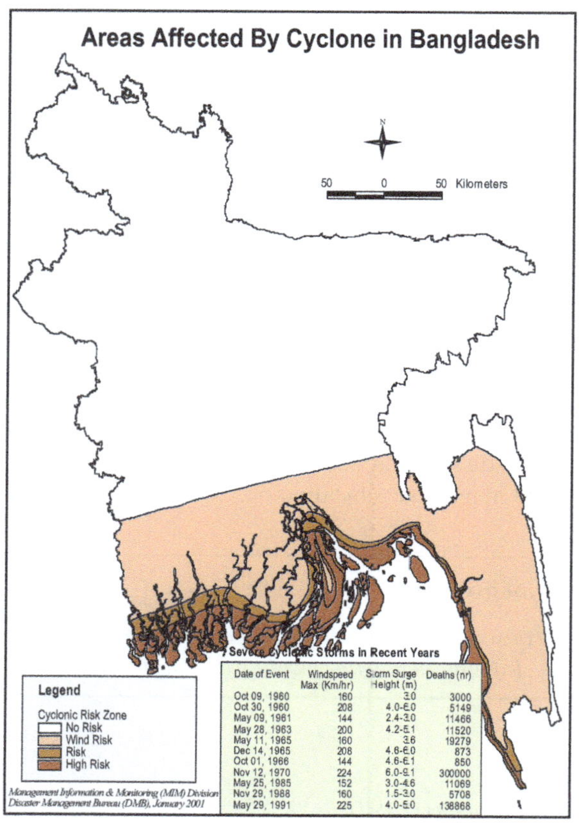

Date of Event	Windspeed Max (Km/hr)	Storm Surge Height (m)	Deaths (nr)
Oct 09, 1960	160	3.0	3000
Oct 30, 1960	208	4.0-6.0	5149
May 09, 1961	144	2.4-3.0	11466
May 28, 1963	200	4.2-5.1	11520
May 11, 1965	160	3.6	19279
Dec 14, 1965	208	4.6-6.0	873
Oct 01, 1966	144	4.6-6.1	850
Nov 12, 1970	224	6.0-9.1	300000
May 25, 1985	152	3.0-4.6	11069
Nov 29, 1988	160	1.5-3.0	5708
May 29, 1991	225	4.0-5.0	138868

Fig. 3.2 Bangladesh's "at risk" coastal regions

Bangladesh is claimed to be the most disaster-prone country in the world. Most of the disasters are meteorological and hydrological in nature, including cyclones and associated storm surges, thunderstorms/tornadoes/hailstorms, floods, droughts, heat waves, cold waves, land slides, earthquakes and tsunamis. The Great Backerganj Cyclone of 1876, the Worst Killer Cyclone of November 1970, the Urichar Cyclone in May 1985, the Killer Cyclone of April 1991, the floods of 1954, 1987, 1988, the Historic Flood of 1998; Demra Tornado of 1969, Manikganj Tornado of 1974, Madaripur Tornado of 1977, Saturia Tornado of 1989, Louhajong Tornado of 1995, and Tangail Tornado of 1996 are a few of the extreme meteorological and hydrological events that have significantly impacted the people of Bangladesh.

3.1.1 Background to the Establishment of EWS in Bangladesh

Disaster Management in Bangladesh had gone through a process of significant reform. After independence, the focus was limited to relief and rehabilitation activities. Following the devastating floods of 1988 and the cyclone of 1991, which created a massive destruction in the economy, the focus has been shifted towards adoption of a holistic approach that embraces processes of hazard identification and mitigation, community preparedness and integrated response efforts. As a result, a short-term project entitled "Assistance to Ministry of Relief in Coordination of Cyclone Rehabilitation: BGD/91/021" was initiated after the killer cyclone of 29 April 1991. In April 1993, the Government of Bangladesh (GoB) established the Disaster Management Bureau (DMB) as the successor to the Disaster Coordination and Monitoring Unit and renaming of the Ministry of Relief and Rehabilitation as the Ministry of Disaster Management and Relief (MDMR). Additionally in 1993, the GoB established Disaster Management Councils and Committees from the national down to the field level.

The Disaster Management Bureau was assigned responsibility to perform specialist support functions, working in close collaboration with District and Upazila level authorities and the concerned line ministries, under the overall authority of a high-level Inter-Ministerial Disaster Management Co-ordination Committee (IMDMCC). The DMB also has the responsibility to create public awareness regarding the severity and risks associated with natural and human-induced hazards and to formulate programmes and projects that will better prepare at-risk communities and public officials to mitigate their consequences. As a technical arm of the Ministry of Food and Disaster management, DMB overviews and coordinates all activities related to disaster management from the national to the grass-roots level. It is also entrusted with maintaining effective liaison with government agencies, donors and non-governmental organizations (NGOs) to ensure maximum cooperation and coordination in all aspects of disaster management.

The main mission of the Government is to bring about a paradigm shift in the disaster management approach from conventional response and relief to a more comprehensive risk reduction culture and to promote food security as an important factor in ensuring the resilience of communities to hazards. In May of 2004, the Ministry of Relief and Rehabilitation was renamed as the Ministry of Food and Disaster Management. The Ministry of Food and Disaster Management has the responsibility for food management and for:

- planning, coordination, monitoring and evaluation of matters relating to relief activities and disaster management;
- coordination among the different organizations for pre-disaster, during-disaster and post-disaster activities including preparation of disaster management guidelines;
- provision of assistance to disaster related ministries and organizations in preparing the action plan on matters related to disaster management;
- formulation of policy and its implementation on matters relating to the preparation, monitoring and evaluation of food assisted projects and programmes;
- coordinating matters relating to the distribution of external food aid and other relief assistance; and
- other relevant matters relating to food, relief and disaster management.

As a continuation of the paradigm shift process, the Comprehensive Disaster Management Programme (CDMP), designed as a long-term programme of the Ministry of Food and Disaster Management with multi-agency involvement, was launched in November, 2003 to optimize the reduction of long-term risk and to strengthen operational capacities for responding to emergencies and disaster situations, including actions to improve recovery from these events.

The idea of a Cyclone Preparedness Programme developed in 1965 when the National Red Cross Society, now the Bangladesh Red Crescent Society (BDRCS), requested the International Federation of Red Cross and Red Crescent Societies (IFRC) to support the establishment of a warning system for the population living in the costal belt. In 1966, the International Federation and the Swedish Red Cross began the implementation of a pilot scheme for Cyclone Preparedness, which consisted of providing warning equipment such as transistor radios, sirens etc. and training the local militia (Ansars). Following the withdrawal of the International Federation as a direct implementation partner in July 1973, the GoB and the Bangladesh Red Crescent Society created a partnership which led to a new programme management structure in a form of joint partner venture. An agreement was signed by both parties spelling out that this programme would be known as the Cyclone Preparedness Programme (CPP) of the Bangladesh Red Crescent Society. Operations would remain, primarily, under the leadership of the Red Crescent Society and there would be increased involvement of local communities. A Policy Committee and an Implementation Board were created to jointly administer and implement the programme. A seven-member Policy Committee, headed by the Minister of the Food and Disaster Management Ministry, was constituted to give policy direction and allocate resources for the programme. The implementation Board is tasked with implementing the policy directives from the Policy Committee. The Director of the Bangladesh Meteorological Department (BMD) is a member of the 15-member Implementation Board.

3.2 Governance and Institutional Arrangements (National to Local Levels)

3.2.1 Policy, institutional and legal Frameworks to support Emergency Planning and Response

Bangladesh has a well-structured framework for disaster risk reduction.

The Bangladesh Government's regulatory framework includes:

- *A Disaster Management Act.*
- *A National Plan for Disaster Management.*
- *A National Disaster Management Policy.*
- *Standing Orders on Disasters (SOD) – guidelines for Government at all Levels.*

The Bangladesh Government's regulatory framework for disaster management provides the relevant legislative, policy and best practice framework under which the activity of Disaster Risk Reduction and Emergency Management is managed and implemented in Bangladesh (Fig. 3.3 for a schematic flow diagram). The framework includes the following components:

3.2.1.1 Disaster Management Act

A Disaster Management Act was enacted with a view to creating the legislative tool under which disaster risk and emergency management actions are undertaken in Bangladesh and providing the legal basis under which activities and actions are identified, undertaken and managed. It also establishes the roles and responsibilities of Ministries, committees and appointments. The objectives of the Act are to:

- Assist communities in mitigation of potential adverse effects of hazardous events;
- Prepare for managing the response to the effects of a disastrous event;
- Assist in effectively responding to and recovering from a disaster or an emergency situation;
- Prepare for and adapt to potential adverse effects of climate change;
- Provide for effective disaster management for Bangladesh;
- Establish an institutional framework for disaster management; and
- Establish risk reduction as a core element of disaster management.

3.2.1.2 National Plan for Disaster Management

The National Plan for Disaster Management is prepared by the Ministry of Food and Disaster Management. This plan points the way to a culture of prevention that should be developed by introducing disaster management in school curricula, including relevant aspects of disaster management in professional courses, enhancing the capacity of disaster managers by better training facilities and by creating massive awareness at all levels. The latter thrust emphasizes involving people at the grassroots, particularly those who are more vulnerable, for better preparedness and response and encouraging community level initiatives for disaster preparedness. Appropriate zoning regulations, design standards, building codes and performance specifications are developed for safe construction. All development schemes in vulnerable areas include a disaster mitigation analysis whereby the feasibility of a project is assessed with respect to the vulnerability of the area. Moreover, disaster mitigation components are built into all development projects and financed under the Plan as part of approved project costs.

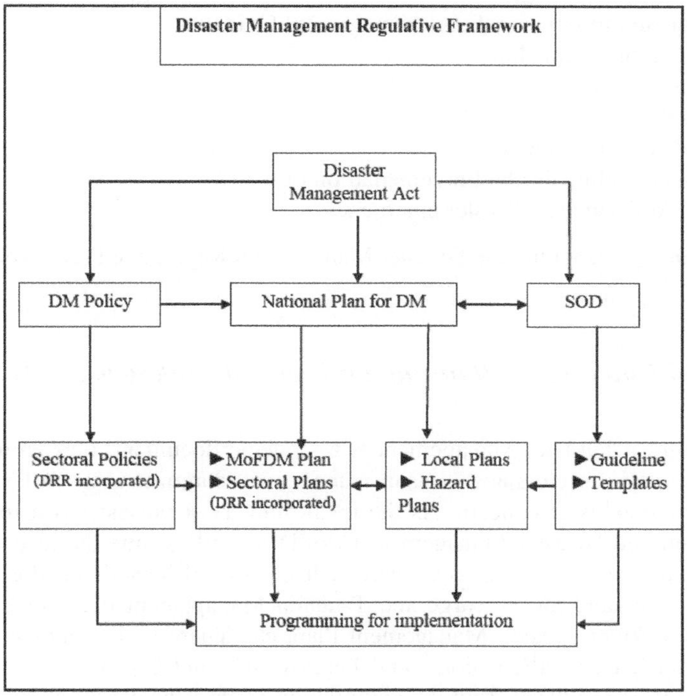

Fig. 3.3 Inter-linkages between various regulative instruments and programming for implementation

3.2.1.3 National Disaster Management Policy

A National Disaster Management Policy has been formulated to define the national perspective on disaster risk reduction and emergency management and to describe the strategic framework and national principles of disaster management in Bangladesh. It is strategic in nature, describing the broad national objectives and strategies in disaster management.

3.2.1.4 Standing Orders on Disasters

Standing Orders on Disaster (SOD) describe, in detail, the roles and responsibilities of committees, Ministries and other organizations involved in disaster risk reduction and emergency management activities and establish the necessary actions required in implementing Bangladesh's Disaster Management Model. An updated SOD is under active consideration by the GoB.

3.2.1.5 Guidelines for Government at all Levels (Best Practice Models)

Guidelines for Government are implemented at all levels, based on best practice models, and are used to assist Ministries, Directorates, Institutions and Divisions under the Ministry, NGOs, disaster management committees and civil society in implementing disaster risk management. The following is a list of Guidelines and Templates prepared to assist in the EWS process:

- Disaster Impact and Risk Assessment Guideline
- Local Disaster Risk Reduction Fund Management Guidelines
- Emergency Fund Management Guidelines
- Indigenous Coping Mechanism Guidebook
- Community Risk Assessment Guidelines
- Damage and Needs Assessment Methodology
- Hazard Specific Risk Assessment Guidelines
- Emergency Response and Information Management Guideline
- Contingency Planning Template
- Sectoral Disaster Risk Reduction Planning Template
- Local Level Planning Template
- National Risk Reduction Fund Management Guideline

- National Disaster Reduction and Emergency Fund Management Guideline
- Local Disaster Management Fund Guide line
- Guideline for road and water safety
- Guideline for industrial safety
- Guideline for Disaster Shelter Management
- Monitoring and Evaluation Guideline for the Implementation of the Plan
- Guideline for international Assistance in disaster emergency

For descriptions of additional plans within the Disaster Management Regulative Framework, please see Appendix 1.

3.2.2 National to Local Emergency Planning and related Linkages to EWS

The Bangladesh National Plan for Disaster Management is a strategic document. It is an umbrella Plan which provides the overall guideline for the relevant sectors and disaster management committees at all levels to prepare and implement specific plans in their respective areas of concern. As the focal ministry for disaster risk reduction and emergency management, the Ministry of Food and Disaster Management (MoFDM) will assume the lead role in disaster risk reduction and emergency management planning. Additionally, there will be several hazard-specific management plans such as a Flood Management Plan, Cyclone and Storm Surge and Tsunami Management Plan, an Earthquake Management Plan, a Drought Management Plan, a River Erosion Management Plan, etc. There will, moreover, be a detailed Disaster Management Plan for each District[17], Upazila[18], Union[19] and Paurashava[20] and City Corporation in the country. A District Disaster Management Plan will be produced from the compilation of the Upazila Disaster Management Plans within the District. Similarly, an Upazila Disaster Management Plan will represent a compilation of the Union Disaster Management Plans of that Upazila, prepared by the Union Disaster Management Committees. The Disaster Management Committees (DMCs) at Union and Paurashava levels will, consequently, be responsible for conducting the risk assessments and preparing the ground-level plans. Once developed these ground-level plans will be sent to the DMCs at the next (higher) level (i.e. Upazila DMCs) whose role will be to verify and compile the Union plans and identify the resource requirements for the Upazila. The disaster management-planning framework in Bangladesh is presented in Fig. 3.4.

The National Plan is a dynamic document, reviewed and evaluated annually to ensure consistency with national initiatives and Government priorities. Key performance indicators are monitored and reported on annually to assess progress with implementation of the framework. The Disaster Management Bureau (DMB) is responsible for carrying out evaluations, in coordination with its stakeholders including the Bangladesh Meteorological Department (BMD). The Inter-Ministerial Disaster Management Coordination Committee (IMDMCC) guides the Ministry of Food and Disaster Management (MoFDM) in monitoring progress with implementation of this Plan at the national level. The MoFDM is responsible for monitoring at the local level through the Disaster Management Bureau.

3.2.3 Organizational Structure for Implementing the Plans

The Ministry of Food and Disaster Management (MoFDM) of the Government of Bangladesh (GoB) has the responsibility for coordinating national disaster management efforts across all agencies. In January 1997, the Ministry issued the Standing Orders on Disaster (SOD) to guide and monitor disaster management activities in Bangladesh. The Standing Orders have been prepared with the objective of making concerned persons understand and accomplish their duties and responsibilities regarding disaster management at all levels. All Ministries, Divisions/Departments and Agencies prepare their own Action Plans with respect to their responsibilities under the Standing Orders, for efficient implementation. The National Disaster Management Council (NDMC) and Inter-Ministerial Disaster Management Coordination Committee (IMDMCC) ensure coordination of disaster related activities at the National level. Coordination at District, Upazila and Union levels is undertaken by the respective District, Upazila and Union Disaster Management Committees. The Disaster Management Bureau renders assistance to them by facilitating the process. At both national and sub-national levels, a series of inter-related institutions (Fig. 3.5) have been created to ensure effective planning and coordination of disaster risk reduction and emergency response management.

[17] Bangladesh is divided into six administrative divisions and further subdivided into 64 Districts.

[18] A District is subdivided into several Upazilas.

[19] An Upazila is subdivided into several Unions; a Union comprises several villages.

[20] Paurashava means the municipality of a small town. The municipality of a city is a City Corporation.

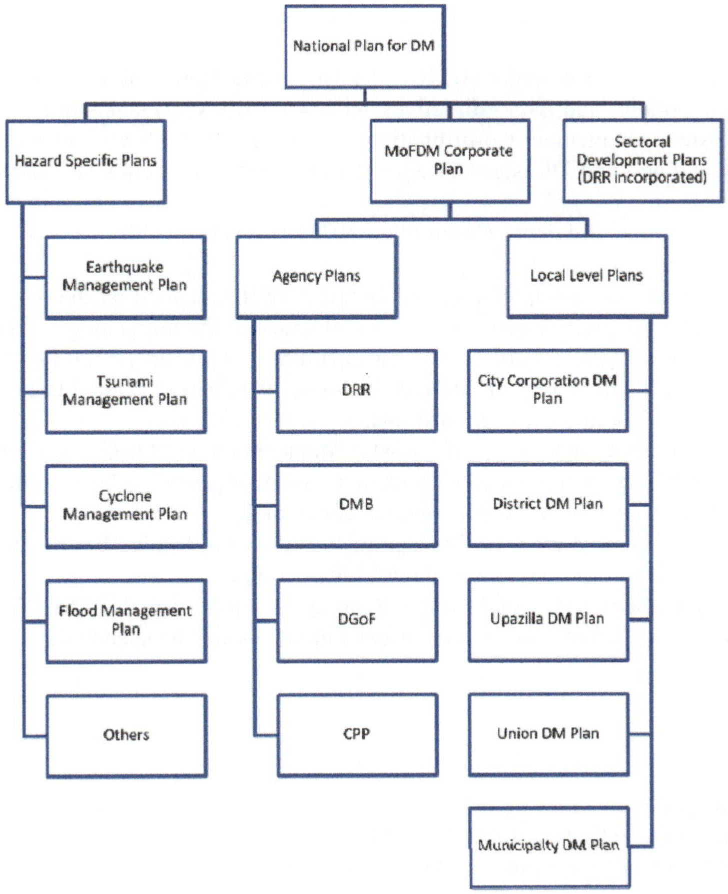

Fig. 3.4 Disaster management planning framework

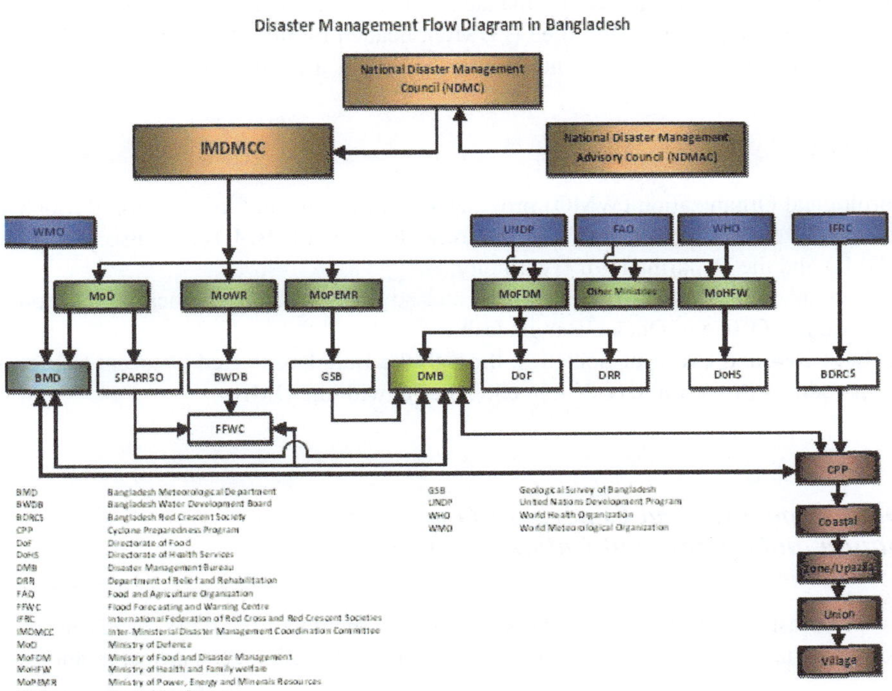

Fig. 3.5 Bangladesh disaster management flow diagram

At the national level:

- The National Disaster Management Council (NDMC), headed by the Honourable Prime Minister, formulates and reviews disaster management policies and issues directives addressing EWS key stakeholder and partner concerns;
- The Inter-Ministerial Disaster Management Co-ordination Committee (IMDMCC), headed by the Honourable Minister in charge of the Ministry of Food and Disaster Management (MoFDM), implements disaster management policies and decisions of the NDMC;
- The National Disaster Management Advisory Committee (NDMAC) is headed by an experienced person nominated by the Honourable Prime Minister;
- The Cyclone Preparedness Programme Implementation Board (CPPIB), headed by the Secretary, Ministry of Food and Disaster Management, reviews the preparedness activities in the face of the initial stage of an impending cyclone;
- The Disaster Management Training and Public Awareness Building Task Force (DMTATF), headed by the Director General of the Disaster Management Bureau (DMB), co-ordinates the disaster related training and public awareness activities of the Government, NGOs and other organizations;
- The Focal Point Operation Coordination Group of Disaster Management (FPOCG), headed by the Director General of DMB, reviews and co-ordinates the activities of various departments/agencies related to disaster management and also reviews the Contingency Plan prepared by the departments concerned;
- The NGO Coordination Committee on Disaster Management (NGOCC), headed by the Director General of DMB, reviews and co-ordinates the activities of concerned NGOs in the country;
- The Committee for Speedy Dissemination of Disaster Related Warning/ Signals (CSDDWS), headed by the Director General of DMB, examines, investigates and ensures ways and means for the speedy dissemination of warnings/signals among the people.

At sub-national levels:

- The District Disaster Management Committee (DDMC), headed by the Deputy Commissioner (DC), co-ordinates and reviews the disaster management activities at the District level;
- The Upazila Disaster Management Committee (UZDMC), headed by the Upazila Nirbahi Officer (UNO)[21], co-ordinates and reviews the disaster management activities at the Upazila level;
- The Union Disaster Management Committee (UDMC), headed by the Chairman of the Union Parishad[22] co-ordinates, reviews and implements the disaster management activities of the Union concerned;
- The Pourashava Disaster Management Committee (PDMC), headed by the Chairman of the Pourashava (municipality), co-ordinates, reviews and implements the disaster management activities within its area of jurisdiction;
- The Pourashava Disaster Management Committee (PDMC), headed by the Mayor of a City Corporation, co-ordinates, reviews and implements the disaster management activities within its area of jurisdiction.

At the international level:

- The World Meteorological Organization (WMO) provides real-time data and information through the Global Telecommunications System (GTS) to the Bangladesh Meteorology Department (BMD) and also provides expertise and guidance materials to BMD and the Department of Hydrology;
- The World Health Organization (WHO) provides help and advice to the governmental and non-governmental health sectors, through its Country Office in Dhaka, Bangladesh;
- The International Red Cross and Crescent Societies, through the Bangladesh Red Crescent Society (BDRCS), administer the Cyclone Preparedness Programme (CPP), in cooperation with the DMB.

3.2.4 Institutional Capacities and Concept of Operations (Coordination and operational Collaboration)

The Standing Orders on Disaster (SOD) document all stakeholders, including the CPP, clarifying their duties and responsibilities regarding disaster management at all levels. All Ministries, Divisions/Departments and Agencies prepare their

[21] Officer who administers the Upazila for the central government.

[22] A Parishad is a Council or Association.

own Action Plans in relation to their responsibilities under the Standing Orders, for efficient implementation. The National Disaster Management Council (NDMC) and Inter-Ministerial Disaster Management Coordination Committee (IMDMCC) ensure coordination of disaster related activities at the National level. Coordination at District, Thana[23] and Union levels is undertaken by the respective District, Upazila and Union Disaster Management Committees. The Disaster Management Bureau renders assistance to them by facilitating the process.

3.2.5 Key Factors considered for Sustainability of different Components of Early Warning Systems with a Multi-Hazard Approach

The national budget allocates funds to government institutions to support their roles in disaster management. These institutions include:

- The Ministry of Food and Disaster Management
- The Disaster Management Bureau
- The Directorate of Relief and Rehabilitation
- The Directorate of Food

Many other Government agencies also contribute to disaster management. In emergencies, these include, principally, the Armed Forces Division, Bangladesh Police, Bangladesh Ansars, Fire Service and the Civil Defense Directorate. In addition a National NGO Coordination Committee on Disaster Management, chaired by the Director General of the Disaster Management Bureau, provides a mechanism for coordination of Government and NGO activities.

The salaries of officers and staff of the Bangladesh Meteorological Department (BMD) and other EWS along with the costs of routine operation and maintenance of the establishments are met directly from the regular budget of the Government of Bangladesh. In addition, the costs of upgrading and improvement of the existing MHEWS and of overseas training are, mostly, covered by WMO, UNDP, JICA, KOICA, CMA etc. as donations through VCP, grant aid and counterpart training. The National Meteorological and Hydrological Service (NMHS) has very limited scope to allocate extra budget for human resources development. Regular professional training courses are conducted using a very small portion of the annual budget (less than 0.5%). All expert training is conducted in foreign countries with the help of international organizations and agencies like WMO, UNESCO, UNDP, ADPC, JICA, KOICA, CMA etc. or of other countries. Some of the training and advocacy programmes in the organizations that are directly involved with disaster management and preparedness are also conducted using foreign aid for human resource development (see Appendix II for details). No systematic mechanism is in place to analyze the costs and benefits in multi-hazard early warning systems. The NMHS is the main source of observational hazard information for impact analysis, however, and the observational data on disasters are always utilized to improve the products and services delivered by the MHEWS, working in conjunction with DMB and other disaster managers.

The Ministry of Food and Disaster Management has established special funds for disaster response and recovery, risk reduction, sectoral planning and disaster management. The following provides additional details regarding these funds.

The National Disaster Response and Recovery Fund

The Government has established a fund known as the National Disaster Response and Recovery Fund from its own resources and donations from home and abroad. This fund is to be used for response, relief and recovery with the allocation and utilization of the fund being governed by rules and guidelines laid down by the Government. The Ministry of Food and Disaster Management, in consultation with Ministry of Finance, takes the initiative to establish the fund by consolidating the existing relief funds.

The National Risk Reduction Fund

The Government has established a fund known as the National Risk Reduction Fund for projects which are designed for the purpose of prevention, mitigation and preparedness. The allocation and utilization of the fund is governed as per rules and guidelines laid down by the Government. The Ministry of Food and Disaster Management, in consultation with Ministry of Finance, takes the initiative to establish the fund by consolidating existing risk reduction funds.

[23] Administrative unit similar to Upazila.

Financing Sectoral Plans

Relevant Ministries/Divisions/Directorates and departments make provisions in their annual budget to fund the activities and programmes set out in the Disaster Risk Reduction component of their Sectoral Development Plans.

District/Upazila/Union/Paurashava/City Corporation Disaster Management Fund

Disaster Management Committees at the District, Upazila, Union, City Corporation and Paurashava levels are responsible for arranging a Disaster Management Fund to implement programmes and activities outlined in the Disaster Management Plans for their respective areas. The expected sources of funding for this mechanism are:

- Funding from the Government;
- Funding from local government; and
- Local donations.

These funds are operated under guidelines formulated by the Government.

3.3 Utilization of Risk Information in Emergency Planning and Warnings

3.3.1 Organizational Responsibilities and Arrangements for the Development of Risk Information

The National Disaster Management Council (NDMC) understands the utility of risk information in the EWS warning process and is developing the risk information capabilities necessary for supporting EWS activities. In the interim, NDMC and its partner agencies are utilizing risk information prepared by external agencies that assist NDMC in this process.

3.3.2 Hazard Assessment, Quantification and Mapping (National to Local)

The NDMC uses all existing risk information in quantifying risk when a hazardous event is likely, through the sharing of risk information and mapping with EWS partners and responsible agencies.

3.3.3 Assessment of Vulnerabilities and Exposure (National to Local)

The NDMC continues to collaborate with its external partners in the assessment of vulnerabilities and exposure, allowing for a better understanding of these elements when threatened. These assessments are shared with all relevant responding and responsible agencies.

3.3.4 Storage and Accessibility of Disaster and national Hazard Risk Information

Both the Bangladesh Meteorology Department and the Bangladesh Water Development Board maintain databases of hydrometeorological and climatological data and information. These data and information are regularly sent (at least yearly) to the Bangladesh Bureau of Statistics, which is mandated by the government to maintain consolidated national data, including population and vulnerability, hydrometeorological and climate data and information.

3.3.5 Development and Utilization of Hazard/Risk Information to support Emergency Planning and Warnings

The NDMC, through its collaboration with external supporting agencies, has developed an interim process addressing how best to utilize hazard/risk information when planning and preparing its response and recovery strategy and SODs.

3.4 Hazard Monitoring, Forecasting, and Mandates for Warning Development

3.4.1 Organizational Responsibilities for Monitoring, Forecasting and Development of Hazard Warnings

Table 3.1 shows the major hazards that affect Bangladesh, ranked by historical figures of casualties and economic losses attributed to the respective hazards, and identifies the national agency mandated to issue warnings and the hazard Type. For Type I hazards, the identified agency has the exclusive mandate for development of warnings for the hazard. In the case of Type II hazards, the identified agencies have a joint mandate for the development of the warnings. Where hazards are classified as Type III, the identified agencies provide information to other agencies that have the mandate for the development of the warning.

Table 3.1 Major hazards that affect Bangladesh and the national agency mandated to issue warnings

Hazard Rank	Hazard	National Agency for Mandate	Hazard Type	Remarks
1	Cyclones	BMD	I	
2	Storm surge	BMD	I	
3	Thunderstorm (Nor'wester), Lightning	BMD	I	
4	Tornado	BMD	I	
5	Hailstorm	BMD	I	
6	River flooding	FFWC (BWDB), BMD	III	
7	Flash flood	FFWC (BWDB), BMD	III	
8	Coastal flooding (due to storm surge/tsunami)	BMD	I	
9	Drought	BMD, DAE	II	
10	Heat Wave	BMD	I	
11	Cold Wave	BMD	I	
12	Dense Fog	BMD	I	
13	Landslide/Mudslide (due to heavy rain)	BMD	I	
14	Earthquake	BMD		BMD monitors earthquakes and issues reports to government and public
15	Tsunami	BMD	III	Tsunami Watch Information (TWI) Bulletins are received from PTWC and JMA
16	Turbulance/Icing	BMD	I	
17	Strong winds	BMD	I	
18	Wind driven surge	BMD	I	
19	Air pollution	DoE	II	
20	Waterborne hazards	DPHE	II	
21	River Erosion	BWDB	I	

(BMD = Bangladesh Meteorological Department; BWDB = Bangladesh Water Development Board; DAE = Department of Agriculture Extension; DoE = Department of Environment; FFWC = Flood Forecasting and Warning Centre; JMA = Japan Meteorological Agency; PTWC = Pacific Tsunami Warning Center; DPHE = Department of Public Health Engineering)

3.4.2 Organizational Collaboration and Coordination for Development of Hazard Warnings

The Bangladesh Meteorology Department collaborates and coordinates closely with all relevant stakeholders that are involved with disaster risk management. This collaboration and coordination is mandated by the Standing Orders on Disaster Management (SOD). Excerpts of the SOD that pertain to tropical cyclones are found in Appendix 3.

3.5 Development of Understandable, Authoritative, Recognizable and Timely Warnings

3.5.1 *Warning Message Development Cycle*

Warning messages and maritime signals are prepared by operational forecasters at the Storm Warning Centre (SWC), a specialized unit of the BMD. Figure 3.6 illustrates BMD's organization.

Thresholds for warning and maritime signals are based on the intensity and impact of the hazard. The lead times for the warning messages are given in Table 3.2:

Table 3.2 Hazard warning products and their issuance times

Warning products		Issued at or before			
		As needed	24 hrs	18 hrs	10 hrs
Cyclone	Alert	X			
	Warnings		X		
	Danger			X	
	Great Danger				X
Storm Surge					X
Heavy Rainfall			X		
Inland River port		X			
Thunderstorms/Squalls		X			
Heat Wave & Cold Wave		X			
Fog			X		

In the previous signal system, two types of signals are used for maritime seaports and inland river ports, respectively. Eleven individual signals were used for different stages of disaster for maritime ports while four separate signals were used for inland river ports. When monitoring the same cyclone formed in the Bay of Bengal, different signals have, as a result, been used for maritime ports and for those on inland rivers. This has created confusion among the general population, maritime port authorities and inland river port authorities, as well as among those associated with disaster management activities. A new warning signal system has been approved by the Government and is currently being implemented to address these issues. The new signal system thresholds are shown in Tables 3.3 and 3.4. In 2009, the number of maritime signals has been reduced to eight and the number of signals used for inland river ports has been increased to six (starting from three), to avoid confusion and to unify the maritime (Table 3.3) and inland river ports (Table 3.4) signal systems. The comparison of these signal systems is given below:

Table 3.3 Signal system for maritime ports, with thresholds

Sl. No.	Signals	Explanation
1	Distant Cautionary Signal Number – I	There is a region of squally weather in which a storm may be forming (well marked low or depression) with surface winds up to) 61 km/h. (33 knots))
2	Distant Warning signal number – II	A storm has formed (cyclonic storm with surface winds 62–87 km/h. (34–47 knots))
3	Local Cautionary signal number – III	The port is threatened by squally weather (cyclonic circulation with surface winds 40–50 km/h. (22–27 knots)) or squalls due to Nor'westers)
4	Local Warning signal number – IV	The port is threatened by a storm, but it does not appear that the danger is as yet sufficiently great to justify extreme measures of precaution (cyclonic circulation) with surface winds 51–61 km/h (28–33 knots))
5	Danger Signal – VI	The port will experience severe weather from a cyclonic storm of moderate intensity (cyclonic storm with surface winds 62–88 km/h (34–47 knots))
6	Great Danger Signal – VIII	The port will experience severe weather from a storm of great intensity (Severe Cyclonic Storm with surface winds 89–117 km/h (48–63 knots))
7	Great Danger Signal – IX	The port will experience severe weather from a storm of very great intensity (Severe Cyclonic Storm with a core of Hurricane winds with surface winds 118–170 km/h (64–119 knots)
8	Great Danger Signal – X	The port will experience severe weather from a storm of very great intensity (Severe Cyclonic Storm with a core of Hurricane winds with surface winds 171 km/h and above (120 knots and above))

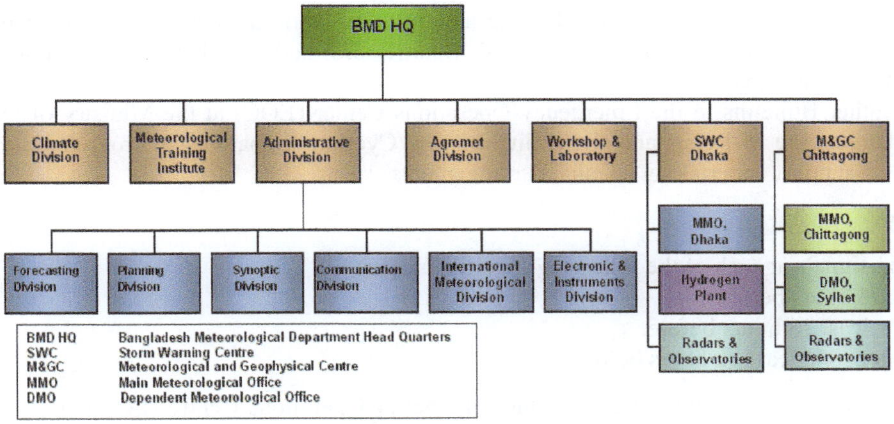

Fig. 3.6 Organizational Structure of the Bangladesh Meteorological Department

Table 3.4 Signal system for inland river ports, with thresholds

Sl. No.	Signals	Explanation
1	Local Cautionary signal number – III	Your area is threatened by squally winds of transient nature (Nor'wester squall of wind speed 40–50 km/h (22–27 knots)). Look out for further development
2	Local Warning signal number – IV	A storm (of depression intensity, associated sustained winds 51–61 km/h (28–33 knots)) or Nor'wester squalls (of wind speed 51–61 km/h (28–33 knots) is likely to strike you (vessels of length 65 feet or less are to seek shelter immediately
3	Danger Signal – VI	A storm of moderate Intensity or Nor'wester squalls, associated sustained winds 62–88 km/h (34–47 knots) may strike you. All vessels are to seek shelter immediately and keep in shelter till further notice.
4	Great Danger Signal – VIII	A violent storm or Nor'wester, associated sustained wind 89–117 km/h (48–63 knots) may strike you. All marine vessels have to keep in shelter till further notice.
5	Great Danger Signal – IX	A very severe cyclonic storm with very high intensity with sustained wind 118–170 km/h (64–119 knots) may strike you. All marine vessels are to be in shelter till further notice.
6	Great Danger Signal – X	A very severe cyclonic storm with the intensity of super cyclone with sustained wind of 171 km/h or more (120 knots or more) may strike you. All marine vessels are to be in shelter till further notice.

3.5.2 Warning Message Improvement Cycle

The warning system is undergoing a continuous process of improvement. There has been an evolution in the development of warning messages based on experience gained over the years. Present-day warning bulletins are very much different from those in the seventies. The advisory component is more elaborate than was the case previously, although the main features remain the same, depicting the nature of the cyclone and including signals.

3.6 Warning Dissemination Mechanisms

There are two stages to the Bangladesh warning system, as follows:

The Alert Stage

- Issue as soon as possible the cyclone alert warning signals, at least 36 hours ahead of formation of a depression in the Bay of Bengal;
- Supply information, through fax/telephone/teleprinter, to the Cyclone Preparedness Programme (CPP) regarding the formation of a depression in Bay of Bengal so as to allow the CPP to take appropriate actions including dissemination of information to all concerned;
- Issue warning signals code 'Whirlwind' to all concerned officials through telephone, teleprinter, telegram, fax, email etc;
- Prepare and submit a Special Weather Bulletin and broadcast/publicize the same through national news media, such as all stations of Radio and Television and national newspapers, for the benefit of the general population. In the case of

Local Cautionary Signal No. 3, arrange for adequate and full-time coordination between the Storm Warning Centre (SWC) of the Meteorological Department, Bangladesh Betar(Radio) and Bangladesh Television for publicity beyond normal broadcasting hours; and

- Send Special Weather Bulletins to the Emergency Operations Center (EOC) at the Ministry of Disaster Management and Relief, the Directorate of Relief and Rehabilitation, the Cyclone Preparedness Programme and Bangladesh Red Crescent Society to support undertaking adequate arrangements.

The Warning Stage

- Publicize warning signals at each of the following specified stages:
  - **Warning** – 24 hours before.
  - **Danger** – At least 18 hours before.
  - **Great Danger** – At least 10 hours before.

The same warning signals are to be repeated to the Emergency Operations Centre (EOC) at the Ministry of Disaster Management and Relief, the Control Room of the Disaster Management Bureau, the Directorate of Relief and Rehabilitation, the Cyclone Preparedness Programme and the Bangladesh Red Crescent Society. The following information should be mentioned in the signals to be disseminated:

- The position of the storm centre;
- The velocity and direction of the storm;
- Mention of the thanas (subdistricts) of the districts likely to be affected, if possible;
- The appropriate time of commencement of gale force winds at different places (Velocity above 32 miles/hour or 51.84 km/hour).

Danger signals and messages are sent to the addressees by fax/telephone/teleprinter or email/Internet, under the code name "Hurricane". In the case of Great Danger signals, messages should be sent to the addressees by fax/telephone/teleprinter or email/Internet under the code name "Typhoon". For the purpose of Inland Water Transportation, appropriate separate messages are sent by fax/telephone/teleprinter or telegraph to addressees mentioned therein under the code name "Water ways and Authority". Alert/Warning signals should be sent to the authorities concerned (Fig. 3.7) for public broadcast through all stations of Bangladesh Betar (Radio) and Bangladesh Television.

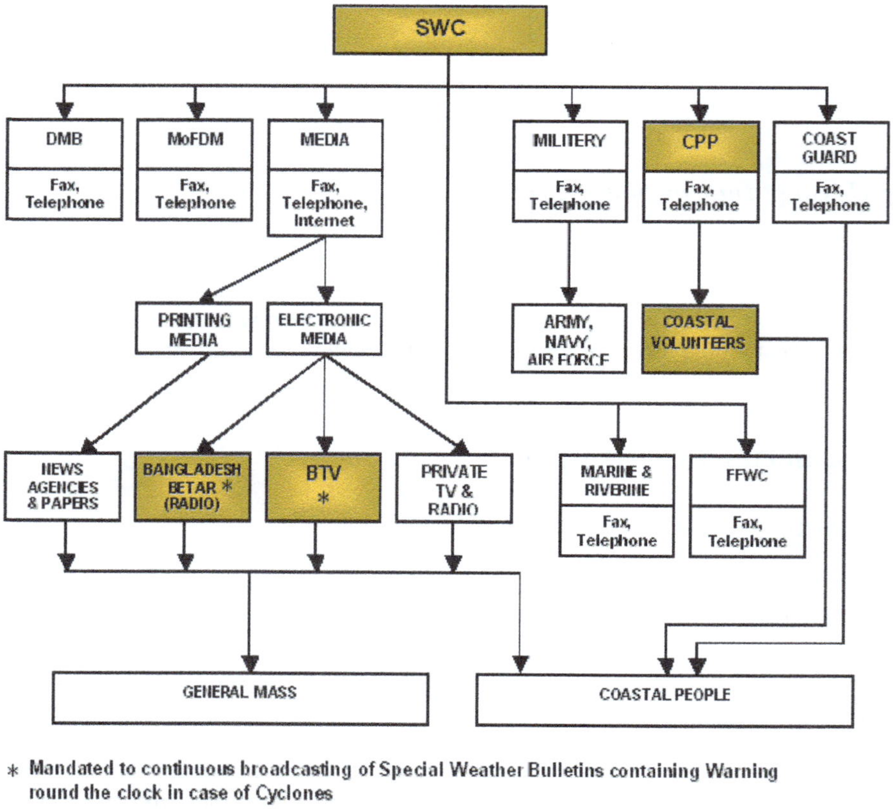

Fig. 3.7 Warning message/bulletin distribution flow chart

In the event of a tropical cyclone, the CPP receives cyclone warning signals from the Storm Warning Centre (SWC) of BMD as soon as a depression has formed in the Bay of Bengal. The information is transmitted to the six Zonal offices over HF radio. The Assistant Directors in turn pass it on to Unions through VHF radio. Where VHF radio has not yet been installed, messengers pass on the message (Fig. 3.8). The Union Team Leaders contact the Unit Team Leaders immediately (Fig. 3.9).

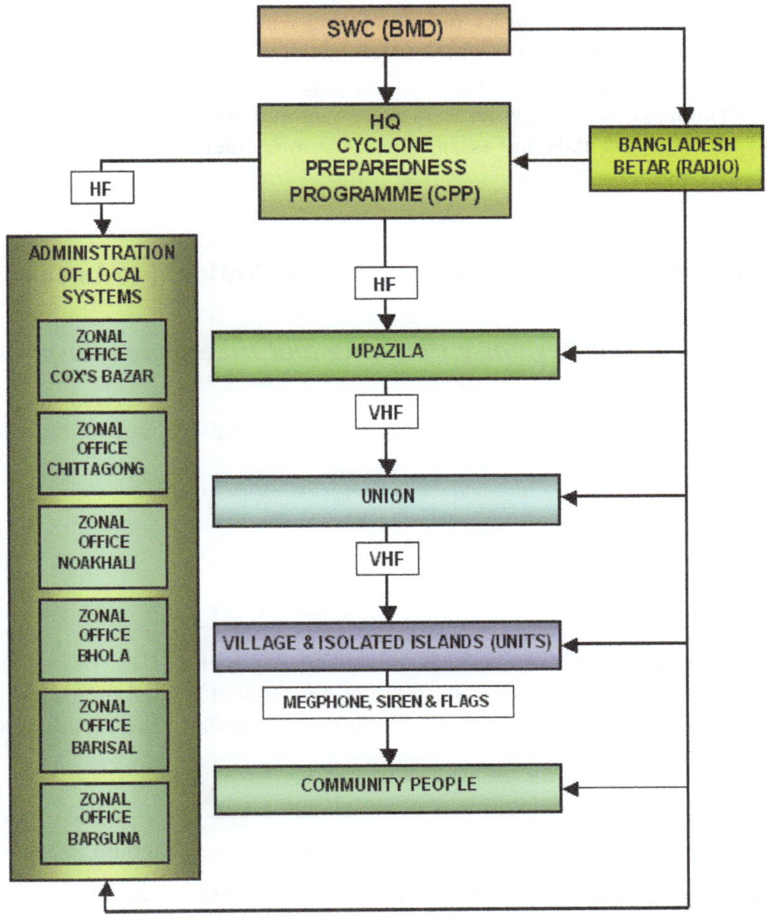

Fig. 3.8 Dissemination of warnings and signals from the Storm Warning Centre (SWC) of BMD through the CPP to the community people

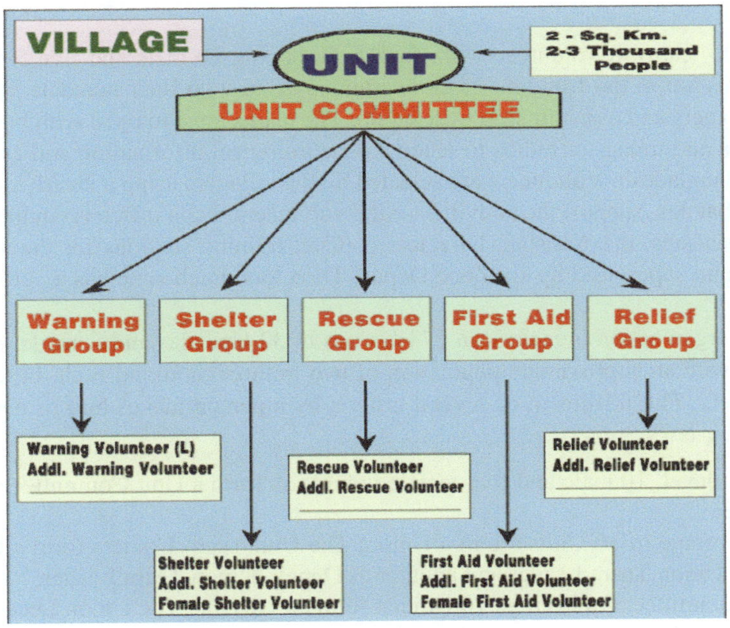

Fig. 3.9 CPP organization from Unit level

The Unit Team Leaders, with their volunteers, spread out in the villages and disseminate the cyclone warnings almost door-to-door, using megaphones, hand sirens and public address systems. The Team Leaders, at the same time, keep track of the approaching cyclone by listening to national radio broadcasts over transistor radios. The Team Leaders are, thus, alerted and start work without losing time. The volunteers keep on announcing the special weather bulletins on the characteristics of the approaching cyclone, as per their action plan.

The Government of Bangladesh passes the order for evacuation when the situation turns serious. The volunteers implement the order and advise and assist people in seeking safety in cyclone shelters or other available safe places. The entire process of communicating a warning issued by the BMD through all levels of the CPP communications network takes about 15 minutes. After the cyclone is over, the volunteers rescue injured and marooned people, provide first aid to the injured, send serious cases to the local hospitals and assist in post-cyclone emergency relief operations.

3.7 Emergency Preparedness and Response Activities (National to Local)

3.7.1 Disaster Preparedness and Response Planning and Emergency Response Activation

In Bangladesh, the planning for disaster preparedness and response is completed according to the disaster management-planning framework mentioned earlier.

3.7.2 Community Response Capacities

The Cyclone Preparedness Programme has a total of 159 full-time personnel and 42,705 trained (28,450 and 14,255 female) volunteers. The Programme office in Dhaka is staffed by 31 personnel headed by the Programme Director with the main function of managing the programme. At field level, the implementation of the Programme is divided into four levels (i.e. Units, Unions, Upazilas and Zones).

> Volunteers make a vital contribution to the Bangladesh Cyclone Preparedness Programme.
>
> *The Cyclone Preparedness Programme utilizes over 42,000 volunteers for dissemination of warning messages and community response. They receive training on response plans, disaster impacts and first aid.*

At unit level there are 2,845 teams of volunteers with each team having 10 male and five female members, headed by a team leader. The Unit teams are in the frontline of the warning system with their mandate being to disseminate cyclone warning signals among villagers and assist in their evacuation. The teams are equipped with basic warning equipment such as hand sirens, megaphones and transistor radios to receive meteorological information and cyclone warning signal bulletins transmitted by Radio Bangladesh. Volunteers are selected by the villagers using a clearly defined a set of criteria. Full-time offices, based in 32 Upazilas, support the activities of the volunteers. Each office is equipped with a transceiver radio. Between the two cyclone seasons, the Assistant Directors conduct training sessions for the volunteers. The work of the Upazila offices is assisted and supervised by six Zone Deputy Directors. Each zone has a permanent office situated at the District level.

The CPP is a volunteer organization organized in 274 Unions of 32 Upazilas and is made up of 2,845 Units. Each unit serves one or two villages with an approximate population of two to three thousand each. Fifteen volunteers are recruited for each Unit by the villagers. The fulfilment of several criteria by a person makes him or her eligible to be a volunteer, according to the volunteer by-laws.

- Unit Committee: At Unit level, 10 male and five female volunteers form a Unit Committee and elect one of their members as Unit Team Leader.
- Union Committee: An average of 10 Units are in a Union. The Unit Team Leaders form a Union Committee and elect one of themselves as the Union Team Leader and another as Deputy Union Team Leader.
- Upazila (Sub-district) Committee: On average 10 Unions form an Upazila. The Union Leaders within the Upazila form the Upazila Committee and elect one member as Upazila Team Leader and another as Deputy Upazila Team Leader.

3.7.3 *Public Awareness and Education*

Public awareness is an integral and very important part of the Cyclone Preparedness Programme (CPP), for its successful implementation. Keeping this in mind, the CPP undertakes the following public awareness activities in the cyclone-prone coastal areas:

- Public Awareness through Volunteers – Local volunteers, through their social contacts, motivate village people in a continuous process. This is an important subject for volunteer training.
- Cyclone Drills and Demonstrations – Cyclone drills and demonstrations are held frequently for volunteers, with local people also participating. These drills and demonstrations are attended by large numbers of villagers.
- Film/video Shows – Documentary films/videos on disaster and preparedness are shown in costal villages, in cooperation with the BMD and the Films and Publications Department of the Government.
- Publicity Campaigns – Immediately prior to the cyclone season, a publicity campaign is organized with government officials, NGO's and local public to familiarize people and create awareness of cyclone preparedness.
- Radio and Television – Arrangements are made with national TV to telecast films relating to disaster preparedness prior to the cyclone season. Special features on disasters and preparedness are also broadcast over Bangladesh Betar (Radio).
- Posters, Leaflets, and Booklets – Posters, leaflets, booklets etc. dealing with preparedness are regularly distributed to coastal inhabitants. Special directives on evacuation, the interpretation of warning signals, instructions for the fishermen etc. have been published in the form of posters and leaflets and are distributed to community people.
- Staging of Dramas – A specially written drama depicting the significance of preparedness is staged in coastal areas to make the village people realize the importance of preparedness. To date, more than 600,000 people have witnessed this drama.

CPP volunteers are given training on the Red Cross and Red Crescent movement, cyclones and their behaviour, warning signals and their dissemination, evacuation, sheltering and rescue, first aid and relief operations, gender issues, humanitarian values and climate change issues, in order to maintain a high level of efficiency. On recruitment, CPP officers give preliminary training to the volunteers at the local level. A three-day basic training course is then given to the volunteers, batch by batch, on the subjects mentioned above. The Team Leaders undergo a five-day training course on the above subjects, as well as leadership training at Upazila or District level.

The officers and Upazila Team Leaders who are engaged in the training of volunteers are given disaster management and leadership training in addition to their normal duties. The training programme is designed to broaden their technical knowledge about disasters in general and cyclones in particular. It aims to enhance their familiarity with all aspects of the CPP, to familiarize them with the principles of disaster management and their application in the CPP, to promote their leadership abilities and to create better understanding of their roles and responsibilities and the need for coordinating efforts with other agencies and local communities. The Bangladesh Red Crescent Society First Aid instructors provide first aid training to First Aid volunteers.

3.8 Development of Warnings and Related Products and Services for Disaster Risk Management User Community

The key users of the products of the MHEWS are different Government organizations, NGO's and community people at risk. The government level recipients who are responsible for major decision making are the Prime Ministers Office, the Ministry of Defence (MoD), the Ministry of Establishment, the Ministry of Water Resources (MoWR), the Ministry of Power Energy and Mineral Resources (MoPEMR), the Ministry of Agriculture (MoA), the Ministry of Health and Family Welfare (MoHFW), the Ministry of Food and Disaster Management (MoFDM) and some other ministries. Among these ministries, the key users in terms of disaster and food security management are MoFDM, MoWR, MoA, MoHFW and MoPEMR. As mentioned earlier, MoFDM is the focal point of the government for all sorts of natural calamities and disaster management. It executes all disaster management activities through the DMB. For cyclone preparedness, a GoB and BDRCS collaboration programme called the Cyclone Preparedness Programme (CPP) is the vital and most effective organization and the key user of cyclone warnings for onward dissemination to threatened communities. Bangladesh Television (BTV), Bangladesh Betar (radio), both state owned, and BSS (the Govt. news agency) are the main recipients of warnings for rapid transmission to the general public. Other private electronic and print media also play an important role in this regard.

The MHEWS delivers products and services generated on the basis of the prevailing weather condition in accordance with the procedure laid down in the SOD. The SOD is updated on a regular basis to meet demands, needs and new requirements arising from the experience with earlier disastrous situations. Moreover, if any GO, NGO or other organization asks for any disaster related product or information for the preparation of their own guidelines, they are provided with same on a priority basis. The MHEWS develops requirements and specifications for certain products or services jointly with the focal points of the related organizations and in conjunction with the DMB.

In Bangladesh, the MHEWS encounters challenges from the users mainly regarding the types and the quality of the products. Today, the demand is increasing for more specific information and warnings in terms of time and space. The current status of the MHEWS is such that it has the capability of generating conventional subjective types of products but operational utilization of objective methods is not in place at present. There is a lack of appreciation by the general public of the existing limitations in forecasting and, at times, this tendency is also seen with the policy makers in that their expectations are too high. Following a major event, a team of experts visits affected areas to obtain first hand information from the local authorities and communities for the assessment of the effectiveness of the warnings. Input from media is also considered in this process. Corrective measures are always taken to avoid the recurrence of identified shortcomings.

3.9 Improvement of Overall Operational Framework of EWS

The improvement of the operational framework is accomplished by incorporating feedback from various sources including the government, media, volunteers and the general public. This feedback is not systematized but instead is a part of the SOD. According to the SOD, each person is responsible for reporting any deficiencies and ideas for system improvement up the chain of command.

3.10 Examples of Previous Events where the Operational EWS Has Led to Improvements in Emergency Preparedness and Prevention

There are many examples of previous events during which the Early Warning System (EWS) led to significant improvements in national emergency preparedness and prevention. Lapses and gaps in preparedness or response during previous disaster events have been studied, solutions have been found and the deficiencies remedied in the operational plan. As an example, forecast bulletins tended to be very technical and did not include impact information during the nineteen seventies and eighties. Today, however, through incorporation of feedback, warning messages are more people-oriented and include the potential impacts of the approaching tropical cyclone.

Over the past 37 years, Bangladesh has been ravaged by three super cyclones (wind speed equal to or over 220 km/h). The Killer Cyclone of 12 November 1970, the Killer Cyclone of 29 April 1991 and Cyclone "Sidr" of 15 November 2007 stand out as having had the greatest impacts on the Bangladeshi people. Some additional details regarding the 1970 and 1991 disasters and their impacts are presented in the following paragraphs.

> The November 1970 super-cyclone killed over 300,000 people.
>
> *The world's conscience was moved by the huge loss of lives.*

- On 12 November 1970, a super cyclone hit the Bangladesh coast near Bhola and took the lives of 300,000 people. Maximum recorded wind speed was 222 km/h and the height of the associated storm surge was about 10 metres. This event happened during the period of high tide. As a result, the duration of standing surge water was long and was responsible for colossal losses of lives and properties. Bio-diversity in the coastal zone was also badly damaged. The world's conscience was moved by the huge loss of lives and property.

In 1971 (16 December), Bangladesh achieved independence from Pakistan and the CPP was revitalized in 1972. A few cyclone shelters were built along the coastal belt most vulnerable to cyclones and associated storm surges. The first Emergency Standing Orders for Cyclones were published by the Ministry of Relief and Rehabilitation in 1984. A severe cyclone hit Urir Char on 25 May 1985. It killed about 10,000 people while another 10,000 people were missing. The second Emergency Standing Orders for Cyclones, containing significant modifications, were published by the Ministry of Relief and Rehabilitation at the end of 1985.

> Over 130,000 people died during the April 1991 super-cyclone.
>
> *Though the intensity of the 1991 cyclone was similar to that in 1970, the inclusion of information on impacts to warning messages contributed to a reduction in the number of fatalities during the 1991 storm.*

- Another super cyclone hit the Bangladesh coast on 29 April 1991. Maximum recorded wind speed was 225 km/h and the maximum storm surge was about 8 metres. Though the intensity of the 1991 cyclone was similar to that of the 1970 system, the death toll (138,882) in 1991 was much less than that in 1970. Cyclone impacts were mentioned in the warning message in this case. And it was effective!

Preparedness and response were further developed through the creation of the Disaster Management Bureau (DMB) in 1992, under the Ministry of Relief and Rehabilitation of the Government of Bangladesh (GoB). Subsequently, the Ministry of Relief and Rehabilitation was renamed as the Ministry of Disaster Management and Relief. In 1998, a great deluge occurred that surpassed all previous floods in terms of duration and magnitude. The total number of cyclone shelters stood at over 2,000. In view of these events, the MoFDM published new Standing Orders on Disasters (SOD) in 1999. In 2004, the Ministry was, once again, renamed as the Ministry of Food and Disaster Management (MoFDM) to reflect a paradigm shift from response and relief to planning and preparedness.

3.11 Overall Lessons Learned and Future Steps for Improving NMHS Contribution in EWS Particularly Focusing on Institutional Coordination and Cooperation

Much of the improvement in Bangladesh's Early Warning System has been achieved through lessons learned from previous experiences. The Cyclone Preparedness Programme was designed and implemented in direct response to the 1970 cyclone and subsequently improved in light of experience gained during the cyclones in 1991 and 2007. In addition, the Comprehensive Disaster Management Programme (CDMP) was launched in November 2003 and implemented under the sponsorship of the United Nations Development Programme and United Kingdom Department for International Development (DFID). The CDMP has been transferred to the Disaster Management Bureau (DMB) and is functioning efficiently. The programme continuously explores ways and means of improving the Multi-Hazard Early Warning System in Bangladesh. It is also playing an important role in institutional coordination among the Governmental Organizations (GOs) and Non-Governmental Organizations (NGOs).

List of Acronyms

BMD	Bangladesh Meteorological Department
BDRCS	Bangladesh Red Crescent Society
BTV	Bangladesh Television
BWDB	Bangladesh Water Development Board
CDMP	Comprehensive Disaster Management Programme
CMA	China Meteorological Administration
CPP	Cyclone Preparedness Programme
CPPIB	Cyclone Preparedness Program Implementation Board
CSDDWS	Committee for Speedy Dissemination of Disaster Related Warning/ Signals
DAE	Department of Agriculture Extension
DDMC	District Disaster Management Committee
DDMP	District Disaster Management Plan
DFID	United Kingdom Department for International Development
DGoF	Director-General of Food
DMB	Disaster Management Bureau
DMC	Disaster Management Committees
DMTATF	Disaster Management Training and Public Awareness Building Task Force
DoE	Department of Environment

DPHE	Department of Public Health Engineering
DRR	Directorate of Relief and Rehabilitation
DRRO	District Relief and Rehabilitation Officer
EOC	Emergency Operations Centre
EWS	Early Warning System
FFWC	Flood Forecasting and Warning Centre
FPOCG	Focal Point Operation Coordination Group of Disaster Management
GO	Governmental Organization
GoB	Government of Bangladesh
GTS	WMO Global Telecommunications System
IFRC	International Federation of Red Cross and Red Crescent Societies
IMDMCC	Inter-Ministerial Disaster Management Co-ordination Committee
JICA	Japan International Cooperation Agency
JMA	Japan Meteorological Agency
KOICA	Korean International Cooperation Agency
MCSP	Multipurpose Cyclone Shelter Programme
MDMR	Ministry of Disaster Management and Relief
MoA	Ministry of Agriculture
MoD	Ministry of Defence
MoFDM	Ministry of Food and Disaster Management
MoHFW	Ministry of Health and Family Welfare
MoPEMR	Ministry of Power Energy and Mineral Resources
MoWR	Ministry of Water Resources
NDMC	National Disaster Management Council
NGO	Non-Governmental Organization
NGOCC	NGO Coordination Committee on Disaster Management
NMHS	National Meteorological and Hydrological Service
PDMC	Pourashava Disaster Management Committee
PPPDU	Policy Programme and Partnership Development Unit (PPPDU)
PTWC	Pacific Tsunami Warning Center
SAARC	South Asian Association for Regional Cooperation
SOD	Standing Orders on Disaster
SWC	Storm Warning Centre
UDMC	Union Disaster Management Committee
UN	United Nations
UNDP	United Nations Development Programme
UZDMC	Upazila Disaster Management Committee
VCP	(WMO) Voluntary Cooperation Programme
VHF	Very High Frequency
WMO	World Meteorological Organization
WHO	World Health Organization

Appendix 1 National Plans

Corporate Plan of the Ministry of Food and Disaster Management (2005–2009): As a long-term strategy the Ministry of Food and Disaster Management (MoFDM) developed the corporate plan a framework for action. The plan also followed by the strategic plan and operational plans of the different directorate of the ministry.

Purpose of the Corporate Plan

- Articulate the Ministry's long-term Strategic Focus.
- Demonstrate a commitment to address key issues: risk reduction, capacity building, climate change, food security, issues of gender and the socially disadvantaged.
- Show the relationship between the government vision, and the MoFDM mission, key result areas, goals and strategies, and to align priorities and strategies with international and national drivers for change.
- Detail a road-map for the development of internal Strategic Plans for the Disaster Management Bureau (DMB), Directorate of Relief and Rehabilitation (DRR), Director-General of Food (DGoF) and the Policy Programme and Partnership Development Unit (PPPDU).
- Guide the MoFDM in the development and delivery of programmes for which the MoFDM is accountable.
- Provide guidance to the MoFDM's Executives in the formulation of programmes, and the allocation of budgets and resources.
- Illustrate to NGOs, other Ministries and civil society how their work can contribute to the achievements of the government vision, and how they can support the MoFDM goals.
- Inform about challenges, opportunities and risks that will have an impact on policies and practices.
- Provide a framework within which to report performance and success in achieving goals and strategies.

Key Result Areas

- Strengthening and Professionalizing the Disaster Risk Reduction System. Mainstreaming of disaster risk management programming through coordination, cooperation and advocacy (partnership development).
- Strengthening of community institutional mechanisms (community empowerment).
- Expanding Mitigation, Preparedness and Response Strategies across a Broader Range of Hazards.
- Strengthening Emergency Response and Recovery (Relief and Rehabilitation) Systems (Operationalising Response and Recovery).
- Maintaining and strengthening the National Food Security System.

A steering committee is established in the Ministry of Food and Disaster Management to implement the comprehensive disaster management approaches in the country. The steering committee is headed by the Honourable Minister/Adviser, Ministry of Food and Disaster Management. The Strategic Partner had also been identified as a part of the medium term strategies. The strategic partners included Government department/agencies, Academic and Scientific Institution as well as Non Governmental Institutions. For immediate strategies implementing partners had also been identified. The implementing partners will implement the programme from community to national level. A coordination committee had also been established to coordinate and monitor the activities of the strategic and implementing partners.

District Disaster Management Plan (DDMP)

There is a District Disaster Management Committee (DDMC) at the District level. The DDMC consists of the Deputy Commissioner of the District as the chairperson and members comprising all District level department heads, NGO leaders and civil society members. District Relief and Rehabilitation Officer (DRRO) acts as member secretary of the committee. Members of Parliament act as advisors of the committees. The committee is required to meet bi-monthly during normal period and as and when necessary during emergency situation.

There is a plan for each District titled "District Disaster Management Plan" (DDMP) comprising both disaster risk reduction and emergency response to be prepared by the District Disaster Management Committee. This is a plan to be prepared by compilation of the Upazila and Paurashava Disaster Management Plans of the district being received from the respective Upazila and Paurashava/City Corporation DMCs. The DDMP should highlight and articulate, among others, the following:

a. The areas in the district vulnerable to different forms of hazards and risks,
b. Total resource requirements and the planned action for the district
    i. to take measures for prevention and mitigation of disasters by government agencies, NGOs, CBOs and the private sector within the district,
    ii. capacity building and preparedness measures to be taken by government agencies, NGOs, CBOs and the private sector
    iii. strengthening emergency response management system plans and procedures in the event of a disaster, providing for –
c. The response plans and procedures in the event of a disaster, providing for –
    i. Allocation of responsibilities to the departments of the government at district level and other DMC members;
    ii. Procedure for mobilization of resources,
    iii. Prompt response to disaster and relief thereof;
    iv. Procurement of emergency supplies;
    v. Operation of disaster shelters,
    vi. Restoration of emergency services, such as water supply, gas supply, power, telecommunication, road links,
    vii. Provision of emergency medical services,
    viii. Burial of dead bodies,
    ix. Trauma counselling, and
    x. The dissemination of information.
d. Recovery plans and procedures delineating damage assessment procedure, restoration of damaged public infrastructure, resumption of educational institutions, restoration of livelihood, rehabilitation of affected people, especially the disabled, and elderly women and children.
e. The DDMP shall be reviewed and updated annually.
f. The copies of the DDMP shall be made available to all district level stakeholders, Divisional Commissioners, etc.
g. A copy of the DDMP will be sent to the Disaster Management Bureau and all relevant ministries and divisions.
h. The DMB/ NDMTI will provide technical advice and capacity building services to all DMCs.

Upazila Disaster Management Plan (UZDMP)

Upazila is an important and vital administrative unit of Bangladesh. There is a Upazila Disaster Management Committee (UZDMC) at the Upazila level. The UZDMC consists of the Upazila Nirbahi Officer as the chairperson and members comprising all Upazila level department heads, NGO leaders and civil society members. The PIO acts as the member secretary of the committee. Members of Parliament act as advisors of the committees. The committee is required to meet bimonthly during normal period and as and when necessary during emergency situation.

There is a plan for each Upazila titled "Upazila Disaster Management Plan" comprising both disaster risk reduction and emergency response to be prepared by the Upazila Disaster Management Committee by compiling all the Union Disaster Management Plans of the Upazila being received from the respective Union DMCs of the Upazila The UZDMP highlights and articulates, among others, the following:

a. The areas in the upazila vulnerable to different forms of hazards and risks
b. Total resource requirements and the planned action for the district
    i. to take measures for prevention and mitigation of disasters by government agencies, NGOs, CBOs and the private sector within the district,
    ii. capacity building and preparedness measures to be taken by government agencies, NGOs, CBOs and the private sector
    iii. strengthening emergency response management system plans and procedures in the event of a disaster, providing for –

c. The response plans and procedures in the event of a disaster, providing for

   i. Allocation of responsibilities to the departments of the government at district level and other DMC members;
   ii. Procedure for mobilization of resources;
   iii. Prompt response to disaster and relief thereof;
   iv. Procurement of emergency supplies;
   v. Operation of disaster shelters;
   vi. Restoration of emergency services, such as water supply, gas supply, power, telecommunication, road links,
   vii. Provision of emergency medical services;
   viii. Burial of dead bodies;
   ix. Trauma counselling, and
   x. The dissemination of information.

d. Recovery plans and procedures delineating damage assessment procedure, restoration of damaged public infrastructure, resumption of educational institutions, restoration of livelihood, rehabilitation of affected people, especially the disabled, and elderly women and children.
e. The UZDMP shall be reviewed and updated annually.
f. The copies of the UZDMP shall be made available to all Upazila level stakeholders and members of DDMCs.
g. A copy of the UZDMP will be sent to the District Disaster Management Committee and DMB.
h. The DMB/NDMTI will provide technical advice and capacity building services to all DMCs.

Union Disaster Management Plan (UDMP)

Union Parishad is the lowest administrative unit of Bangladesh. There is a Disaster Management Committee at the Union level. The elected Chairman of the respective Union Parishad chairs the UDMC. The Union Disaster Management Committee consists of the Union Parishad Chairman as the Chairperson and members comprising all the Government department head at Union level, members of Union Parishad, NGO leaders working in respective union and civil society members. Secretary of the respective Union Parishad acts as the member secretary of the committee. The committee is required to meet bimonthly during normal period and as and when necessary during emergency situation.

There will be a plan for each Union titled "Union Disaster Management Plan" comprising both disaster risk reduction and emergency response to be prepared by the Union Disaster Management Committee following a proper community risk assessment procedure to be provided by MoFDM with the participation of vulnerable groups and the communities. The UDMP should highlight and articulate, among others, the following:

a. Defining and redefining community risks to hazards utilizing both traditional and scientific knowledge,
b. Total resource requirements and the planned action for the district

   i. to take measures for prevention and mitigation of disasters by government agencies, NGOs, CBOs and the private sector within the district,
   ii. capacity building and preparedness measures to be taken by government agencies, NGOs, CBOs and the private sector
   iii. strengthening emergency response management system plans and procedures in the event of a disaster, providing for –

c. The response plans and procedures in the event of a disaster, providing for –

   i. Allocation of responsibilities to the departments of the government at district level and other DMC members;
   ii. Procedure for mobilization of resources;
   iii. Prompt response to disaster and relief thereof;
   iv. Procurement of emergency supplies;
   v. Operation of disaster shelters;
   vi. Restoration of emergency services, such as water supply, gas supply; power, telecommunication, road links;
   vii. Provision of emergency medical services;
   viii. Burial of dead bodies;
   ix. Trauma counselling, and
   x. The dissemination of information.

d. Recovery plans and procedures delineating damage assessment procedure, restoration of damaged public infrastructure, resumption of educational institutions, restoration of livelihood, rehabilitation of affected people, especially the disabled, and elderly women and children.
e. The UDMP shall be reviewed and updated annually.
f. The copies of the UDMP shall be made available to all Union level stakeholders, UNOs and DCs.
g. A copy of the UDMP will be sent to the Upazila Disaster Management Committee.
h. The DMB/NDMTI will provide technical advice and capacity building services to all DMCs.

Paurashava/City Corporation Disaster Management Plan

Paurashava is at the bottom of the urban administrative tier of Bangladesh. There is a Disaster Management Committee at the City Corporation/Paurashava level. The Paurashava Chairman is the head of the committee. The members of the Committee are all Paurashava commissioners, representatives from all the Government departments, NGOs and CBOs. Chief Executive Officer of the Pourashava is the member secretary of the committee. The committee is required to meet monthly during normal period and as and when necessary during emergency situation. Besides, metropolitan cities in Bangladesh have City Corporation Disaster Management Committees with the Mayor as the Chairman and comprising members as it is in case of Pauroshavas. There will be a plan for each Paurashava/City Corporation titled "Paurashava/City Corporation Disaster Management Plan" to be prepared by the "Paurashava/City Corporation Disaster Management Committee having linkages with the National Plan for Disaster Management. The PDMP/ CCDMP should highlight and articulate, among others, the following:

a. The areas in the Paurashava/city corporation vulnerable to different forms of hazards and risks,
b. Total resource requirements and the planned action for the district
   i. to take measures for prevention and mitigation of disasters by government agencies, NGOs, CBOs and the private sector within the district,
   ii. capacity building and preparedness measures to be taken by government agencies, NGOs, CBOs and the private sector
   iii. strengthening emergency response management system plans and procedures in the event of a disaster, providing for –
c. The response plans and procedures in the event of a disaster, providing for –
   i. Allocation of responsibilities to the departments of the government at district level and other DMC members;
   ii. Procedure for mobilization of resources;
   iii. Prompt response to disaster and relief thereof;
   iv. Procurement of emergency supplies;
   v. Operation of disaster shelters;
   vi. Restoration of emergency services, such as water supply, gas supply, power, telecommunication, road links;
   vii. Provision of emergency medical services;
   viii. Burial of dead bodies;
   ix. Trauma counselling, and
   x. The dissemination of information.
d. Recovery plans and procedures delineating damage assessment procedure, restoration of damaged public infrastructure, resumption of educational institutions, restoration of livelihood, rehabilitation of affected people, especially the disabled, and elderly women and children.
e. The PDMP shall be reviewed and updated annually.
f. The copies of the PDMP shall be made available to all Paurashava/city corporation level stakeholders, UNOs and DCs.
g. A copy of the PDMP will be sent to the District Disaster Management Committee and Disaster Management Bureau.
h. The DMB/ NDMTI will provide technical advice and capacity building services to all DMCs.

Plan for Women, Children, Elderly and Disabled

The gender dimension of vulnerability suggests that risks and impacts of environmental crises and natural disasters are experienced by women and men differently and are mediated by their differential access to and control over resources and familial relationships. Natural disasters also result in reaffirmation and re-genesis of traditional gender patterns in communities as the need for care giving and daily household maintenance in a difficult disaster situation expands their workload dramatically. Where life-saving skills are concerned, women are at a disadvantage due to social confinement and are more vulnerable to violence, personal injury, drowning, and health hazards. Furthermore, break up of marriages and desertion of wives, children and the elderly often increase during and after disasters following migration of men in search of employment. Female-headed households have less access to social, political and financial resources and are more likely to experience a more difficult recovery process than similarly poor male-headed households.

Age differentials play a critical role in creating greater vulnerability to disease, discomfort and dependency during a disaster. Particularly the children and the elderly suffer more and mortality rates are found to be much higher among the very young and the very old. People with disabilities are one of the most neglected and alienated sections of the society and as such remain more vulnerable to natural hazards. In the absence of any special provisions, they face particular difficulties during evacuation and in post evacuation shelters. Human casualties during cyclones and tornadoes comprise proportionately more people with disabilities. On the other hand, many who sustain physical injury become disabled due to the lack of appropriate medical facilities. Every concerned Ministry of the GoB will prepare their own Disaster Management Plans with a view to safeguarding women, children, elderly, and the disabled elderly from disaster.

Casualty Information Management Plan

There are thousands of people casualties, which are not officially reported and documented in every disaster. This plan is planed to make the provision for the DMCs to document all relevant information regarding casualty at the time of disaster including the victim age, gender and profession. It is also giving emphasis to ensure attendance of doctors to the every disaster victims. Post mortem report at list death certificates have to be issued for the victims of disaster by the local government/non-government doctors.

Involvement of Ansar-VDP in the Programme

This plan provides the opportunity to develop Ansar VDP personnel to maintain an effective emergency deployment skill during disaster situation for search, rescue and relief purpose and manage and provide information on various disasters, loss and damage assessment as required. The Ansar VDP will be the helping force for the Local Government agencies, local administrative authorities and disaster management committees for reducing the impact of natural disaster.

Sectoral Development Plans incorporating Disaster Risk Reduction

Every Ministry/Division of the GoB prepares its own Sectoral Development Plans. MoFDM with the participation of sectoral experts will prepare a general guideline to incorporate disaster risk reduction agenda for the sectors. MoFDM will also be responsible for overall monitoring and follow-up of the process to ensure that disaster risk reduction agenda are mainstreamed within the sectoral policies, plans and programmes. The development plans addresses, among others, the following:

a. Defining and redefining risk environment through hazard analysis, vulnerability assessment, risk evaluation, risk treatment options, and risk treatments.
b. Managing the risk environment by developing programmes and strategies that eliminate, or reduce the level of risk. Traditionally mitigation programmes were viewed as engineering solutions to eliminate risk, but it is now accepted that all activities undertaken to eliminate or reduce risk are "mitigation" strategies (e.g. community education and awareness, planning activities, development of warning systems). This includes activities previously described as the PPRR Model-Prevention, Preparedness, Response and Recovery.

c. Regularly review and update the plan; and
d. Submit a copy of the plan, and of any amendment thereto, to appropriate authority including the MoFDM.
e. Submit a copy of its disaster management plan, and of any amendment thereto, the concerned authority.

Hazard Specific Multi-sectoral Disaster Management Plans

In addition to area specific disaster management plans and sector specific disaster risk reduction plans, it has been proposed that there will be a few hazard-specific management plans, such as earthquake management plan. These type of plans will be multi-sectoral and will be divided into two components: risk reduction and emergency response and will address specific hazard related issues.

Cyclone and Flood Shelter Management Plan

To face associated with cyclone, flood and tidal surge, different government and non-governmental organizations have constructed about 2033 cyclone shelters in the coastal belts of 15 districts of the country. These shelters are insufficient in terms of capacity to provide shelter to all the affected people as well as their domestic animals. A survey team captioned as Multipurpose Cyclone Shelter Programme (MCSP) headed by Prof. Dr. Jamilur Reza Chowdhury recommended in its report of 1993 to construct 1250 new cyclone shelter as priority no. 1 and 1250 as priority no. 2 for providing shelters to the affected people during disasters. The report also mentioned that the total number of 2500 cyclone shelters including primary school, madrasha and secondary schools were proposed to be constructed.

These shelters will be constructed on the government khas land/institution`s land/purchased lands. There is a provision of separate latrine facilities for women. One tubewell for each shelter will set-up for supplying pure drinking water. In normal period these shelters will be used as educational institution.

It was decided that Bangladesh's plan of action should be inclusive to multi-hazard, all risk, and all sector approach. Therefore, following technical options are considered as critical element of the plan of action.

- Comprehensive Risk assessment (Hazard Assessment and Vulnerability Assessment), including tsunami inundation modelling and evacuation mapping;
- Warning Guidance, including seismic and sea level monitoring, data evaluation, processing and interpretation, forecasting methods and warning dissemination (a detailed plan of action is prepared);
- Mitigation and Preparedness, including education and awareness programmes, structural and non-structural mitigations, and government policy and emergency management procedures.
- Development of Rescue, Relief and Rehabilitation Plan of Action based on Comprehensive Risk Assessment.
- Existing Cyclone Preparedness Programme (CPP) should be strengthened in a way that they can prepare the community for tsunami as well as cyclone.

Management Aspects of Shelter Center in Coastal Zones

DMB proposes the facilitating role of local Disaster Management Committee in forming the Cyclone Center Management Committee for each centre. The committee will have the following types of representation:

- A member of local Disaster Management Committee
- Locally Elected Representative (UP Member)
- Head Master of local Primary School
- Imam of Local Mosque
- NGO representative
- Women representative

DMB also propose for multipurpose use of the Cyclone Centers by local NGOs, Civil Society Groups and community people for public functions like marriage ceremony, meeting, training session and other social functions under the supervision of CC Management Committee. The users will pay a minimum fee for using CC as maintenance charge. The Management Committee will be responsible for keeping financial statement of CC.

Implementation and Follow-Up

1. Disaster management which includes disaster risk reduction and emergency response management will be viewed holistically and the strategic goals and priorities for action presented in Section XII should be addressed by various stakeholders in a multi-sectoral approach, including the development sector.
2. All relevant organizations and agencies will integrate disaster risk reduction considerations into their sustainable development policy, planning and programming at all levels. MoFDM as the focal agency for disaster management will provide the guidelines to be developed with the participation of sectoral experts to ensure its compatibility.
3. Civil society, including volunteers and community-based organizations, the scientific community and the private sector are vital stakeholders in disaster management. MoFDM/DMB will ensure contribution of all actors in the development and implementation of the plan to build disaster resilient nations and communities.
4. Following the broader guideline described in the National plan and the specific guidelines to be provided by MoFDM, DMB, the district, upazila, union, City Corporation and paurashava disaster management committees will prepare their respective plans for the area of responsibility of the committee.
5. MoFDM with the assistance from sectoral experts will develop a general guideline to mainstream disaster risk reduction within their respective sectoral development plans.
6. The planning process at all levels should be completed by December 2010.

Budget for MHEWS Operation

Name of Organization	Fiscal Year	Expenditure (in million US $)	Source	Remarks
BMD	2008–2009	1.88	GoB	Yearly salary and other allowances for the officers and staffs.
		0.80		Cost for operation, and maintenance of the equipment
		4.34	WMO, JICA, KOICA, ADPC, UNDP, UNESCO	Overseas training/Counterpart training
		28.94	JICA, UNDP	Up-gradation/Improvement of existing M-EWS systems.
FFWC	2009–2010	0.22	GoB	Yearly salary and other allowances for the officers and staffs.
	2009–2010	0.21	GoB	Cost for operation, and maintenance of the equipment.
	2001–2010	6.51	Foreign Aids	Up-gradation/Improvement of existing H-EWS systems & Training.
DMB	2009–2010	0.21	GoB	Yearly salary and other allowance.
		0.04	UNDP, DFID, EU	Training & public awareness.
		0.04	UNDP, DFID, EU	Mainstreaming risk reduction issue in development process.
CPP	2009–2010	0.32	GoB	Yearly salary and other allowances for staff.
		0.002	IFRC	Operation cost
		0.23	Save the Children-USA, CDMP(DMB)	Training & equipment.

(GoB = Government of Bangladesh; DMB = Disaster Management Bureau; CPP = Cyclone Preparedness Programme; BMD = Bangladesh Meteorological Department; FFWC = Flood Forecasting & Warning Center; CDMP = Comprehensive Disaster Management Programme)

Appendix 2 Standing Orders for Relevant Cyclone Preparedness Programme

Disaster Management Bureau

The Disaster Management Bureau was created under the Ministry of Disaster Management in 1992. Its responsibilities are noted below:

General Responsibilities

a. Advise the government on all matters relating to disaster management.
b. Maintain liaison with different government agencies, aid-giving agencies, NGOs and Voluntary Organizations and ensure their maximum cooperation and coordination in all matters of disaster management.

Normal Times

a. Undertake various activities for creating awareness among the people, government employees and people of other professions for reducing risks during disaster.
b. Provide secretarial support to the National Advisory Committee on Disasters.
c. Prepare guidelines for mitigation of disaster with the assistance of Planning Commission and other agencies for devising steps to reduce disaster risk.
d. Arrange publication and distribution of Standing Orders on Disasters, National Disaster Management Plan and other related guidelines.
e. Propose legislation on Disaster, Preparedness and Mitigation and other instructions.
f. Assist in the preparation and implementation of framework for Action Plan on disaster management at District, Thana and Union levels.
g. Impart training to the government employees, elected representatives and others on disaster management in cooperation with different Ministries, Local authorities, Training Institutions and NGOs.
h. Establish an Emergency Operations Centre (EOC) with improved communication facilities at national level and to disseminate inputs/information to government and private agencies.
i. Supply books, maps and other information to those associated with disaster management.
j. Monitor obstacles to the Action Plan or project implementation for probable disaster risks to lives and properties, preparedness, disaster response and mitigation and inform the Ministry o Food and Disaster Management.
k. Collect and preserve lists with location, condition and ownership of Cyclone shelters, embankments, platforms at higher than flood level (flood proofing)
l. Coordinate the activities of NGOs.
m. Educate all about weather signals.
n. Arrange public awareness programmes at regular intervals in radio and television on various disaster related issues.
o. Organize meetings/seminars/workshops at national, District, Thana and Union levels to increase awareness about disasters.
p. Ensure liaison with the Ministry of Education for inclusion of disaster topics in the curricula of schools and colleges.
q. Publicise cyclone signals in the villages through posters, cultural functions, documentary films etc.
r. Arrange research on disaster management issues.

Alert and Warning stage

a. Ensure receipt of warning signals of imminent disasters by all concerned officials, agencies and mass communication media.
b. Assist the Ministry of Food and Disaster Management for undertaking emergency programmes in risk areas by different agencies and the people in the structured areas.
c. Activate the Emergency Operations Centre (EOC) and keep touch with other agencies for making their Action Plan effective and also to activate the Control Room.
d. Publish daily bulletins during disaster period for foreign embassies and UN Missions.
e. Instruct local authorities for assessment of loss and damage and requirement of relief.

Disaster Stage

a. Keep Emergency Operations Centre (EOC) open on non-stop basis (24 hours).
b. Assist the Ministry of Food and Disaster Management for formation of groups for primary assessment of loss and damage.
c. Assist the Inter-Ministerial Disaster Management Coordination Committee (IMDMCC) for ensuring coordination among government, NGOs and different agencies for relief and rehabilitation activities.
d. Monitor progress of rescue, relief and rehabilitation operations, identify the problems and needs and to draw the attention of the proper authority.
e. Assist the Ministry of Food and Disaster Management for supply of required information to the Economic Relations Division, Ministry of Information, Foreign agencies, NGOs etc.
f. Supply information to foreign missions and UN agencies regularly through daily news bulletins.
g. Keep the Ministry informed about the progress of use of materials in relief operations.

Rehabilitation Stage

a. Supply information/input to concerned authority for the preparation of rehabilitation plan.
b. Ensure adoption of steps for minimizing future disaster risks in the rehabilitation plans.
c. Undertake post-mortem of steps taken in the overall management in the emergency response of disaster and to prepare reports on the basis of experience/knowledge acquired in this respect and publish the same and to bring required changes accordingly in the training programmes and future policy.

Cyclone Preparedness Programme (CPP)

Cyclone Preparedness Programme (Central Office, Dhaka)

Normal Times

a. Organize preparedness programmes in disaster prone areas on continuing basis and in order to assess the state of preparedness, hold drills in the months of April and September every year.
b. Ensure recruitment and training of volunteers before the month of April every year.
c. Ensure formation of Unit, Union and Thana Committees.
d. Ensure establishment of Union and Thana Officers and holding of meetings at Union and Thana level.
e. Ensure issue of warning signal kits to the leaders of volunteers and use by them.
f. Ensure wireless communication between CPP Headquarters and Thana Office and between Thana Office and Union Office.
g. Instruct local level Disaster Management Committees for selection of shelters, and high rise safe places and inform people about evacuation plans.
h. Make public awareness about cyclone signals and with the help of Disaster Management Bureau (DMB) popularize preparedness plans, through discussion meetings, posters, pamphlets, films and drama etc.
i. Operate Fax machines for ensuring non-stop communication with meteorological department.

Alert Stage

a. Establish Control Room at CPP headquarters and in regional offices and assist District, Thana and Union authorities in the establishment of Control Rooms.
b. Maintain close liaison with Meteorological Department.
c. Receive special weather bulletins from Meteorological Office and send them to Thana and regional offices and to instruct the Thana offices to send them to Union offices as quickly as possible.
d. Instruct the volunteers to listen to normal radio news/CPP radio news. To communicate with Union group leaders through radio or via liaison volunteers and to advise Union leaders and their colleagues to instruct the volunteers for starting work according to guidelines of cyclone instructions booklet published by CPP.
e. Keep the CPP Implementation Board Chairman and the Members informed about the cyclone.
f. Alert the Deputy Commissioners, Thana Executive Officers, Union Parishad Chairman and Members and local non-govt. organisations.

Warning Stage

a. Propose calling of emergency meeting of the Implementation Board.
b. Verify whether the volunteers of the cyclone preparedness programmes have started work at field level.
c. Advise people about evacuation under proper protection of animals, poultry and other domestic animals to raised land and Killas.
d. Instruct CPP Development Officer for advising Chairman of the Thana Disaster Management Committee to call meeting.
e. Instruct Development Officer for calling a meeting of Union Disaster Management Committee.
f. Implement the decisions taken in the emergency meeting of the implementation Board.
g. Instruct Development Officer for cooperation in the implementation of the decisions of the Thana and Union Disaster Management Committees.
h. Send the special weather bulletins through wireless to Regional, Thana and Union Offices.
i. Ensure that the people are properly warned (about the disaster) through the Development Officer and volunteers.
j. Keep the Chairman, Implementation Board; Director General, Disaster Management Bureau; Relief and Rehabilitation Department and Bangladesh Red Crescent Society informed at regular intervals about the local situation.

Disaster Stage

a. Coordinate its own activities with those of all concerned agencies and non-govt. agencies.
b. Ensure rescue operations and first aid by Development Officers and volunteers in times of need.
c. Instruct Development Officer to maintain wireless communication and to keep touch with the Headquarters at regular intervals and to transmit loss and damage report as soon as it is received.
d. Instruct the Union Group leaders to maintain wireless communication and to keep touch with the Development Officer frequently and to instruct the Development Officer to send the reports of loss and damage received from fields to Regional/Central Office.

Rehabilitation Stage

a. Instruct every Union leader to send a primary report of loss and damage in their respective areas to the Development Officer.
b. Instruct the field office to carry out the following tasks:
c. To assist in the programmes of vaccination, prevention medicines and other health measures.
d. To participate in the rehabilitation programmes.
e. (ii) To assist in coordinating the activities of Non-Govt. organizations (NGOs)

Field Level CPP

Normal Times

a. Organize simulated drills continuously in disaster prone areas and monitor the status of preparedness drill participated by the people in April and September every year.
b. Select volunteers and complete their training according to rules of the CPP before April, and organize grouping of families of evacuation in times of need.
c. Make physical verification of equipment to facilitate allotted to the leader of volunteers and their condition and undertake their repairs/replacement, if needed.
d. Examine the wireless system and keep it running in order to maintain communication between CPP Headquarters, Thana and Union level offices.
e. Educate and create awareness amongst the local people about the cyclone preparedness programme and the understanding of different warning signals.
f. Earmark shelter places, Killas and safe raised land and to keep them fit for use and to publicise the people about the evacuation plan.
g. Make the people and the volunteers aware of the dangers of resistance to disaster preparedness activities, in coordination with Thana and Union Parishad administration.

Alert Stage

a. Set up Control Room and maintain contact with Thana, Union offices and CPP Headquarters
b. Maintain close link with the Meteorological Department and other offices and collect information about cyclone.
c. Receive special weather bulletins from CPP Headquarters and to communicate the same to local offices.
d. Alert the chairman of District/Thana/Union Disaster Management Committees, members, religious leaders, local elites, teachers and persons connected with different organizations.
e. Advise the CPP volunteers to listening to radio broadcasts in order to ensure that work of the volunteers specified in the booklet published by CPP by the union group leaders and their colleagues is started.

Warning Stage

a. Request Deputy Commissioner/Thana Nirbahi Officer/Union Parishad Chairman to call emergency meeting of the respective Disaster Management Committee.
b. Assist in the implementation of the decisions taken in the emergency meeting.
c. Depute CPP volunteers for shifting the livestock, poultry and other domestic animals to raised land, Killas under proper safety measures.
d. Instruct every field office to inform all offices under its control about the special weather bulletins.
e. Warn the people about the disaster.
f. Advise the people to go to shelter place in the event of evacuation order and to assist them in the process.
g. Give the final warning to the people by using megaphone, light signals and flash lights.
h. Keep Thana and District administration, CPP Central Headquarters and Disaster Management Bureau and others concerned informed about the field level disaster situation report.
i. Assist other agencies and non-government organizations in the implementation of their programmes.

Disaster Stage

a. Keep wireless in operation and to maintain communication with CPP Central Office and to send the loss and damage statement to concerned authority as soon as it is received.
b. Conduct rescue work in coordination with Union and Thana authority and provide first aid according to need.
c. Assist local authority in relief distribution.

Rehabilitation Stage

a. Collect data on loss and damage due to cyclone, prepare report and send it to CPP Headquarters, Union Disaster Management Committee, Thana and District administration.
b. Assist local administration in burying the dead bodies and dumping the carcasses under the earth.
c. Assist in inoculation-vaccination drive and other health measures.
d. Participate in the rehabilitation programme with Non-Government Organizations (NGOs) and other agencies.

Bangladesh Betar (Radio, Bangladesh)

The role of Bangladesh Betar (Radio) at every stage of pre-disaster, disaster and post disaster periods is very important. On government orders Bangladesh Betar (Radio) would, specially at the stage of the pre-disaster period, alert the people and motivate them and inform them about the disaster position and their duties in this respect. The responsibility of Bangladesh Betar about disaster will be the following:

Normal Times

a. Establish and maintain quickest communication channels with Bangladesh Meteorological Department, Flood Forecasting and Warning Centre and T&T Board and ensure continuous operation of these channels. Establish non-stop communication arrangement with Disaster Management Bureau and Meteorological Department through fax (and email).
b. Publicize special action plans in cooperation with Bangladesh Meteorological Department and the Ministry of Food and Disaster Management for the information of the people and taking up activities for disaster prevention, mitigation and preparedness.

c. Publicise programmes relating to the meaning of warning signals in cooperation with the Meteorological Department/Disaster Management Bureau/CPP.
d. Ensure audio briefing from the cyclone forecaster of the Storm Warning Centre (SWC) based on weather charts, radar and satellite pictures while broadcasting special weather bulletins.
e. Publicise short talks on the government disaster management policy and work plans.
f. Ensure stand-by arrangement to continue round the clock transmission if so ordered.

Alert Stage

Broadcast weather forecast from all Broadcasting Stations on receiving warning signals 1, 2, 3 from Bangladesh Meteorological Department (BMD) and announce it at regular intervals.

Warning Stage

a. On receipt of warning signal No. 4, arrange to broadcast the same along with the explanations received from the Bangladesh Meteorological Department on hourly basis and on the advice of the Ministry of Food and Disaster Management continue the broadcast without any interval after the normal broadcasting hours.
b. Broadcast the announcement of the Ministry of Food and Disaster Management from Dhaka Radio station. If required the Radio Stations at Chittagong, Rajshahi, Sylhet, Rangpur and Khulna will broadcast the instructions of the local authority. It will be desirable to use the local dialect in the broadcast.
c. Broadcast the Flood Warning on receipt of the same.
d. Broadcast Danger Signal in case of cyclone every 30 minutes and Great Danger Signal every 15 minutes and continue the broadcast after normal broadcasting hours if so advised by the Ministry of Food and Disaster Management.
e. Broadcast the evacuation order for the population living in vulnerable areas to the local administration/authority if so instructed by the Ministry of Food and Disaster Management and Relief.

Disaster Stage

a. (In case of cyclone) Broadcast Danger Signal every 30 minutes and Great Danger Signal every 15 minutes and continue to do so even after normal broadcasting hours on the advice of the Ministry of Food and Disaster Management.
b. Broadcast information relating to disaster prevention.
c. Broadcast safety programmes on matters relating to survival, rescue and domestic belongings and drinking water, health etc.

Rehabilitation Stage

a. Broadcast programmes to keep the morale of the people in affected area high.
b. Broadcast short and long term rehabilitation measures.

Bangladesh Television

Bangladesh Television will perform the following duties at different stages of disaster.

Normal Times

a. Establish contact with Bangladesh Meteorological Department and ensure proper functioning of telephone and teleprinter. To prevent possible disconnection, arrange, if necessary, non-exchange Magneto telephone line of T&T Board. Arrange fax communication (e-mail) with Bangladesh Meteorological Department and Disaster Management Bureau at all times.
b. Telecast special programmes as authorised by the Bangladesh Meteorological Department and the Ministry of Food and Disaster Management for information and action of the public.
c. Telecast special precautionary signals of the Bangladesh Meteorological Department along with their meanings.
d. Ensure audio briefing from cyclone forecaster of the Storm Warning Centre (SWC) of BMD and video representation of radar and satellite pictures along with verbal information while telecasting Special Weather Bulletins.
e. Show short films and video on disaster management.

Alert Stage

Soon after receipt of signal Nos. 1, 2 and 3 from the Bangladesh Meteorological Department, Bangladesh Television shall immediately telecast them through all its sub-centres as well as from Dhaka Centre and continue to do so at regular intervals. Also telecast Flood Warnings issued by the Flood Forecasting and Warning Centre (FFWC).

Warning Stage

a. Bangladesh Television must telecast on receipt of Warning Signal No. 4 along with explanation given by Bangladesh Meteorological Department every one hour and continue such telecast if advised by the Ministry of Food and Disaster Management without any interval even after normal broadcasting hours. As soon as signal No. 3 is hoisted, Bangladesh Television will establish contact with the Ministry of Food and Disaster Management on full time basis for non-stop broadcast beyond normal broadcasting time.
b. Dhaka Television shall telecast all announcements issued by the Ministry of Food and Disaster Management and the Bangladesh Meteorological Department (BMD).
c. Announce instruction relating to precautionary measures issued by the Ministry of Food and Disaster Management and the Bangladesh Meteorological Department (BMD).
d. Telecast in details the Flood Warnings when received.

Disaster Stage

a. In case of cyclones, telecast Danger Signals every 30 minutes and Great Danger Signal every 15 minutes and continue to do so beyond normal broadcasting hours if advise by the Ministry of Food and Disaster Management.
b. Telecast the directions of the Ministry of Food and Disaster Management issued to Local Administration/Authority for evacuations of population from risk areas to safer places.
c. Telecast survival devices and advice regarding rescue and protection of domestic belongings and drinking water.

Rehabilitation Stage

a. Telecast programmes to keep the morale of the affected people high
b. Telecast short and long term programmes regarding rehabilitation

Bangladesh Meteorological Department

In addition to normal duties, the Bangladesh Meteorological Department will perform the following duties:

Normal Times

a. Keep ever careful watch over weather conditions, and ensure improvement of cyclone forecast procedures and supply of information on regular basis.
b. Ensure full time effectiveness of the quickest channel of communication for disseminating weather warnings to all concerned. Fax arrangement must be established between Storm Warning Centre (SWC) of BMD and Bangladesh Betar (Radio), Bangladesh Television and the Ministry of Food and Disaster Management.

Alert Stage

a. Issue as soon as possible the alert Warning Signals of cyclone, at least 36 hours ahead of formation of Depression in the Bay of Bengal.
b. Supply information through Fax/telephone/teleprinter to Cyclone Preparedness Programme (CPP) about the formation of depression in Bay of Bengal so as to allow CPP to take appropriate actions including dissemination of information to all concerned.

c. Issue warning signals code 'Whirlwind' as per Appendix-A to all concerned officials through telephone, teleprinter, fax, email/internet.
d. Prepare and submit Special Weather Bulletin and broadcast/publicise the same through national news media such as the all stations of Bangladesh Betar (Radio) and Bangladesh Television and in national newspapers for the benefit of the general people. In case of Local Cautionary Signal no. 3, arrange for adequate and full time coordination between SWC of the Meteorological Department, Bangladesh Betar (Radio), Bangladesh Television for publicity beyond normal broadcasting hours.
e. Send Special Weather Bulletins to EOC at the Ministry of Food and Disaster Management, the Directorate of Relief and Rehabilitation, the Cyclone Preparedness Programme and Bangladesh Red Crescent Society for undertaking adequate arrangements.

Warning Stage

Publicize warning signals at each of the following specified stages.

a. Warning: 24 hours before
b. Danger: At least 18 hours before
c. Great Danger: At least 10 hours before

The same warning signals are to be repeated to the EOC at the Ministry of Food and Disaster Management, Control Room of the Disaster Management Bureau, the Directorate of Relief and Rehabilitation, the Cyclone Preparedness Programme and the Bangladesh Red Crescent Society.

The following information should be mentioned in the signals to be disseminated.

a. Position of the storm centre;
b. Velocity and direction of the storm;
c. Mention of the thanas of the districts likely to be affected, if possible;
d. Appropriate time of commencement of gale wind at different places (Velocity above 32 miles/hour or 51.84 km/hour).

In the case of Danger Signals, messages are to be sent to the addresses under code name "Hurricane" through Fax/telephone/teleprinter or e-mail/internet. In the case of Great Danger Signals, messages should be sent to the addresses under code name "Typhoon" through Fax/telephone/ teleprinter or e-mail/internet. For the purpose of Inland Water Transportation, appropriate separate messages should be sent to addresses mentioned therein through Fax/telephone/teleprinter or e-mail/internet under code name "Water ways and Authority". Alert/Warning signals should be sent to the concerned authorities for publicity through all centres of Bangladesh Betar (Radio) and Bangladesh Television.

Rehabilitation Stage

Work in collaboration with the Disaster Management Bureau to perform the following tasks.

a. Compare the severity of cyclone with that of warning signal.
b. Collect data from affected areas for research purpose
c. Obtain opinion of the people about the signals issued.

Chapter 4
The French Vigilance System. Contributing to the Reduction of Disaster Risks in France

Catherine Borretti

Abstract Managing natural risks has been a major concern for France ever since the 1950s. The relevant mechanism, which initially focused on operational response on-site then on prevention, relied on specific weather forecasts tailored to the needs of each user. Following the storms of December 1999 whose severe impact could have been lessened by provision of better information to the public, it was decided to introduce a new mechanism that utilized a common language – a four-colour scheme reflecting different risk levels. This would be applied at the scale of the French département, the administrative division of the national territory which is best suited to planning and crisis management. It would also be applied on a 24-hour time frame that combined satisfactory forecasting reliability with sufficient advance warning for action.

Following the storms of December 1999, whose severe impact could have been lessened by better information of the public, the Ministry of the Interior, the *Ministère de l'Ecologie, de l'Energie, du Développement Durable et de la Mer*[24] (MEEDDM) introduced a new mechanism relying on a common language: a four-colour scheme reflecting different risk levels, used at the scale of the French *département*, the administrative division of the national territory which is best suited to planning and crisis management, along with a 24-hour time frame combining satisfactory forecasting reliability with sufficient advance warning for action.

This simple common language allows the French weather service to disseminate widely its "potential risk" forecasts linked to meteorological criteria in the form of a "Vigilance map". Information is transmitted simultaneously to civil defence authorities, national operators, the media and the general public. Even though this information is basically a meteorological forecast, this clear and unified signal triggers a preventive response which is accompanied, as need be, by status reports, personalized briefings and imagery in order to enable each decision-maker to react accordingly. As the general public receives clear risk information from the authorities accompanied by safety guidelines, it can also play a role in ensuring its security.

This mechanism, originally designed for five types of weather hazards, namely high winds, heavy precipitation, storms, snow/ice and avalanches, was broadened in 2003 to include heatwaves and intense cold episodes. In 2007, the parameter "heavy precipitation" was changed to "rains/flooding", thanks to the operational cooperation with the flood prediction network. There are also some specific features in the overseas territories, mainly in relation to tropical cyclones, as presented in the Appendix. These developments reflect a commitment to improve the system, which has demonstrated its value in the implementation of the French policy for reducing the risk of natural disasters in France, and is now well known to French citizens: in 2008, a survey revealed that the Vigilance was known by 86 per cent out of a statistically significant sample of 2000 people.

4.1 Overview

4.1.1 Background to the Establishment of the Vigilance System

The French national meteorological service[26] has always played a key role in relation to all of the laws, regulations and procedures concerning the prevention of risks and the management of crises linked to hydrometeorological hazards that

Catherine Borretti, Météo-France[25]

[24] Ministry of Ecology, Energy, Sustainable Development and Spatial Planning
[25] The significant contributions of several French government organizations are acknowledged at the end of the document.
[26] Directorate for Meteorology, a Central Administrative Directorate within the Ministry for Transportation until 1993, then, since 1994, Météo-France, a public administrative body.

M. Golnaraghi ed., *Institutional Partnerships in Multi-Hazard Early Warning Systems*,
DOI 10.1007/978-3-642-25373-7_4, © Springer-Verlag Berlin Heidelberg 2012

have been in force since the end of the Second World War. A particular contribution has been making aware and warning the governmental authorities in charge of implementing security measures and actions in response to emergency situations. Until the present Vigilance system was introduced, however, the French national meteorological service was only involved within the government information system. Its role consisted of sending messages, primarily by fax, to addressees selected on the basis of the nature of the severe weather events, though interactions via dedicated phone and secondment of staff from the French weather service to event or crisis management centres were also available options.

The performance of this information system was reviewed after France was struck by terrible storms on 26 and 28 December 1999, with strong winds up to 170 kph and very severe impacts on the country, including the loss of nearly 100 lives. The review concluded that Météo-France had accurately predicted the phenomenon and alerted the governmental services through ad hoc bulletins. It also concluded that provision of earlier warnings and better information to mayors, the media and the general public, together with information on risks and safety guidelines, would undoubtedly have limited the impacts of this disaster. In response, Météo-France's proposal to draw up a Vigilance map, accessible on the Internet and simultaneously available to the authorities, the media and the public at large, was accepted. The proposal was enhanced through institutional dialogue between the main ministries and bodies involved, particularly with respect to the definition of follow-up procedures and safety guidelines for the public. Operational implementation began in October 2001.

4.2 Governance and Institutional Measures from National to Local Level

4.2.1 Policy, Institutional and Legal Frameworks in Support of Emergency Relief Planning and Implementation

The legal framework governing disaster risk reduction in France has been updated.

In recent years, French legislation on natural risks has been updated by the passage of two new laws:

1. *The "natural risks" component of Law No. 2003-699 of 30 July 2003 Relating to the Prevention of Technological and Natural Risks and Repairing the Damage Caused was designed to remedy shortcomings identified during floods in the south of France in September 2002.*
2. *The organizational structure of the nationwide civil defence system was updated by means of Law No. 2004-811 of 13 August 2004 on the Modernization of Civil Defence.*

Over the past few years, existing legislation on natural risks has been updated by two new laws, one relating to risk prevention and the other to the modernization of civil defence:

- The "natural risks" component of *Law No. 2003-699 of 30 July 2003 Relating to the Prevention of Technological and Natural Risks and Repairing the Damage Caused* was designed to remedy the shortcomings and malfunctioning with regard to the prevention of natural risks that had been seen on the occasion of the torrential floods in the south of France in September 2002. Its provisions hinge on four guiding principles – strengthening information and coordination in the field of major risks, controlling urbanization in exposed zones, prevention targeting the root causes of risks, and improving terms for the compensation of disaster victims. This text was guided by two main goals – developing risk awareness, memory and ownership at all levels of preventive action (individual and joint, territorial), and adjusting or establishing the necessary frameworks for intervention to tackle the root causes of risks and reduce vulnerability.
- The organizational structure of the nationwide civil defence system was updated by means of *Law No. 2004-811 of 13 August 2004 on the Modernization of Civil Defence.* Article 1 stipulates that the State guarantees the coherence of civil defence nationwide but that citizens are the key players in this respect. This Law reflects an ongoing shift with regard to risk management policy in France towards enhanced protection at all levels. It emphasizes the importance of awareness of severe weather phenomena and their consequences, the legal allocation of responsibility, risk anticipation, and the preparedness of the public and emergency services. The Law gives a new, territorial dimension to the ORSEC (Organisation de la Réponse de Sécurité Civile) disaster contingency plans aimed at organizing the civil defence response through coordinated mobilization of all actors. At the local level, this response is structured by means of communal safeguard plans which determine immediate protection and security measures and stipulate modalities for the dissemination of warnings. The public is mobilized via the creation of communal civil defence reserves that help build awareness, inform the public and intervene in post-crisis phases.

The preceding two laws round out French legislation for the prevention and management of natural risks that has been in existence for nearly 60 years (for example, the creation of the ORSEC disaster contingency plans dates back to 1952), in particular:

- Law No. 82-600 of 13 July 1982 on the Compensation of Victims of Natural Disasters, which introduced the Plans d'Exposition aux Risques[27] (PER) to prevent natural risks;
- Law No. 87-565 of 22 July 1987 on the Organization of Emergency Relief, which introduced the necessary legal bases for a global approach to the prevention of technological and natural risks and regulated preventive information as well as cyclone-proof and earthquake-resistant construction;
- Law No. 95-101 of 2 February 1995, which instituted the Plans de Prévention des Risques naturels prévisibles[28] (PPR), designed to replace the PER and the other prevention mechanisms and funds and authorizing expropriation on grounds of major natural risk;
- The "Seveso II" directive No. 96/82/CE of 9 December 1996 requiring operators of all high-risk installations to draw up an emergency plan that must be revised and tested at least every three years.

At the institutional level, natural risk management mobilizes the services of the *Ministère de l'Ecologie, de l'Energie, du Développement durable et de la Mer*[29] (MEEDDM) and the Ministry of the Interior in a coordinated, complementary and balanced fashion. The *Direction de la Sécurité Civile*[30] (DSC) of the Ministry of the Interior is the central body responsible for risk management. It brings together various groups:

- Firemen and emergency services officials who are responsible for nationwide guidance of *Services Départementaux d'Incendie et de Secours*[31], approval of civil defence associations and the promotion of the communal civil defence reserves.
- Risk management drives and underpins the work of State representatives in the defence zones and the *départements*. Inter alia, it calls for the analysis of risks, regardless of their origin and, therefore, requires cooperation with a great many specialized bodies. It is also linked to civil defence monitoring in cooperation with the headquarters of the defence zones.
- Nationwide operational services that comprise air back-up, reinforced if need be by the army's civil defence training and intervention units, which can be dispatched abroad in the form of air mobile disaster intervention detachments.

The MEEDDM was recently restructured further to *Decree 2008-608 of 9 July 2008*. This reorganization took the form of the establishment of six major units including the *Direction Générale de la Prévention des Risques*[32] (DGPR), which is responsible for coordinating prevention of all types of risks by ensuring greater coherence. The MEEDDM focuses on activities to promote resilience not only by reducing the vulnerability of property but also by helping exposed persons and societies to become more proactive. The idea is to combine sustainable development with protection of persons and property.

Within the DGPR, it is primarily the Service for Natural and Hydraulic Risks which draws up, coordinates and implements policies relating to the awareness, assessment and prevention of risks linked to human activity or natural risks (in particular flood prevention). It also drafts policies relating to the safety of hydraulic and hydroelectric installations. The DGPR cooperates with the other ministries concerned to increase knowledge of risk prevention. In addition, it works to promote greater awareness of problems linked to emerging risks and, therefore, prepares the contribution of the Ministry for Health Policy whenever environmental issues are involved. In addition, the MEEDDM has an Advisory Board for the Prevention of Major Natural Risks. This body was established in December 2003 via *Decree 2003/728*. It is tasked with issuing advice, in particular in the report of the delegate for major risks, and is responsible for making proposals on the prevention of natural risks in three areas: (i) risk awareness – monitoring – prediction – preventive information; (ii) land management and vulnerability reduction; (iii) analysis and expertise. Parliament is associated with its work.

On the territorial level, civil defence has a specific relationship to the organizational structure of the French State between the *Préfets de département*[33] and the mayors at the communal level. The *commune*[34] constitutes the first organizational level in terms of risk, in particular in the phases of preparedness, public information and vulnerability reduction. As the highest level governmental authority in a *département*, the Prefect plays a key role in each of France's 96 metropolitan *départements* both (i) in preparedness – by validating the various preparedness plans at the local level and the departmental plans;

[27] Exposure to Hazards Plans.
[28] Plans for the Prevention of Foreseeable Natural Risks.
[29] Ministry of Ecology, Energy, Sustainable Development and Spatial Planning
[30] Directorate for Civil Defence.
[31] Departmental Fire Brigade and Rescue Services
[32] Directorate General for Risk prevention
[33] Prefects of Departments
[34] There are 36,000 communes in France, the smallest administrative units.

and (ii) in disaster management – being responsible not only for alerts but also response and the mobilization of public or private resources. The departmental operational centres (permanent bodies that can activate crisis management modules as required) and the operational command posts (inter-service command bodies set up in the field) interact with the communes concerned. At a higher level (i.e. when response capabilities at the level of the *département* are insufficient to deal with a crisis), the *Centre Opérationnel de Zone*[35] (COZ) of the defence zone[36] in question coordinates activities, with national level support, if required, from the *Centre Opérationnel de Gestion Interministérielle de Crise*[37] (COGIC) (Fig. 4.1).

The French national meteorological service, Météo-France, was established as a public administrative body by decree of 18 June 1993 tasking it with State authority to ensure the meteorological safety of persons and property. As such, Météo-France must meet the needs of civil defence and risk prevention authorities in the fields of risk forecasting and prevention through the provision of data and assessments to the services concerned. Consequently, its role is mentioned in various official texts (Table 4.1) as an authoritative provider of meteorological and climate information products and services. Nevertheless, it is primarily the contracts which Météo-France signs every four years with the State that detail its missions, establish its goals and allocate its resources accordingly. As a particular example, the second major goal for Météo-France established by the State in the 2005 – 2008 contract was – improving prevention of meteorological risks. The "Vigilance" Early Warning System (EWS) for meteorological and hydrological risks was introduced in 2001 and subsequently updated in 2004 and 2007 by means of circulars issued jointly by the two Ministries responsible for risk management, the MEEDDM and the Ministry of the Interior.

Table 4.1 lists the main texts specifying the role of Météo-France with regard to the prevention of natural risks.

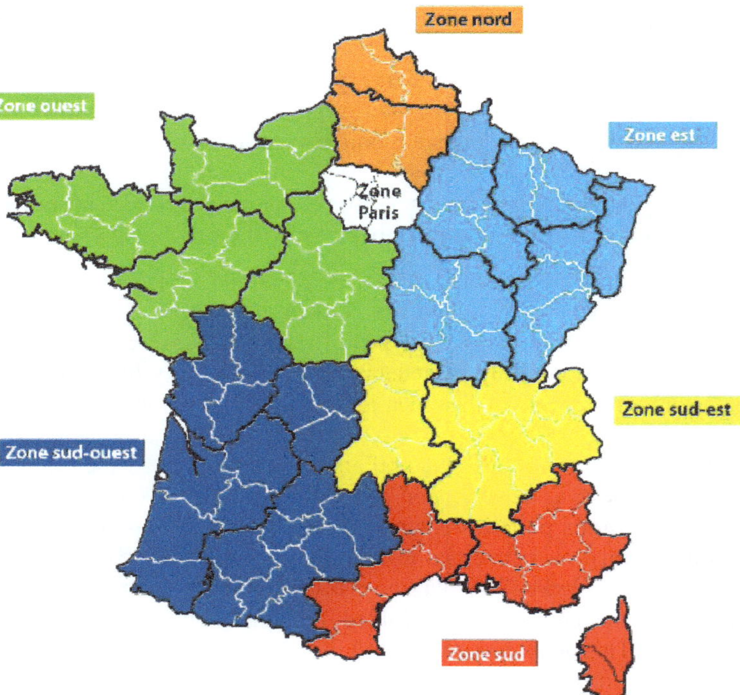

Fig. 4.1 The seven defence zones and 96 départements of metropolitan France

Table 4.1 Main texts specifying Météo-France's roles with regard to natural risk prevention

Sector	Type	Title	Date
General	Decree	Creation of the Public Administrative Body "Météo-France"	June 18, 1993
Hydrology	Circular	Operational Production of Flood Monitoring	Dec 6, 2007
Health and social	Circular	Heat waves	May 13, 2008
Civil defence	Circular	ORSEC Planning for *Départements*	Dec 29, 2006
Weather monitoring	Circular	Procedure for Meteorological Monitoring and Warnings	Oct 15, 2007

[35] Defence Zone Operational Centre.

[36] France is divided into seven Defence Zones.

[37] Inter-Ministerial Crisis Management Centre.

4.2.2 *National to Local Planning of Emergency Relief and Related Links With the EWS*

The 2004 Law on the Modernization of Civil Defence and its enabling decrees have led to a radical change as far as doctrine for planning emergency relief in France is concerned. As mentioned above, it is primarily the mayors at the communal level and the prefects at the level of *départements* who are responsible for the safety of the population. Consequently, emergency relief planning focuses on these levels.

The ORSEC mechanisms at the level of *départements* describe the steps taken to cope with disaster events that affect the public. They are designed to coordinate and mobilize, under the prefect's sole authority, civil defence actors at the level required above their normal, day-to-day services. Their aim is to enhance the preparedness of all actors who can play a role in protecting the public and to develop a civil defence culture. ORSEC stresses four key points: (i) establishment of a civil defence network; (ii) definition of an operational doctrine; (iii) implementation through exercises; and (iv) continuous improvement through feedback and lessons learnt. By doing so, it covers preparedness with regard to crisis management, implementation during crises and feedback. The breakdown of the system is as follows:

- Identification and prior assessment of risks;
- An operational mechanism providing for a general organizational structure for managing all types of events as well as specific arrangements by type of event;
- Preparedness, exercise and training phases.

With regard to risk analysis and awareness, ORSEC documents are based on:

- The *Dossiers Départementaux des Risques Majeurs*[38] (DDRM), drawn up by the prefectural authorities;
- The component of the Departmental Plan for Risk Analysis and Coverage of the Departmental Emergency Service dedicated to "specific risks and critical sites"; and
- The *Plans de Prévention des Risques*[39] (PPR) which the communes listed in the DDRM are required to draw up.

In practical terms, an ORSEC disaster contingency plan is composed of a series of general provisions valid for all sorts of crises, expanded upon by specific provisions by foreseeable risk types such as flooding, cyclones, etc. Specific intervention plans are incorporated for technological risks that apply to certain industrial installations such as large dams or nuclear power plants. Finally, organizational plans specific to certain civil defence actors now form part of the new ORSEC corpus – for example, Communal Safeguard Plans for the communes, Blue Plans for elderly people communities, Customized Safety Plans for schools and White Plans for hospitals. Moreover, when the crisis extends beyond the level of the *départements*, ORSEC zonal plans can be activated for one or more of the seven defence zones. These, primarily, specify the arrangements for organization, mobilization and coordination between the services of the different exposed *départements*. In the case of a nationwide weather hazard event, the appropriate means can be mobilized via the Interministerial Crisis Management Operational Centre (COGIC) or, if further escalation is required, even at the level of the European Union (Fig. 4.2). National plans have been drawn up to deal with specific risks that do not fall within the usual crisis management

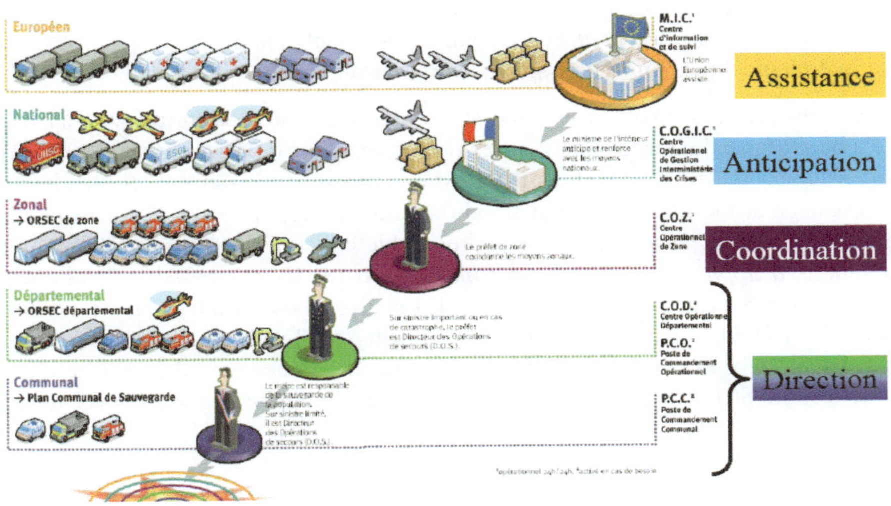

Fig. 4.2 Roles and responsibilities of the different levels in the management of natural risks

[38] Departmental Files for Major Risks.

[39] Plans for the Prevention of Risks.

framework. This is the case, for example, with the Heat Wave Plan which relies on other mechanisms involving the services of the Ministry of Health.

Prefects are responsible for implementing the ORSEC plans. They are also the individuals who decide to activate the plan and who alert the communes and the various services concerned as well as the general public by means of the *Gestion de l'Alerte Local Automatisée*[40] (GALA) system. In the case of meteorological and hydrological disasters, as the recipients of information from the Météo-France Vigilance early warning system and the flood Vigilance early warning system of the SCHAPI (*Service Central d'Hydrométéorologie et d'Appui à la Prévision des Inondations*)[41], prefects rely on these early warning products when deciding whether to activate ORSEC and, if so, within which areas. Finally – and we shall revert to this point later –, the Vigilance system is the essential tool used to generate pre-alerts aimed at the general public and to monitor events linked to any meteorological or hydrological hazards.

Role of Météo-France

Among the thematic cards of the ORSEC plans is a "weather" card which describes the products and services provided by Météo-France, operational contacts, means of accessing or disseminating this information, etc. In addition, Météo-France is involved at all stages of meteorological risk management:

- *Prevention* – the provision of specific weather data and assessments of exposure to meteorological risk is spelled out in the framework conventions which Météo-France has signed with the various prevention actors in France;
- *Crisis preparedness* – Météo-France helps define its role and responsibilities in the various intervention plans relating to hydrological and meteorological risks, such as the weather hazard plans of the defence zones or the national heat wave plan;
- *Warning and Pre-alerts* – Météo-France's Vigilance system is the key component warning the public of meteorological or hydrological risks;
- *Implementation of interventions* – risk monitoring and the production and circulation of risk status reports and weather safety guidelines for the general public are part of the response phase;
- *Lessons learnt* – by participating in the various "lessons learnt" mechanisms foreseen within the Vigilance system and more broadly by the various civil defence mechanisms, Météo-France is, constantly, able to refine the Vigilance system and continuously improve civil defence preparedness and response.

4.2.3 Organizational Structure of the Vigilance Early Warning System

The actors in the Vigilance System who bear risk management responsibilities can be classified according to:

- Responsibility/field of action – detection, alert, communication, response;
- Level of action – national, zonal, *département*, communal, general public; and
- Type of weather or hydrological risk – meteorological, hydrological, health.

Upstream, are the various services involved in forecasting, detecting and monitoring weather hazards. These are primarily Météo-France's decentralized services, the *Centres départementaux de Météorologie*[42] (CDM) and the *Centres interrégionaux de Météorologie*[43] (CMIR), as well as the *Centre National de Prévision*[44] (CNP) based in Toulouse which pools all information obtained at the local level. For flood risk monitoring, there are 22 Flood Prediction Services (FPS) which divide up the national territory according to the hydrographic limits of the main catchment areas. The *Service Central d'Hydrométéorologie et d'Appui à la Prevision des Inondations* (SCHAPI), based in Toulouse, centralizes information from the Flood Prediction Services at the national level. Finally, where health risks are concerned, the *Institut National de Veille Sanitaire*[45] (INVS), which comes under the Ministry of Health, provides special monitoring, particularly with regard to heat waves and related health monitoring signals. In accordance with their respective mandates, these three main actors (Météo-France, the SCHAPI and the INVS) decide or agree on the Vigilance level suited to the forecast phenomenon and its potential implications.

40 Automated Local Alert Management System.
41 Central Service for Meteorology and Flood Prediction Support.
42 Departmental Weather Centres.
43 Inter-regional Weather Centres.
44 National Forecasting Centre.
45 National Health Monitoring Institute

As soon as a detected severe weather phenomenon meets a number of criteria, these bodies assess the risk and activate the appropriate warning levels of the Vigilance system. They transmit the early warning simultaneously and directly to the prefects of the *départements* concerned, to the prefects of the defence zones concerned, to COGIC and to the general public through the media (TV, radio, Internet, press agencies, national and local media), as well as to the other recipients registered in the Vigilance information dissemination system, such as the main network operators (Energy, Railways), certain NGOs (Red Cross/Red Crescent) and the private sector. The prefects of the *départements* concerned then contact the weather forecasting services to decide what follow-up measures to take in response to this warning and, if need be, to implement the ORSEC disaster contingency plan, alert the communes concerned and the general public, and mobilize the available services and resources available, that is:

- Firefighting and Emergency Relief Service of the *Département*;
- Directorate for Equipment of the *Département*;
- Regional Road Information Centre; and
- Police and constabulary services.

In case of need, prefects can request additional resources by alerting the prefecture of the defence zone(s). COGIC is responsible for nationwide mobilization of resources, as required.

4.2.4 *Institutional Capacities and the Operational Concept (Coordination and Operational Cooperation)*

All of the working relationships between the institutions represented in Fig. 4.3 are spelled out in conventions that establish arrangements for the provision of necessary data, assessments or specific assistance. These conventions are tailored to fit each of the working relationships between the institutions involved in the Vigilance system. Two specific documents explain the overall system and define the responsibilities of each body of actors involved in the operational phase of risk management. These are circulars relating to the Vigilance procedure and weather alerts and to internal Météo-France instructions for the application of the conventions. The circular, which is signed jointly by the Ministry of the Interior and the Ministry in charge of sustainable development (MEEDDM) is intended for local risk management officials and, consequently, targets prefects. It details the procedure to be followed depending on the Vigilance warning issued by Météo-France. The first circular issued in 2001 was amended in 2004 to include heat wave risks and in 2007 to incorporate "rain-flooding" as an integrated hazard. The latter hazard replaced "heavy precipitation", involving the introduction of a procedure coordinated by Météo-France and the flood forecasting network of the SCHAPI.

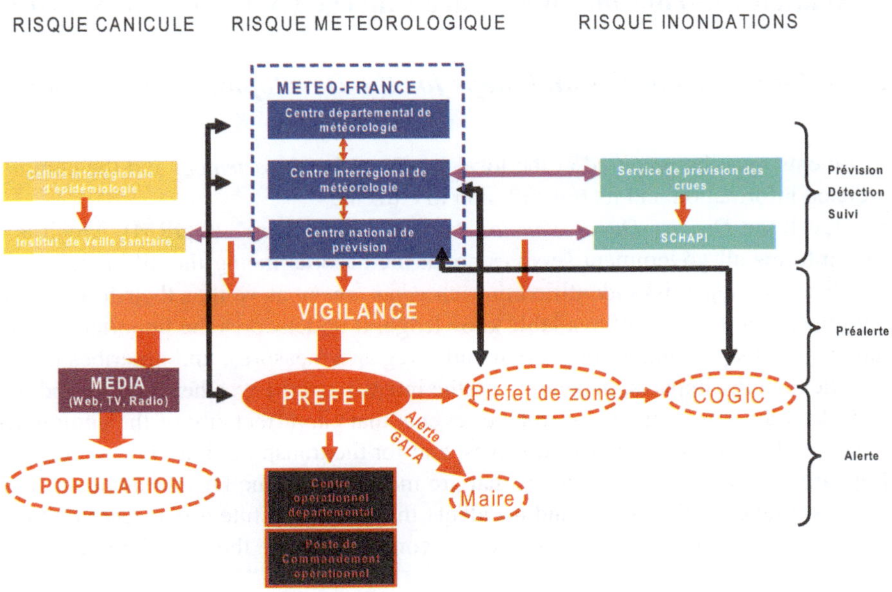

Fig. 4.3 Diagram showing the Vigilance structure and link with crisis management mechanisms

The regulatory provisions contain recommendations or requirements relating to the following:

- Bottom-up information and the recipients of a consolidated final report after each crisis (for both the governmental authorities and elected officials);
- Identification of an evaluation body (usually the body of inspectors of an administrative department);
- A process for regular audits, including an initial audit shortly after the introduction of a new mechanism;
- The choice of a follow-up committee/body, possibly ad hoc, to act in a coordination and steering role.

Since the Vigilance System was originally designed by Météo-France, the national meteorological service has been instrumental in defining the system and the relevant operational procedures. In consequence, Météo-France is consulted prior to the publication of relevant regulatory texts.

4.2.5 Financial and Budgetary Aspects

At the national level in France, the annual Finance Law specifies the budgets allocated to the various ministries. Within the framework of the Organic Law relating to the Finance Laws (LOLF in French), various programmes or missions include a reference to risk prevention responsibilities. The LOLF includes a "meteorology" programme under the "Transport" mission. Thanks to contracts signed with the government (MEEDDM and the ministry of budget), the French national meteorological service receives an overall governmental subsidy covering some 55 per cent of its expenditures. The balance of its budgetary requirement must be financed by cost recovery of services to air navigation and by revenues generated from contracts for specific products and services that Météo-France is free to sign with private or public customers. With regard to the financing of some infrastructure (e.g. weather radar facilities), specific arrangements make it possible to raise funds from local authorities, the State and European Union on a case-by-case basis.

A special feature of the LOLF is that it establishes both the objectives to be achieved and the corresponding performance indicators. An example, Objective No. 2 of the "Meteorology" programme consists of "Ensuring the prevention of weather risks" and two indicators are provided to evaluate whether this goal has been reached, specifically:

- Performance levels (percentages of false alarms, non-detections, and relevant warnings with less than three hours' advance notice) for the Vigilance procedure and maritime safety bulletins;
- Degree of user satisfaction with the Vigilance procedure by the general public.

These indicators are used as a basis for drafting new Finance Laws and contracts with the government.

4.3 Use of Risk-Related Information to Prepare Emergency Relief and Warnings

4.3.1 Organizational Responsibilities and Steps for the Development of Risk-Related Information

Several regulatory documents must be prepared at the local levels of the *département* and the commune in relation to the development of risk-related information and its transmission to citizens.

At the departmental level, the *Dossier Départemental des Risques Majeurs*[46] (DDRM), which is developed under the prefect's responsibility, informs all government services about the hazards facing the different communes within the *département*. The DDRM lists the major risks identified in each *département* as well as their foreseeable impact on persons, property and the environment, on the basis of available knowledge. It underscores the importance of exposed critical sites, particularly in built-up areas; lists prevention, protection and safeguard measures; and describes the mitigation modes that can be implemented to alleviate the impacts, depending on the intensity of the weather hazards and the vulnerability of the exposed critical sites. It does the same for severe weather events that can affect any of the communes in the *department*, such as storms, heavy snowfalls, intense cold or heat waves, and for the transportation of dangerous goods. Prefects are responsible for annual updating of the list of communes that are mentioned in the DDRM as being subject to specific risks. The DDRM provides background on the events and accidents that can constitute a risk record and summarizes the main studies, Internet sites or reference documents available for consultation by those seeking complete information. This

[46] Departmental File on Major Risks.

document, which is updated every five years, is public and must be accessible to citizens within administrative bodies of the *département* and on the Internet.

The State may require the preparation of a *Plan de Prévention des Risques Naturels*[47] (PPR) in the communes mentioned in the DDRM. The PPR, introduced by Law No. 95-101 of 2 February 1995, now represents one of the essential tools for preventing natural risks or reducing the vulnerability of persons and property. Based on up-to-date knowledge of weather hazards and critical sites at the local level, the DDRM makes it possible to define regulatory prescriptions in the communes, particularly with regard to urbanization and spatial planning. Its primary goal is to delimit zones exposed to risks, more particularly to floods, landslides, earthquakes, avalanches or forest fires. It produces maps (at least one map providing information on natural phenomena, a map of weather hazards, and a map of critical sites) drawn up to medium scale at 1 : 25,000, possibly blown up to 1 : 10,000 to make it easier to read, or even more detailed if available documents so allow. The MEEDDM has introduced specific mechanisms for financing their preparation and, as a result, 7,000 PPRs have now been validated from the 36,000 communes in France while around 500 new PPRs are prepared every year (in the long run, 1 out of 2 communes should be covered by a PPR). The preparation of meteorological guides has also facilitated their development. The completion of a PPR provides an opportunity to encourage ownership of risk awareness by means of in-depth discussions with various administrative officials, elected community officials, local associations, the private sector, etc. Subsequently, risk-related information is transmitted and circulated to citizens via the *Document d'Information Communal des Risques Majeurs*[48] (DICRIM) which can be consulted at city hall.

The preparation of these documents is the end result of the work done by all of the decentralized administrative bodies of the State, as mobilized by the prefect, including, in particular, those of the MEEDDM such as the Departmental Directorates for Equipment, the Regional Directorates for the Environment, and the Regional Directorates for Industry and Research, or the Departmental Directorates for Agriculture. The National Geographic Institute (IGN), the National Statistical Institute (INSEE) and Météo-France work together to provide mapping, statistical, economic and meteorological data.

4.3.2 Assessment of Risks, Quantification and Mapping (National to Local)

Those parts of metropolitan France that are not situated in cyclone-prone regions or areas with high earthquake activity are, nevertheless, subject to numerous, unevenly distributed, weather hazards and, thus, to disasters caused by natural hazards. There are also some specific features in the overseas territories, mainly in relation to tropical cyclones, as presented in the Appendix. The most serious risk in metropolitan France is that of flooding. To the north (e.g. the Seine, Somme, Meuse and Moselle), water levels rise slowly. Near the Mediterranean, however, flooding occurs rapidly with heavy rains, especially in the autumn. Other weather hazards stemming from natural risks are storms, drought, heat and cold waves and snowfalls/icing. In the mountains, avalanches are a deadly form of weather hazard. On mountain slopes, hillsides and cliffs, there are also mudslides, landslides and rock falls that can pose a threat to houses or other facilities. Table 4.2 presents the estimated cost of the main disasters caused by natural hazards that have occurred in France, viewed from the perspective of the National Reinsurance Fund.

> December 1999 storms led to the Vigilance system
>
> *The deaths and property damage caused by storms on 26 and 28 December 1999 led to the introduction of the Vigilance system in France.*

Over the past ten years, metropolitan France has experienced several large-scale disasters caused by natural hazards, with significant impacts in terms of loss of life and/or property damage. Each of these events has led to the improvement of the relevant parts of the early warning system. The following three events are of particular relevance.

- In 1999, storms on 26 and 28 December brought havoc across three-quarters of metropolitan France with winds gusting from 130 to 170 km/h. Eighty-eight persons lost their lives, 60 million trees – 140 million m^3 – were uprooted and 200 electricity pylons were blown down, leaving over three million households without electricity for several days. In January 2000, the physical damage caused by the storms was estimated at over 75 billion francs (11.4 billion euros), prompting the introduction of the Vigilance system.

[47] Prevention Plans for Natural Risks.
[48] Communal Information Document on Major Risks.

Table 4.2 Cost impact estimates for the main disasters caused by natural hazards experienced in France in recent years (National Reinsurance Fund). (These estimates do not represent the total cost of the disasters as they only concern goods covered by the legal recognition of natural disaster status)

Exercices de survenance Years of occurrence	Désignation - Type	Estimation coût Marché Estimated cost
De 1989 à 2002	Subsidence - Subsidence[1]	2 700 M€
1995	Inond. Août/Sept., cyclone (Antilles) Flood August/Sept., hurricanes in the Carribeans[2]	110 M€
1996	Séisme Juillet (Annecy) - Earthquake July (Annecy)	60 M€
1997	Inond. Juin (Normandie) - Floods June (Normandy)	< à 50 M€
1998	Inond. Octobre (Meurthe et Moselle) - Floods October (Meurthe et Moselle)	< à 50 M€
1999	Inond. Novembre (Grand Sud) - Floods November (South)	260 M€
1999	Ouragans José et Lenny (Antilles) - Hurricanes José and Lenny (Carribeans)	50 M€
1999	Tempête Lothar et Martin - Storm Lothar et Martin[2]	270 M€
2000	Inondations Septembre (Marseille) - Floods September (Marseille)	60 M€
2000	Inondations Décembre (Bretagne) - Floods December (Britain)	80 M€
2001	Inond. Janvier (Bretagne, Normandie) - Floods January (Britain, Normandy)	< à 50 M€
2001	Inondations Avril (Somme) - Floods April (Somme)	100 M€
2002	Cyclone Dina Janvier (Réunion) - Cyclone Dina January (Reunion)	95 M€
2002	Inondations Septembre (Sud) - Floods September (South)	715 M€
2003	Inondations Décembre (Sud) - Floods December (South)	790 M€
2003	Subsidence - Subsidence[3]	1 000 M€
2005	Inondations Septembre (Gard/Hérault) - Floods September (Gard/Hérault)	95 M€
2007	Cyclone Dean Août (Guadeloupe)	200 M€

The 2003 heat wave led to the inclusion of heat wave risk in Vigilance.

Following the disastrous heat wave in 2003, coordinated action by the Ministry of Health, INVS, the National Health and Medical Research Institute and Météo-France resulted in the addition of heat wave risk to the Vigilance system.

- Europe was hit by an unprecedented hot spell between June and August 2003. Both the duration of this heat wave and the temperature records established during the first half of August were particularly significant in France. Based on existing records, Météo-France estimated that "*this period of intense heat far exceeded everything reported since 1873 in terms of intensity and length as well as minimum, maximum and average temperatures*". The human impact of this disaster was especially severe. An abnormally high death rate, estimated at nearly 15,000 persons, occurred during the month of August, primarily among the elderly. This represents a 55 per cent increase in mortality and led to a drop in life expectancy in France for the first time in over 50 years. Hospitals and funeral parlours were soon swamped by the flood of victims. State authorities were slow to grasp the magnitude of this human tragedy. Consequently, there was a great deal of criticism concerning the delay in implementing an emergency plan. The overhaul of the disaster prevention and warning services and the relief and emergency services that was undertaken following this disaster led to the inclusion of heat wave risk in the Vigilance system.

Severe flooding in September 2005 resulted in enhancement of Vigilance flood warnings

As a result of the flooding in the départements of Gard and Hérault in September 2005, coordination was reinforced between the hydrological and weather services. A new parameter "rains-flooding" was included on Vigilance maps, based on a consolidated procedure involving Météo-France and the flood prediction network.

- Between 5 and 9 September 2005, flooding in the *départements* of Gard and Hérault claimed two lives and caused property damage in 242 communes. Yet these were not by any means the worst floods to hit this part of France in recent years. The tolls for the floods in 1999, 2002 and 2003 were even worse with 35 dead and €609 million in property damage, 25 dead and €1 billion worth of damage and 7 dead and €1.1 billion worth of damage, respectively. Nevertheless, the events of 2005 are worthy of note because they resulted in improved coordination between the hydrological and weather services in the early warning field, following confusion that had arisen from the juxtaposition of two systems. Météo-France's Vigilance system, which was purely meteorological in nature and focused on the phenomenon of heavy precipitation, did not address the impact of flooding. Following the 2005 floods, a decision was taken to replace the parameter "heavy precipitation" by the parameter "rains-flooding", based on a consolidated procedure involving cooperation and coordination between Météo-France's weather forecasting services and the flood prediction network (SCHAPI and the flood prevention services) (Box 4.1).

4.3.3 Assessment of Vulnerability and Exposure

The *Plan de Prévention des Risques Naturels* (PPR) takes a close look at vulnerability and exposure to the various natural risks, in particular flooding, at the communal level. Consequently, the 1 : 25,000 scale PPR maps that identify critical sites mention schools, establishments for the elderly or the disabled, hospitals, prisons and tourist infrastructure, especially campsites, for which specific disaster contingency plans exist. In addition to these types of facilities, special studies have been conducted on specific risks that affect other population groups. In the case of heat waves, for example, isolated elderly people are the most vulnerable. However, the idea of advance registration to identify this group clashes with the notion of protection of privacy. Cold spells hit the homeless hardest. At night, roving teams, often organized by NGOs, scour the streets of large cities to provide this vulnerable population group with a place to sleep, a meal and a hot drink.

During the vulnerability assessment process, Météo-France may take part in studies focussing on correlations between meteorological parameters and measurements that are representative of impacts (e.g. the link between temperatures and excess mortality during heat waves). In addition, Météo-France provides local authorities with historical weather data to facilitate risk assessment studies. Furthermore, Météo-France participates in various exercises to test crisis management mechanisms, providing support in terms of real-time weather information delivery. The meteorological service is also invited to help prepare and evaluate these exercises that are, primarily, aimed at reducing vulnerability and risk exposure.

4.3.4 Storage and Availability of Information on Disasters and Related Risks at the National Level

Under the aegis of MEEDDM, the body of prevention documents, including in particular the completed PPRs and DDRMs, has been brought together in a database that is available on the Internet at www.prim.net. By accessing this web site, users can readily obtain information on the identification and mapping of risks and vulnerabilities on a commune by commune basis. In addition, the map of avalanche-prone zones is updated on a relatively regular basis.

Météo-France is tasked with helping to establish and safeguard climate-related records. For this purpose, it publishes reference atlases every decade on climatic variables calculated over the previous 30 years, making it possible, inter alia, to detect long-term trends (variations in intensity or frequency). Over the past decade, it has, In addition, established and updated, for its own needs, a Database on Extreme Events that comprises archives of meteorological elements along with expert assessments and other forms of public information pertaining to such events. When a major weather event occurs, Météo-France calibrates the event in terms of climatology and comparison to recent events that are still fresh in people's minds, in order to help the various actors (authorities, media, and general public) understand the significance of the phenomenon.

4.3.5 Development and Use of Information on Dangers/Risks to Support Emergency Intervention Planning and Alerts

Prevention documents identifying weather hazards and vulnerabilities (DDRM, PPR, DICRIM) serve as a basis for planning emergency interventions both at the level of the *département*, through ORSEC, or at the communal level with the *Plan Communal de Sauveguarde*[49] (PCS) initiated by the 2004 Law on the Modernization of Civil Defence, For example, it is obligatory to develop a PCS if a PPR is applicable. The risk-related information produced via above-mentioned procedures is not, however, solely intended for the planning of emergency intervention. It also provides a means for incorporating specific risk messages intended for the public in the Vigilance products distributed by Météo-France. The Vigilance map, for example, contains safety guidelines for those facing voluntary or involuntary exposure while the status reports provide more precise and specific, pre-formatted, advice. Prefects can tailor their advice, particularly their guidelines to

[49] Communal Safegard Plans.

mayors, to fit the specific characteristics of their *département* when alerts are disseminated. All of these measures are set out in the inter-ministerial circular establishing the Vigilance warning mechanism.

Box 4.1 Severe flooding in September 2005 resulted in enhancement of Vigilance flood warnings

In the Météo-France Vigilance map, risk information is included with specific safety guidelines as soon as an orange or a red warning level is reached in any of the 95 French metropolitan départments. Follow-up bulletins with even more detailed and updated risk information are also frequently released in these.

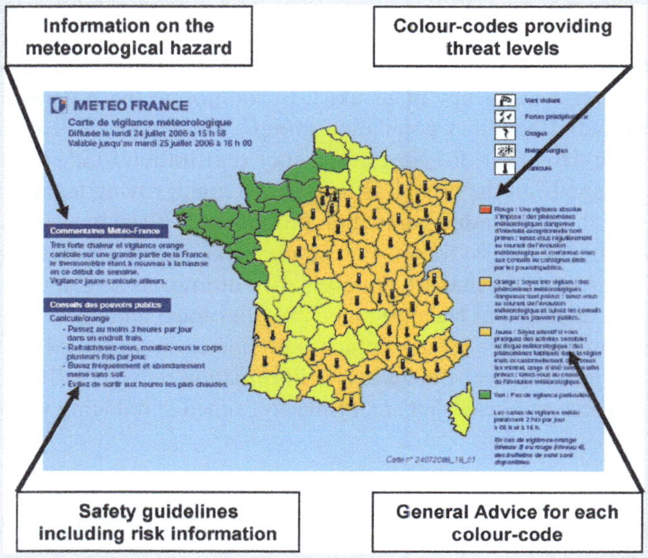

4.4 Observation and Forecasting of Weather Hazards and Mandates for the Development of Warnings

4.4.1 Organizational responsibilities to observe, predict risks and issue warnings

The following outlines the organizational responsibilities of key disaster risk reduction agencies in France:

- Météo-France is responsible for observing, forecasting, detecting and warning for all meteorological hazards. These include, in particular, those associated with the seven main types of risk covered by the Vigilance system (high winds, rains-flooding (in cooperation with the flood prediction network), storms, snow and ice, avalanches and heat and cold waves) in addition to tropical cyclones in the overseas territories and maritime risks.
- Where flooding and high-water risks are concerned, the 22 *Services de Prévision des Crues*[50] (SPC) monitoring the main catchment areas, along with their lead body, SCHAPI, bear this responsibility, having their own Vigilance flood system covering waterways. In this field, coordination with Météo-France has been developed both for the provision of meteorological information and data and for decision-making concerning Vigilance rains-flooding warnings at the *département* level.
- Responsibility for monitoring health risks and issuing health warnings lies with the *Institut National de Veille Sanitaire* (INVS). This mandate applies, in particular, to the risk associated with heat waves though Météo-France has particular responsibility in this area as well.
- The MEEDDM is the responsible ministry with regard to more diffuse risks that require the aggregation of several sources of data (drought, landslides). It relies on its various services (e.g. Departmental Directorates of Equipment, Regional Directorate for the Environment), those of other Ministries (for example the Departmental Directorates for Agri-

[50] Flood Prediction Services.

culture in the case of droughts) and specialized agencies and partners (for example BRGM[51] for anything pertaining to the subsoil), CEMAGREF[52] a research centre that plays a lead role with regard to land management in the mountains, and Electricité de France (EDF) for monitoring water levels in hydroelectric dams).

- Finally, the civil defence authorities deal with everything concerning forest fires, the Ministry of the Interior handles risks stemming from hazardous airborne substances while the General Directorate of Civil Aviation (DGAC) covers hydrometeorological risks specific to the field of aviation. Météo-France helps to assess and monitor risk factors for each of these hazards.

Table 4.3 sums up the different responsibilities for the main natural risks faced by France, distinguishing between:

- Type I risks – those for which the national meteorological service is solely responsible for developing warnings;
- Type II risks – those for which Météo-France is jointly responsible with other agencies for developing warnings; and,
- Type III risks – those for which Météo-France merely provides information to other agencies that are responsible for developing warnings.

4.4.2 *Organizational Cooperation and Coordination for the Development of Warnings*

With regard to Type I hazards, Météo-France operates a weather observation network and a permanent forecasting mechanism composed of its local meteorological centres (Centres Départementaux de Météo France (CDM) and Centres Interrégionaux de Météo-France (CMIR)) in metropolitan France and overseas and the national forecast centre (*Centre National de Prévision* (CNP)) in Toulouse. The CMIRs at the interregional level are responsible for triggering the Vigilance system for Type I hazards, subject to validation by the CNP at the national level. The arrangements between Météo-France and SCHAPI for rains-flooding risk, between Météo-France and INVS for heat waves, and between Météo-France and the civil defence authorities for forest fires, illustrate the institutional cooperation and coordination that takes place in developing warnings for Type II and Type III risks, providing a clear picture of the French system.

Where flood risk is concerned, the SCHAPI network provides risk level information on all waterways for which it has measurement networks and hydrological models. Météo-France supplies data on observed or forecast precipitation accumulation. Based, inter alia, on meteorological data, SCHAPI produces a national flood monitoring map for the main waterways. Météo-France carefully monitors rainfall measurements and has defined indicators for heavy rains by *département*. Combining these two types of information makes it possible to produce the "rains-flooding" Vigilance warning. At the level of their local structures (CMIR for meteorology and SPC for hydrology), prior coordination between meteorological and hydrological experts takes place as early as possible in relation to awareness of observed or predicted phenomena (Fig. 4.4). Such coordination is not systematic but is initiated following a triggering event that foreshadows the occurrence or persistence of a hydrometeorological phenomenon that requires monitoring. Coordination between Météo-France's National Forecasting Centre and SCHAPI ensures the consistency of the Météo-France map from a spatial perspective.

Table 4.3 Natural hazards and institutional monitoring responsibilities

Hazard	Warning	Responsible authority	Contributors	Type
Strong gales – high winds	YES	Météo-France		I
Rains – flooding-high water	YES	SCHAPI – Météo-France	Météo-France, EDF (dams)	II
Storm	YES	Météo-France		I
Snow and ice	YES	Météo-France		I
Tropical cyclones	YES	Météo-France		I
Intense heat – heatwave	YES	Météo-France, INVS	INVS	II
Intense cold	YES	Météo-France		I
Drought	YES	MEESDSP	MF, BRGM, DIREN, DDA, CEMAGREF, EDF…	III
Landslide	YES	MEESDSP	CEMAGREF, MF, BRGM, RTM	III
Maritime risk	YES	Météo-France		I
Avalanche	YES	Météo-France	Ski areas	I
Forest fire	YES	DSD	MF	III
Tsunami	YES	Prefect	MF, CEA, IOPG	III
Hydrometeorological risks for aviation	YES	DGAC	MF	II

[51] France's major public institution in Geosciences.

[52] Public research institute focused on sustainable development in non-urban areas.

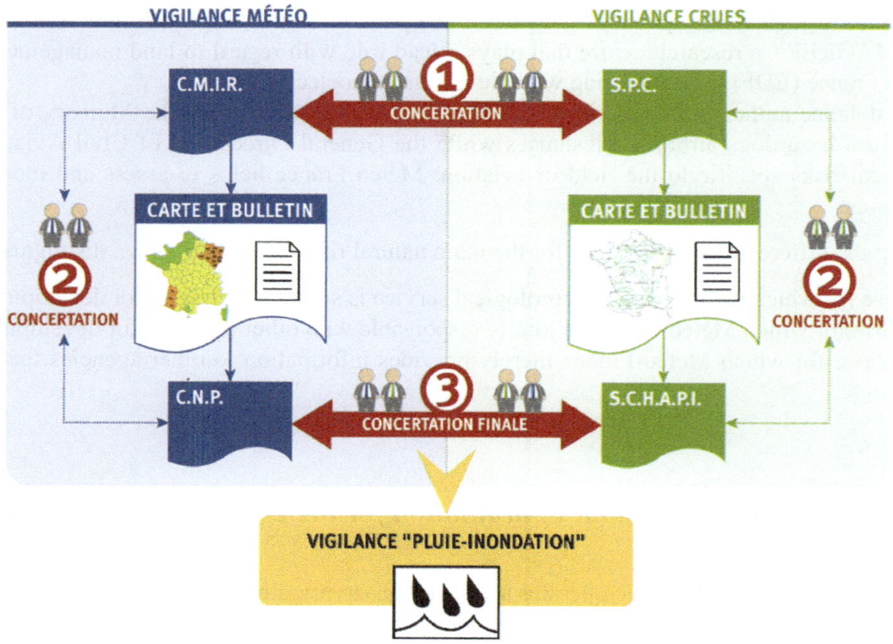

Fig. 4.4 Cooperation between SCHAPI and Météo-France for the Vigilance "rains-flooding" warnings

In the case of the Vigilance heat wave warning, forecasting and detection fall within the purview of Météo-France which proposes a Vigilance warning level to the Health Monitoring Institute (HMI) within the Ministry of Health. INVS, which monitors health-related feedback and possible aggravating factors, then provides an expert opinion on the choice of warning level and drafts an "Alert Sheet" for the Ministry of Health. Finally, the Ministry validates this document and distributes it to the actors in the health sector, civil defence authorities and the general public while Météo-France circulates the Vigilance map.

For Type III hazards, Météo-France supplies contractually defined information to the various agencies and institutions in charge of managing such hazards. In the case of forest fires, for example, the civil defence zonal structures (COZ) introduce a special management mechanism every summer season in the zones located in southern France. Météo-France provides all of the requisite meteorological data (wind speed and direction, temperature, hygrometry, etc.) to these civil defence structures that bear direct responsibility for mobilizing resources and alerting the public using specific mechanisms.

4.5 Development of Authoritative, Understandable, Recognizable and Timely Warnings

4.5.1 Development Cycle for Vigilance Products

Weather monitoring in metropolitan France is built around a Vigilance map showing the required warning level for each département in terms of meteorological and hydrological dangers for the upcoming 24 hours. Warning levels are identified by means of four colours that correspond to the following risk thresholds:

- Green – No particular vigilance required;
- Yellow – Be careful if you engage in activities that are sensitive to meteorological risk or vulnerable to flooding; forecasts call for phenomena that are usual in the region but may be dangerous occasionally and locally; keep informed of any developments in the situation;
- Orange – Be extremely vigilant; forecasts call for dangerous phenomena; keep informed of any developments in the situation and follow the safety guidelines issued by the authorities;
- Red – Utmost vigilance is required; forecasts call for exceptionally intense dangerous phenomena; keep regularly informed of any developments in the situation and make sure you follow the safety guidelines issued by the authorities.

This procedure pursues three basic goals – focusing on major phenomena; anticipating these events as well as possible; and providing the public with ever broader and increasingly effective information.

The Vigilance Map and Status Reports are forwarded to several State bodies operating on national, zonal, regional or departmental level, to local authorities and to the media. They can also be consulted at any time at Météo-France's Internet site: *www.meteo.fr*. The Vigilance Map is prepared twice daily, at 6 a.m. and 4 p.m., but can be updated at other times as dangerous phenomena evolve. It covers seven different meteorological risks: high winds, rains-flooding, storms, snow and ice, intense cold, avalanches and heat waves.

In the event of a Vigilance Orange or Red level, pictographs display the predominant dangerous phenomenon on the map. The second part of the procedure is activated at this point. Status Reports on the dangerous phenomenon in question are issued frequently, enabling all actors to monitor the changing weather situation. The Status Reports cover two geographic scales with nationwide bulletins being issued by the National Forecasting Centre and regional bulletins being issued by the interregional forecasting services concerned by the phenomenon. These bulletins are updated as often as necessary – that is as soon as new information of interest to users is available. Such updates are undertaken, at a minimum, every three hours for regional bulletins and every six hours for nationwide bulletins, with more frequent updates in the case of fast-moving phenomena. Their content is defined according to the needs of users (civil defence authorities, the media, the general public) using a simple text format. Regardless of the meteorological risk being addressed, Status Reports provide the following information:

- The date and the issuing centre, the time of the next report, the time predicted for the end of the event;
- The location and chronology of the event;
- Its classification – to give a qualitative idea of the probability that the event will be more or less severe;
- New facts – to indicate whether the forecast is confirmed or draw attention to a change (change in chronology, path, intensity);
- A description of the phenomenon; and
- Possible consequences and weather safety guidelines issued by the authorities (civil defence officials and the MEEDDM; Ministry of Health for heat waves and cold spells).

Météo-France warns its contacts by telephone when weather forecasts make it possible to provide more than 24 hours' advance warning of the occurrence of a hazardous weather phenomenon. The Vigilance map, however, always operates on a 24-hour time frame. Whenever possible, forecasters predict the arrival of the phenomenon in advance of its occurrence. Twelve hours advance warning is an ideal time frame for dealing with a critical event, especially if there is a need to mobilize emergency resources, whereas a three-hour notice is tight but adequate.

Notification of the end of a Vigilance Orange or Red level warning is by means of a Final Status Report that describes the ongoing meteorological situation, justifies the termination of the alert and provides an initial summary of the phenomena that have been observed during the episode. As a rule, safety guidelines are, however, maintained since they are still of use because the consequences of the phenomenon are still dangerous even though it has weakened or ceased. In many instances, the prefecture takes over to provide a status report on the departmental impact and explain the instructions to be followed.

For its part, SCHAPI operates a Vigilance flood warning system using the same colour-coded scheme. Here, however, the colour-code is applied to a national map representing the main water courses and not to *départements*. In the case of Yellow, Orange or Red levels, the procedure for flood warnings also includes the issuance of status reports on the hydrological situation.

Coordination between the weather and hydrological services and risk prevention and disaster management services is an integral part of the Vigilance System. At the pre-crisis stage, this relates to design and during crises and in the post-crisis stage concerns feedback for fine-tuning purposes. In the pre-crisis phase, the development of systems and procedures takes due account of the requirements of the civil defence authorities. In particular, Status Reports include a list of possible consequences and safety guidelines for the general public. These messages are developed in an orderly fashion by all risk management actors.

The choice of a four-colour scheme meets an operational requirement of the emergency services. An uneven number would have led to a situation of median risk that would have been difficult to translate into action and would not have reflected the danger as far as the public was concerned. At the same time, colours are translated into meteorological criteria solely for the internal use of the hydrological and weather services, with criteria grids drawn up by region depending on the situation. "Aggravating phenomena" criteria are also taken into consideration when choosing a colour. However, as will be seen below, the refinement of criteria relies increasingly on feedback from both forecasters and emergency services.

In a crisis period, the first step is internal cooperation at Météo-France (and with SCHAPI) on the basis of the technical tools that have been developed. The internal software for producing the Vigilance Map and the Status Reports enables sharing of event monitoring experience between the CNP and the CMIR. It is possible to determine the chronology of the phenomenon, risk by risk, assign a colour using predetermined criteria grids, and check for consistency between regions and with Status Reports. The software also makes it possible to start producing maps, circulate them and place them on the

Internet. It has been customized to incorporate expertise from SCHAPI, refinements in status reports and ergonomic improvements to reflect more advanced forecasting techniques. Once the Vigilance system is activated, ongoing dialogue is established between Météo-France and the civil defence authorities. In response to the needs of the emergency services, customized maps and satellite imagery can be provided via a dedicated Internet site to support the experts' input. This process takes place on all levels:

- National – Météo-France's CNP and COGIC maintain a constant dialogue;
- Zonal – the Interregional Weather Centres (CMIR) are in touch with the Zone Operational Centres (COZ) at the administrative centre of each of the defence zones;
- Département – the *Départemental* Weather Centres (CMD) are in contact with the prefects, decentralized State bodies and the CODIS.

Vigilance maps are widely recognized and understood.

In its most recent survey, Météo-France found out that the Vigilance map is known and understood by 86 % of the French people.

The Vigilance products are now well known to the general public (the mechanisms for their dissemination will be discussed in greater detail later in this report). In the 2008 annual survey conducted by Météo-France, the recognition rate for the Vigilance map was 86 per cent based on a survey of 2,000 persons. Knowing that a Vigilance Map exists and becoming familiar with it are one thing, however, but changing one's behaviour accordingly is something more. Nevertheless, French citizens are gradually becoming relatively disciplined as over three-quarters of them (79 per cent) say that they follow safety guidelines in the event of a hazardous meteorological situation.

4.5.2 Cycle for Improving Vigilance Products

Given the high recognition rate for the Vigilance Map and its colour-coded scheme in France, Vigilance products are, clearly, now well established in terms of overall form. However, regular improvements have been made to the system since its introduction in 2001. Firstly, the Vigilance System has gradually been broadened to cover new risks, with the development of the Vigilance "heat wave" and "intense cold" warnings in 2004 being followed by the replacement of the Vigilance "heavy precipitation" by the "rains-flooding" warning in 2007 and the introduction of extensive cooperation arrangements between Météo-France and SCHAPI. Secondly, on the technical level, the criteria defining the four Vigilance levels and the accompanying status reports are constantly being refined. A follow-up group, with representation from all categories of risk management actors, meets three times a year for this purpose. The relevance rate, a guarantee of the procedure's credibility, is analysed jointly. Every Vigilance Orange or Red is documented by Météo-France (description of the phenomenon and meteorological assessment in the light of internal criteria), civil defence authorities (impact in the field, activities of emergency teams, atmosphere at the local level) and the other partners (actions and impacts observed in their sector). They rate each Vigilance Orange or Red episode (number of *départements* actually affected, false alarms, non-detection, attenuated phenomena, advance warning, etc.) and suggest changes. For example, if emergency managers feel that an episode deserved a Vigilance Orange, Météo-France will study the meteorological criteria making possible the detection of the phenomenon then explore the database to determine whether that would have led to other changes to Orange level. Météo-France also checks these criteria for spatial consistency and predictability. Depending on the findings, a proposal may be made to modify the meteorological criteria for a shift to Orange level or to add cyclical aggravating factors. These conclusions are presented at the inter-ministerial steering committee which meets at least once a year to decide on improvements to the procedure and is co-chaired by Météo-France and the civil defence authorities. The analysis by the follow-up group gives rise to an annual Vigilance assessment that is circulated by Météo-France's local services at the regional and departmental levels of the civil defence system. Whenever possible, it is hand-delivered and accompanied by a local assessment. This "post-event" dialogue provides an extremely useful means of ensuring lasting contacts and guaranteeing that improvements are, in fact, implemented. For intense or large-scale events, a feedback report is drafted by the Forecasting Directorate.

The following is one example of the value of feedback. When the Vigilance "rains-flooding" warning was being tested, it became clear from discussions with the services and from feedback (for instance, the water level of the Nivelle River on 4 May 2007 and other storm-related events in May–June 2007) that it was necessary to reinforce the operational dialogue in some Vigilance Orange warning situations. While the stakes of a Vigilance Yellow warning are, by definition, relatively low for the general public and, hence, the media, there remains a probability of damage to the safety of property and even people, in the event of a localized thunderstorm, for example, or to persons who are exposed owing to their activities. Consequently, a decision was taken in 2007 to further strengthen links between the civil defence authorities, Météo-France and the flood prediction network and to arrange for systematic telephone contacts in situations involving significant safety issues.

As a final note, Météo-France also strives to improve the graphics and visual impact of Vigilance products, in particular the Vigilance section at the site www.meteo.fr, with a view to enhancing the visibility of status reports.

4.6 Mechanisms for the Dissemination of Warning Messages

At this point, it is worth stressing two key characteristics of the Vigilance System. First, Vigilance products are circulated simultaneously to the operational services of the civil defence authorities and to the general public. Second, this task falls to the national weather and hydrological services. Accordingly, Météo-France circulates Vigilance Maps and Status Reports, including those for jointly assessed parameters (i.e. "heat wave" and "rains-flooding"), simultaneously to all Vigilance addressees:

- Operational actors – the prefectures, COGIC, the MEEDDM Ministerial Operational Monitoring and Alert Centre, the directorates for road equipment and safety, and the health services – by e-mail;
- The media – audiovisual (radio, TV) and press agencies – by e-mail or fax;
- The general public via the Internet site www.meteo.fr ;
- Customers – private sector (energy, industry, building, etc.), town councils, NGOs, etc. – by mail, site or dedicated Internet site (these services are provided free of charge).

SCHAPI circulates Vigilance "flood" warnings directly. On the Internet, links have been reinforced to ensure easy access to "flooding" information from Météo-France's site and vice versa.

As noted earlier, Vigilance Maps are published daily at 6 a.m. and 4 p.m. in order to ensure maximum impact – 6 a.m., when stations are gearing up for the morning news, and 4 p.m., when TV stations are preparing for the evening news broadcast. Since Vigilance information is likely to be updated at any moment, a specific mechanism has, in addition, been introduced for the media. In the event of a shift to a Vigilance Orange or Red level warning, Météo-France's Media Unit combines the dissemination of the broadcast with a briefing call to major media. The main nationwide television and radio stations have signed the "Vigilance red protocol" which stipulates that "*stations consider that this information has priority and agree to circulate a short message of a shift to Vigilance Red Alert as soon as possible via the most appropriate means – unwinding of a stream on TV screens, circulation in the next information flash, interruption of the programme for a "breaking news" flash, etc.*". In general, circulation is also combined with telephone calls from Météo-France's National Forecasting Centre to major customers who are likely to be affected (for example managers of electricity supply networks).

4.7 Activities to Prepare and Implement Emergency Intervention

4.7.1 Planning of Disaster Preparedness and Implementation and Activation of Emergency Intervention

The various emergency arrangements described above, the ORSEC disaster contingency plan at the departmental level and the PCS at the communal level, are developed under the lead of the prefects and mayors who are primarily responsible in terms of civil defence for the territories under their jurisdiction. In the definition phase, these plans are based primarily on risk and vulnerability assessments (PPR and DDRM) and are implemented in conjunction with all civil defence actors, including citizens, who can mobilize within the framework of the Communal Civil Defence Reserves. Specific plans exist for all sites hosting particularly vulnerable population groups, such as the Blue Plans for old people's homes, the Customized Safety Plans for schools and the White Plans for hospitals. Similar arrangements exist for prisons, campsites and certain tourist attractions. All of these plans call for regular exercises and post-crisis evaluations making it possible to constantly refine mechanisms and ensure appropriate information for citizens, the public's involvement in exercises, and the circulation of specific information tools such as evacuation routes at campsites. Météo-France is involved in the preparation and development of emergency plans through exercises based on mock weather scenarios, training for emergency staff in meteorology and hazard awareness, and participation in joint brainstorming sessions on risk management. In addition, Météo-France experts assist at crisis or support centres during disasters as required.

4.7.2 Awareness-Building and Education Activities Aimed at the General Public

Educating the public and building risk awareness are cornerstones of modern risk management policies in France. A reference has already been made to public information provided through the various regulatory documents (PPR, PCS, DICRIM, DDRM, etc.) but other types of activities are also carried out. In real estate transactions, for example, purchasers must be informed of the risks they may run. In addition, communes now have an obligation to report the levels reached during the heaviest flooding, recorded by means of precise, clearly visible, indicators. Furthermore, numerous communications campaigns have been and are organized to increase awareness of the Vigilance system among the general public and officials. For example, measures to accompany the introduction of the Vigilance “rains-flooding” warning included an inter-ministerial press briefing, a file for the 10,000 mayors of communes with flood-prone areas, a Vigilance brochure and a small poster for the general public as well as a Météo-France briefing for weathermen to roll out the new pictograph and supply explanations. Météo-France also contracts a polling institute to undertake an annual survey on recognition of and feelings towards the Vigilance map in order to obtain a more accurate reading of public awareness. In 2008, the Vigilance recognition rate was 86 per cent, based on a poll of 2000 people. Even though this awareness rate varies depending on lifestyle and region, it is at least 75 per cent regardless of social group.

4.8 Development of Warnings and Related Products and Services for the Disaster Risk Management User Community

Météo-France maintains contact with meteorological warning system partners by means of a circular signed by the Ministry of Ecology (the entity in charge of our agency) and the Ministry of the Interior that defines operational relationships as well as assessment and guidance procedures. Météo-France has also signed bilateral agreements with these partners that have made it possible to define the necessary collaborative efforts with regard to warning systems and, beyond that, to determine if specific products are needed. Implementation of the agreements is monitored at an annual meeting chaired by the directors or their representatives. This allows for regular assessment of the collaboration as a whole and for decisions regarding possible further developments.

There are frequent opportunities to meet with operational partners in order to identify the requirements of the user community. These arise during the drafting of new plans, participation in exercises and follow-up meetings to monitor the warning system. As well, telephone contacts or participation by meteorologists in crisis teams enable them to get a sense of the situation and to gather ideas for making improvements right on the spot. In addition, after a major crisis, lessons learned from the experience are shared by different services affected by the crisis. Requests regarding improvements or shortcomings that have been identified are handled at the national level – within Météo-France by the institutional missions unit, within the flood forecasting services network by the SCHAPI, and within the Civil Security services by the Direction de la Sécurité Civile (DSC), in the process of preparing for the follow-up meetings mentioned later.

Warning maps and bulletins are delivered simultaneously to the disaster risk management user community. The e-mail delivery mechanism is supervised by the Météo-France transmission service and a back-up mechanism allows for delivery by fax in the event of problems. Information is also available on the “meteo.fr” web site and on a back-up Internet site reserved for public safety authorities. In the event of a crisis, a dedicated Internet site is made available for public safety service in order to provide a more detailed explanation of the meteorological situation. It also provides support to local meteorologists who are working on site or by telephone so that their forecasts can be illustrated with images that are updated regularly (for example, tracking heavy rain cells in south-eastern France).

Working with users from different perspectives has been challenging for Météo-France. In overcoming challenges, the first step was to find a common language to define the levels of risk. These levels of risk cannot simply be meteorological thresholds. Using four colours, was developed through correspondence between the risk level and meteorological criteria which was supplemented with criteria to fit the current situation. The second step was to provide information with greater precision without becoming too technical in the process. The consultative work on the items in the follow-up report is a first level of response, supplemented by editorial advice to forecasters, with suggestions such as, for example, that defining the magnitude by making comparisons with other previous crises. Finally, as mentioned earlier, many follow-up mechanisms have been established to improve Vigilance including regular follow-up reports and meetings between Météo-France and the user community.

4.9 Improvements to the Operational Framework of the Early Warning System

Two other processes designed to improve the system as a whole are worthy of note, in addition to the above mechanisms for refining Vigilance products. First, the *Conseil Supérieur de la Météorologie*[53], chaired at ministerial level, is a nationwide body for coordination between the top management of Météo-France and its public or private contact partners or users. Within this body, a "civil defence" commission, chaired by the Civil Defense, is responsible for the following:

- Evaluating the services provided;
- Drafting requests or recommendations;
- Proposing ways of responding to new needs which have emerged;
- Making proposals to secure the corresponding funding; and
- Following up on action taken.

Second, Météo-France has implemented a quality management system that meets ISO 9001 standards. Two processes have been identified in this connection:

- "Meeting the State's needs with regard to civil defence, prevention of major risks, nuclear safety and ensuring the safety of persons and property, on land and sea, in the meteorological field"; and
- Matters pertaining to "warning".

These two processes make it possible to follow and report on actions taken, from identifying customers' needs to guaranteeing the reliability of operational production.

4.10 Examples of Prior Events Where the Operational EWS Has Led to Improvements in Emergency Preparedness and Prevention

The following illustrate three cases where the Vigilance system provided warnings of severe events, assisting in emergency preparedness and prevention.

> Vigilance delivered nationwide warnings during the 2006 heat wave.
>
> Three successive Vigilance "heat wave" Orange Warnings were issued during the hot spell from 30 June to 1 August 2006.

- Much of France was affected by a hot spell from 30 June to 1 August 2006 that gave rise to three successive Vigilance "heat wave" Orange Warnings during which 66 *départements* were placed on Orange level at some point. Consistency was good with the activation of the nationwide heat wave warning levels as 68 *départements* triggered the warning and action level. A study of the meteorological characteristics of the summer of 2006 revealed that the heat wave was one of the most severe observed in France since the 1950s, following the one in 2003. The 2006 episode lasted longer but was less intense and less widespread than in 2003. The abnormal mortality rate observed (estimated at 2000 extra deaths) though significant was considerably less than expected. A mathematical model based on the relationship between the daily fluctuations in mortality observed in France and temperature variations during the summer period made it possible to provide forecasts of the number of deaths to be expected on a given day depending on the temperature. When applied to the summer 2006 heat wave, this model indicated that excess mortality of some 6,400 deaths would normally have been observed with the temperatures recorded. A hypothesis for the population's reduced vulnerability to summer heat

[53] Higher Meteorology Board

waves can be put forward on the basis of the findings which have been published. Such a reduction can be explained by a general awareness of the health risks linked to extreme summer heat and by the introduction of prevention, surveillance and alert mechanisms for this type of extreme weather phenomenon.

> Vigilance warned of the 2008 All Saints' Day floods in the Loire.
>
> Vigilance Red and Orange Alerts for "rains-flooding" were issued for affected départements during the 2008 floods in the upper Loire.

- Heavy rains lashed the Cévennes mountains on the evening of 31 October 2008. A "Mediterranean"-style episode was then triggered during the night of 1 to 2 November, stagnating in the catchment area until dawn. Local rainfall exceeded 300 mm in 48 hours. This 50-year record rainfall provided grounds for placing the stretches of the Loire concerned on a Red Level Warning, from the source up to where the river meets the Allier. Accordingly, five départements were placed on Red Alert for "rains-flooding" and 12 were placed on Orange Level for heavy related rains. Numerous villages were flooded and road and rail traffic was disrupted in the Rhone-Alps region and on several major roads in the south. The information was given broad coverage by the media. This event has resulted in valuable feedback, in particular, with regard to the need to beef up Internet servers (which were overwhelmed) and to re-think certain flood monitoring criteria.

> A storm on January 24, 2009 caused a Red Alert.
>
> Météo-France placed nine départements on Red Alert, the maximum danger level, during a storm in south-western France on 24 January 2009.

- An exceptionally strong storm swept over south-western France on 24 January 2009. Winds reached speeds comparable to those recorded in December 1999, with gusts peaking at 190 km/h on the Mediterranean coast. During this episode, Météo-France placed nine *départements* on Red Alert, the maximum danger level, marking the first time since the introduction of the Vigilance system that a Red Alert was triggered for a storm. As the event was accurately predicted by forecasting models, it was possible to give civil defence officials and the general public two days' advance warning via a press release. With confirmation of the model results on each run and observations from satellite imagery which showed the storm forming and developing over the Atlantic, the shift to a Red Alert was activated more than 12 hours in advance. Crisis centres were set up, extra staff was laid on for each service, and emergency resource deployment and traffic restriction measures were activated. Starting on the morning of January 24, a system of conference calls between Météo-France's National Forecasting Centre, COGIC, SCHAPI and all major operators provided an opportunity to share expertise and feedback from the field. The impact of the storm was dramatic, with eight deaths due directly to the storm and four deaths from carbon monoxide poisoning caused by back-up generators or heating facilities in areas hit by power cuts. Nearly 400 people were injured or suffered from carbon monoxide poisoning, including almost 50 from the storm itself, and considerable property damage occurred. Though overall deaths bore no comparison to the 1999 storm during which nearly 100 people lost their lives, there was massive impact on forests, even though the storm affected a smaller area than in 1999.

4.11 General Experience Acquired and Future Steps to Improve the Contribution of the Weather and Hydrological Services to the EWS, With a Focus on Institutional Coordination and Cooperation

The key principle of the Vigilance system is that of simultaneous circulation by Météo-France of early warnings for natural weather and hydrological risks to civil defence authorities and the general public. As we have seen, the system was introduced by Météo-France following the catastrophic storms of 1999 where insufficient public information resulted in considerable loss of human life. Once this principle had been noted and accepted by all civil defence actors, cooperation mechanisms were put in place, primarily between Météo-France and civil defence authorities, with the aim of constantly improving the system via ongoing feedback and dialogue. As the system proved its value, it has been gradually improved and broadened, as we have seen, to cover other types of risks, in particular heat waves and rains-flooding. In this connection, it was found necessary to establish new types of partnerships with technical bodies such as SCHAPI and INVS. In the future, plans call for the inclusion of coastal risks in the Vigilance system through the establishment of new cooperative ties with other public operators, based on agreements similar to those signed in the past.

Acknowledgements:
The author wishes to acknowledge the significant contributions to the preparation of this paper made by the following French organizations:
- The Ministry of the Interior, Overseas Territories and Departments and Territorial Authorities – Directorate for Civil Defence; Bureau for Crisis Warning, Planning and Preparedness.
- The Ministry of the Ecology, Energy, Sustainable Development and Spatial Planning – General Directorate for Risk Prevention; Bureau for Meteorological Risk; Central Service for Hydrometeorology and Flood Prediction Support.
- Météo-France – Conseil Supérieur de la Météorologie; Forecasting Directorate; Directorate for Institutional Missions and International Affairs.

List of Acronyms

BRGM	France's major public institution in Geosciences
CDM	Centres départementaux de Météorologie (Departmental Weather Centres)
CEMAGREF	Public research institute focused on sustainable development in non-urban areas
CMD	Départemental Weather Centres
CMIR	Centres interrégionaux de Météorologie (Inter-regional Weather Centres)
CNP	Centre National de Prévision (National Forecasting Centre)
COGIC	Centre Opérationnel de Gestion Interministérielle de Crise (Inter-Ministerial Crisis Management Centre)
COZ	Centre Opérationnel de Zone (Defence Zone Operational Centre)
DDRM	Dossiers Départementaux des Risques Majeurs (Departmental Files for Major Risks)
DGAC	General Directorate of Civil Aviation
DGPR	Direction Générale de la Prévention des Risques (Directorate General for Risk prevention)
DICRIM	Document d'Information Communal des Risques Majeurs (Communal Information Document on Major Risks)
DSC	Direction de la Sécurité Civile (Directorate for Civil Defence)
EDF	Electricité de France
EWS	Early Warning System
FPS	Flood Prediction Services
GALA	Gestion de l'Alerte Local Automatisée (Automated Local Alert Management System)
IGN	National Geographic Institute
INSEE	National Statistical Institute
INVS	Institut National de Veille Sanitaire (National Health Monitoring Institute)
LOLF	Organic Law relating to the Finance Laws
MEEDDAT	Ministère de l'Ecologie, de l'Energie, de Développement durable et de l'Aménagement du territoire (Ministry of Ecology, Energy, Sustainable Development and Spatial Planning)
NGO	Non-Governmental Organization
ORSEC	Organisation de la Réponse de Sécurité Civile
PCS	Plan Communal de Sauveguarde (Communal Safegard Plans)
PER	Plans d'Exposition aux Risques (Exposure to Hazard Plans)
PPR	Plans de Prévention des Risques naturels prévisibles (Plans for the Prevention of Foreseeable Natural Risks
SCHAPI	Service Central d'Hydrométéorologie et d'Appui à la Prévision des Inondations (Central Service for Meteorology and Flood Prediction Support)
SPC	Services de Prévision des Crues (Flood Prediction Services)

Appendix An Adaptation of the French "Vigilance" System in the French Antilles and French Guiana

Jean-Noel Degrace and Philippe Cova

The territories of French Antilles and French Guiana are *«départements»* governed in the same way as those of mainland France.[54] However their geographical location in the tropical belt and their specific geographic features as islands or Amazon plateau imply that major natural hazards have to be handled somewhat differently from the French mainland. This section outlines how the French Vigilance System has been adapted for the French Antilles and French Guiana.

1. Background

The Lesser Antilles archipelago lies at latitudes between 10° and 20° North, forming an arc between 60 and 62° West, approximately 6,500 kilometres from mainland France.

The Antilles are small islands, a few dozen kilometres long and wide (1,128 km² for Martinique, 1,638 km² for the islands of the Guadeloupe archipelago and a mere 54 and 24 km² for St-Martin and St-Barthélémy), but have a highly variable and rapidly changing topography, sometimes transitioning from the sea level to 1,000 metres in altitude in less than 10 km. French Guiana is in the North of the Amazonian plateau, with altitudes from 500 to 800 metres; it is by far the largest French "département", with an area of 83,500 km² (1/7 that of mainland France of which 94% is covered by the Amazon forest). It is crossed by many rivers, the main two of which from the borders with Brazil to the East and south Surinam to the West. (Fig. 4.A.1)

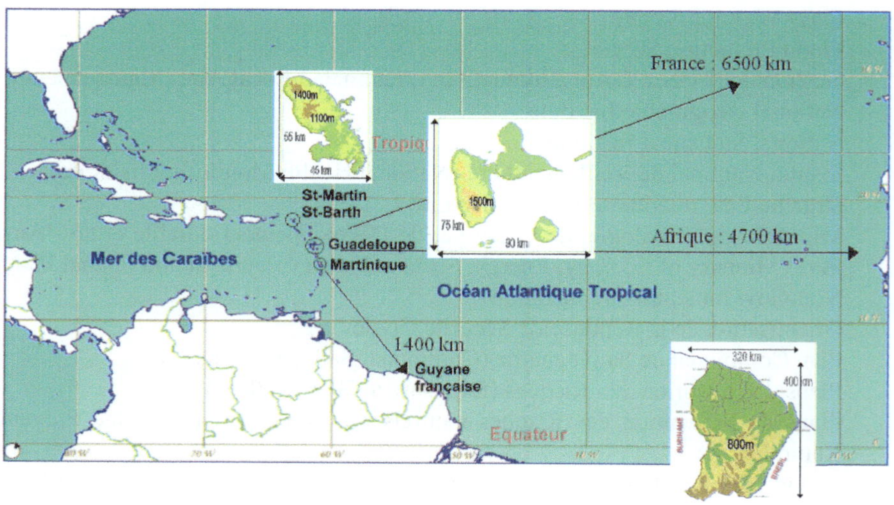

Fig. 4.A.1 Map of the Antilles Islands and French Guiana in relation to France

Jean-Noel Degrace, Head of the Regional Service of Météo-France, Martinique, French Antilles-French Guiana and Inter-Regional Coordinator for forecasting, vigilance and risk management
Colonel Philippe Cova, Chief of Staff for the Antilles Zone for the civil protection aspects

[54] Mainland France = the French part of the European continent, including the coastal islands and Corsica.

The French Antilles islands have a high overall population density (approximately. 350 inhabitants/km² in Martinique and Grande-terre, Guadeloupe) despite many areas of mountain and steep slopes.

- *Martinique:* approx. 400,000 inhabitants, 135,000 of whom live in the Fort de France agglomeration (seat of the regional authority) and 40,000 in Lamentin.
- *Guadeloupe:* a little over 400,000 inhabitants, 16,500 of whom live on the small islands of the archipelago (Marie Galante: 12,000, Les Saintes: 3000, Désirade: 1500)

More people live on Grande-terre (180,000 inhabitants in Pointe à Pitre) and Basse-terre (45,000 inhabitants in Basse-Terre, the seat of the departemental government).

- *St-Martin* (French part): 30,000 inhabitants
- *St-Barthélémy:* 9,000 inhabitants
- *French Guiana:* has a population of approx. 230,000 inhabitants spread along the coastal fringe (90%), 55,000 of whom live in Cayenne (seat of the regional authority).

French Guiana, Martinique and Guadeloupe are French overseas territories (and hence outermost regions of the European Union) falling in the category of "départements d'Outre-mer" which are run almost like the French mainland. The law applicable in the mainland regarding prevention of natural hazards (eligibility for the Major Natural Hazards Prevention Fund for example) and civil protection, also applies throughout the French Antilles (Martinique, Guadeloupe, Saint-Barth and Saint-Martin). The Prefect of each "département" is responsible for civil protection, assisted to that end by the Interministerial Defence and Civil Protection Service (SIDPC).

There is also an Antilles defence zone (under the responsibility of the Prefect of Martinique) and a French Guiana defence zone, each with its own Zone Headquarters (EMZ). French Guiana, Martinique and Guadeloupe are single "département" regions and combine the powers of a *«département»* and a region in two distinct but complementary local authorities. The authorities of Saint-Martin have the powers of a municipality, but also those of the *«département»* and region of Guadeloupe.

Meteorology

Tropical Islands

The Antilles have an oceanic tropical climate with prevailing easterly trade winds. The tropical climate is characterized by more random phenomena, on a smaller scale but more intense and of shorter duration than in temperate regions, which makes them more difficult to forecast. There are hardly ever any disturbances with stable and widespread clouds; all the phenomena are convective and formed of cumuliform clouds whose presence and intensity varies very quickly over only a few kilometres.

The islands are subject to tropical storms from Africa, some of which bring heavy rain, thunder storms and gales. Of the sixty tropical waves that cross the Antilles on average between June and November, a few (5–10 on average) become tropical cyclones. These can cause high winds, heavy rain and coastal hazards due to storm surges (very rare but devastating) or more often by the cyclonic swell that can hit the coast even if the cyclone is offshore and does not make landfall.

During the dry season, the boreal winter, temperate disturbances can also affect the islands, mainly those in the North, with a lot of heavy rain and high winds, especially during the transitional months of May and November. Additionally, long period waves from the far North formed by large temperate low pressure areas off the coast of the United States and which can hit the highly exposed and vulnerable coast.

Equatorial French Guiana

French Guiana is right on the equator and has a climate typical of such regions, very humid, hot and often rainy. The weather in French Guiana is highly dependent on the located activity of the inter-tropical convergence zone (ITCZ), which is where the trade winds from each hemisphere meet and often give rise to rainy and stormy areas which are sometimes very intense. The main hazards in French Guiana are heavy rains and flooding from the major Amazonian rivers. French Guiana never experiences cyclones, as it is too close to the equator, and the wind and sea are rarely hazardous. In some cases the long period waves from the North may break on the coast and cause damage, particularly when combined with high tides.

Organization of Forecast in the French Antilles and French Guiana

The Antilles-French Guiana Directorate (hereinafter DIRAG) is responsible for Météo-France operations throughout the French islands of the Lesser Antilles. DIRAG (based in Martinique) has the following operational forecasting institutional structure:

- Three short-range forecasting centres, responsible for managing crises and warnings at "département" level, one in French Guiana, one in Martinique and one in Guadeloupe.
- An interregional synoptic forecasting centre, based in Martinique, providing real-time coordination and harmonization of forecasting for the whole region.

2. *Involvement of Météo-France in Early-Warning Systems (EWS) in the Antilles and French Guiana*

In the context of its quality management system, Météo-France-DIRAG has highlighted the importance of its risk-management mission by setting up a specific process known as "GesRisk". Its purpose is to contribute to safeguarding persons and property, by satisfying the requirements of institutional customers and other players in civil protection. Météo France can be asked to take part in specialized emergency plans by making appropriate forecasts, and supporting prevention or communication by issuing special bulletins and products. "GESRISK" is supported by the following support and processes:

- Basic expertise (observations, forecasts),
- Telecommunications and data processing (IT systems),
- Design and development (studies, tools, etc.)

GESRISK relies on the process of listening to customers, focusing DIRAG's activities on the customer. Most hazards are covered by specific regulations and are managed in different ways depending on the type of hazard or event and whether weather conditions have any impacts.

3. *Differentiation of Hazard Types in Relation to the Role of Météo-France*

Hazards Managed by the "Vigilance"

Currently there are four main hazards managed by the "Vigilance", forming the core of the weather hazards for the Antilles and French Guiana region, i) cyclone (including its global effects such as cyclonic swell, high winds, heavy rain and/or thunderstorm and storm surges); ii) heavy rain and/or thunder storms; iii) hazardous coastal sea; and iv) high winds (non-cyclonic). In contrast to practice in mainland France, the hazards of heavy rain and thunder storms are grouped together.

Risks Managed in Regulated International Contexts, Outside the "Vigilance" System

Aviation

Aviation risks are managed in a separate process, for aircraft in flight and air navigation, risks to airport installations and aircraft on the ground. However, the triggering of certain levels of vigilance for hazards such as cyclones, heavy rain/thunder storms or high winds requires coordination of special messages so that aviation procedures and the Vigilance System are linked.

Marine hazards

As the Antilles are located in Metarea IV (United States responsibility) and French Guiana is in Metarea V (Brazil's responsibility), the official safety and warning bulletins for open-ocean navigation are produced by the authorities of those countries. However, to supplement this international regulatory framework, the DIRAG meteorological services produce special marine bulletins in the event of danger for open-ocean navigation (open sea) or at less than 25 nautical miles from the coasts (coastal area); these warnings are produced in accordance with the international regulations for open ocean Special Weather Bulletins (open-ocean SWBs) and depend more on local conditions for coastal Special Weather Bulletins (coastal SWBs).

Decision Support Information Provider

The French Antilles and French Guiana are no different from mainland France. Météo-France plays the part of a data provider, using the procedures of national framework agreements, for example with CEDRE (Centre for Documentation, Research and Experimentation on accidental water pollution), for marine pollution plan (POLMAR) or the National Institute for the Industrial Environment and Hazards (INERIS) for the management of Individual Action Plans (PPIs) for business in the context of environmental hazards. Météo-France also acts as an expert in many local governmental plans where there is a particular need for meteorological information.

4. *Application of the French "Vigilance" System to the French West Indies*

The "Vigilance" procedure is information with three primary parts: hazard type, risk level (colour) and an instruction. On this basis, individual behaviour measures are recommended and collective or societal measures are decided. It is an interregional procedure harmonized and coordinated by DIRAG (Martinique) but risk management takes place at *«départemental»* level. Risk levels are allocated for the whole *«département»* for Martinique, the Guadeloupe archipelago (Grande-Terre, Basse-Terre, Marie-Galante, Désirade, Les Saintes) and the northern islands (Saint-Martin and Saint-Barthélémy). French Guiana, which is much larger, is divided into four zones. The value of this procedure is that it is:

- Simple to implement and communicate, covering everyone's needs,
- The same whatever the hazard, season or «département» of the French Antilles and French Guiana,
- Based on the production of an interregional warning map and follow-up reports giving details of the event,
- Intended for everyone, at the same time, by all available means (fax, e-mail, internet, SMS).

Risk Levels

The procedure is based on four colour coded graduated risk levels including:

– Situation normal
– Be alert
– Be prepared / Be careful
– Protect yourself / Be extremely careful

There are two additional levels for cyclones only, including

– Stay indoors
– Stay careful

Assessing Danger Levels in "Vigilance"

The hazard threshold values (rainfall, wind, seas, etc.) give an indication but are not fixed and allow for assessment of the environmental conditions, the recent past and the specific geographical location. The overall procedure of "Vigilance" allows flexibility in the choice of anticipation lead-times. It may vary from several days that the authorities need to prepare the services and the public for cyclones to just a few hours for phenomena that are difficult to forecast such as localized heavy rain. The procedure is based on a gradual escalation of risk level and actions to be taken but it also allows for the possibility of downgrading as the hazard becomes clearer and the risk abates. The crisis team, chaired by the Governor or his representative, identifies in real time the collective measures to be taken among those defined for a given risk level. An illustration of identification of the assessment of danger is presented in Fig. 4.A.2 and Fig. 4.A.3.

Given the need to anticipate, sometimes by several days (yellow vigilance sometimes announced 72 hours in advance), and the risk of making too many false alarms which could lead to losing credibility with the authorities and the public, Météo-France DIRAG has decided to determine the danger using a probabilistic rather than a deterministic approach. This is also in line with civil protection's request to have a system for escalating (or downgrading if the risk abates) the means deployed as the danger evolves.

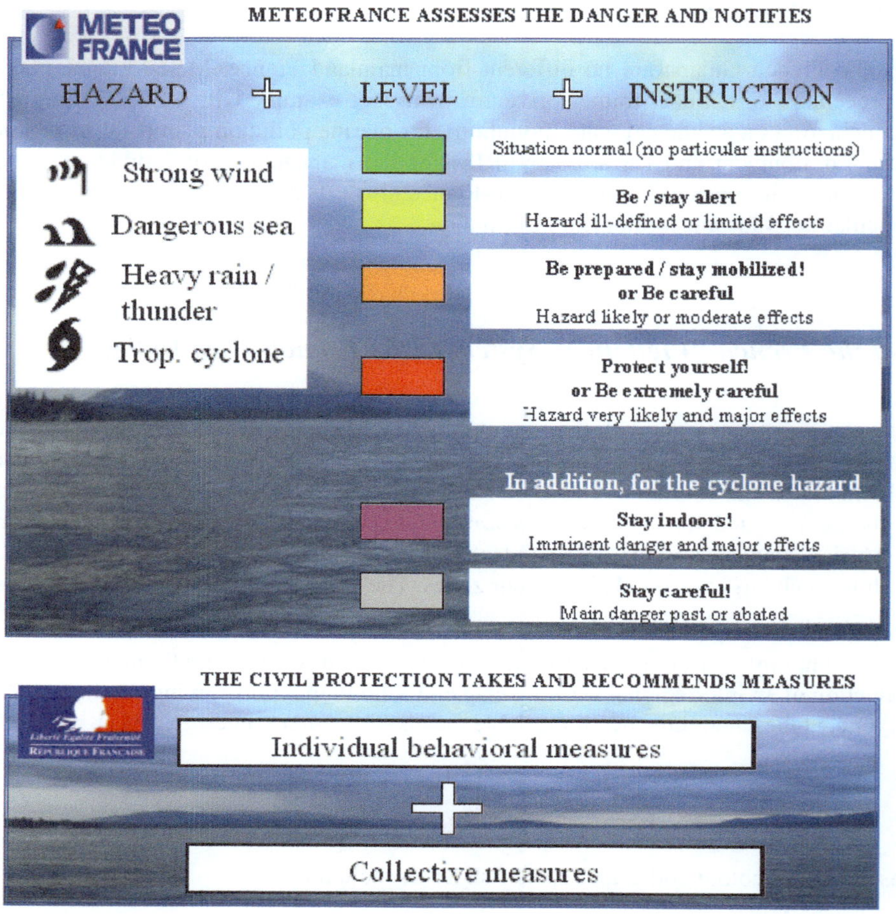

Fig. 4.A.2 Assessment of danger and actions

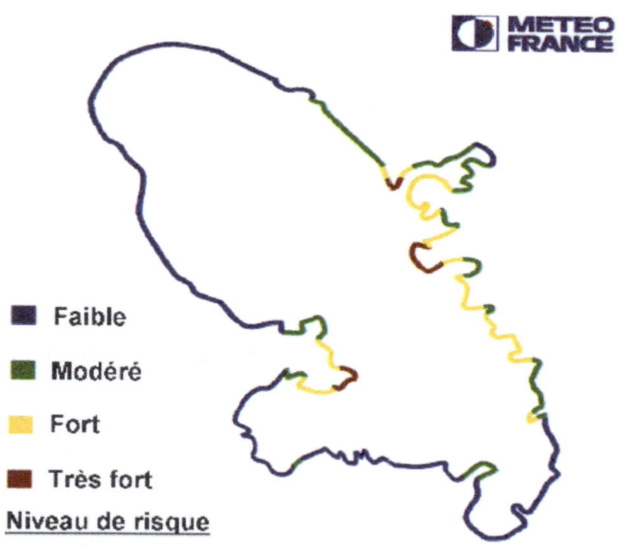

Fig. 4.A.3 "Vigilance" map of storm surge for Martinique

The level of danger corresponds to the current level at the time of broadcast and not to the maximum level expected for the event; otherwise (e.g. more than 48 hours in advance) the number of red alerts would be too high and would not reflect observed reality. For example, with a 48-hour lead-time, the average trajectory error for a tropical cyclone is about 150 to 200 km for the Lesser Antilles, which has to be put in the context of the length of the islands, which is often less than 100 or even 50 km!

The probability is closely linked to the lead-time, though it also depends on the structure of the hazard itself. For instance, for a tropical wave (a classic tropical Atlantic disturbance), the probability of heavy rain will also be linked to the extent and compactness of the rainy area.

Another important parameter to be taken into account in managing and anticipating risks is the expected impact on the *«département»* or island in terms of potential disruption or damage.

Determining the impact requires:

- Knowledge and quantification of the meteorological-hydrological hazard;
- Understanding of the consequences in terms of repercussions on the weather, hydrology and marine features; and
- Quantifying the vulnerability of the areas at risk.

For example, for a given hurricane (hazard) generating a cyclonic swell of 3–3.5 metres (marine consequences/repercussion) arriving on the Atlantic seaboard of islands protected by a coral reef (vulnerability), the impact will be low and the danger level no higher than yellow. If the same swell reaches the Caribbean coast which is much more exposed and vulnerable, the danger level might be red. As shown in Fig. 4.A.4 and Table 4.A.1, vulnerability for the same hazard could be different based on the extent of the hazard and other geographical characteristics.

Cross-Tabulating Probabilities With Impacts

Cross-tabulating these two approaches determines the danger level (1 to 4 corresponding to the colours listed above) as per Table 4.A.2. This corresponds to sentences that are simple to communicate, such as:

"there is a very high probability of a moderate impact on Guadeloupe" → orange
"there is a high probability of a severe impact on Martinique" → red

Table 4.A.1 Quantifying the risk of coastal storm surge in Martinique (values in centimetres)

Cyclone type Vulnerability	Storm	Hurricane Cat. 1	Hurricane Cat. 2	Hurricane Cat. 3	Hurricane Cat. 4	Hurricane Cat. 5
Low	5 to 20	15 to 40	25 to 60	35 to 80	45 to 90	50 to 100
Moderate	15 to 40	30 to 60	50 to 100	70 to 150	90 to 170	100 to 200
High	25 to 60	50 to 110	80 to 170	120 to 250	170 to 270	200 to 300
Very high	50 to 100	90 to 130	150 to 250	200 to 350	270 to 400	Over 300

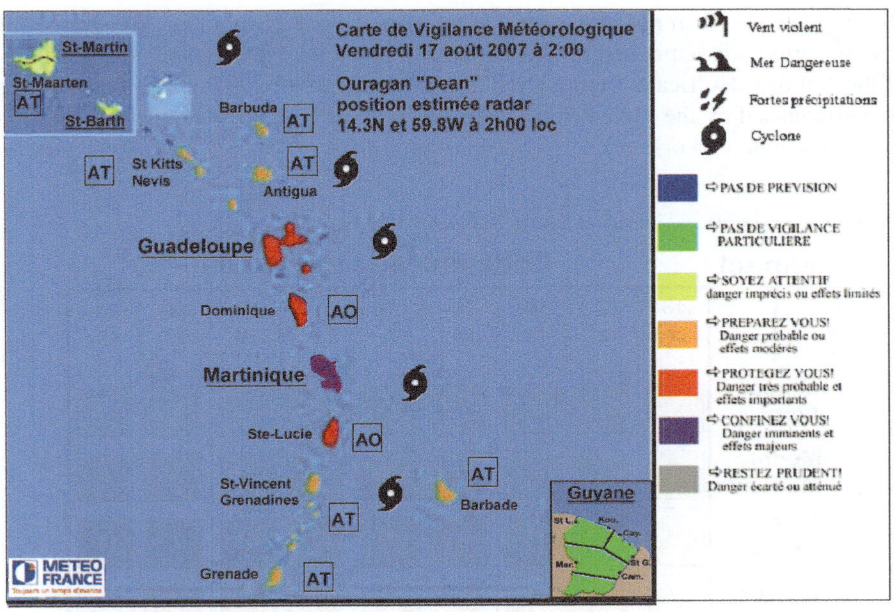

Fig. 4.A.4 Weather Warning Map, Friday 17 August 2007 at 02:00 Hurricane Dean, estimated radar position 14.3 N and 59.8 W at 02:00

Table 4.A.2 shows the lead-time verses impact with the probability colour coding which gives an estimation of the timeline of hazard risk

Impact / ***Probability***	Low (imprecise)	Moderate	High	Major
Low (plausible)	1	1	1 / 2	2
High (probable)	1	1 / 2	2 / 3	3
Very high	1	2	3	3
Almost certain	1	2	3	4

For certain cross-tabulations; two danger levels are possible, which allow flexibility for fine-tuning to match it to the real situation.

Table 4.A.3 below shows how risk gradually increases as a function of time, where highest risk corresponds to the last hours before the hazards makes its impacts (e.g., hurricane making landfall).

Vigilance Production

Vigilance products are based on an interregional warning map, follow-up reports for each *«département»* concerned, and Vigilance SMS (text) messages. The warning map is produced by the CPS in Martinique (production by emergency services in Guadeloupe) after discussion and coordination with the various forecasting centres, at least twice a day. Follow-up reports are the responsibility of each *«départemental»* CPP. The minimum vigilance information (hazard + risk level) is given at the top of all bulletins and regular forecasting products.

Interregional Vigilance Map

The vigilance map covers the Lesser Antilles, from Grenada to Saint Martin, with an enlargement of the northern islands and an insert showing French Guiana (see below), illustrating any weather hazard in the area, using pictograms (hazard type) and colours (vigilance level).

The vigilance map shows the risk level which includes a colour for each *«département»* of the Antilles a single colour for;

- The whole island of Martinique
- Guadeloupe archipelago
- St-Martin and St-Barth as a group
- A colour for each of the four zones of French Guiana

Special case of foreign islands when not in danger and for non-cyclone phenomena, foreign islands are coloured blue, meaning that there is no warning disseminated by these islands. For cyclone phenomena, the colour is allocated on the basis of the criteria of the National Hurricane Centre (NHC) in Miami, based on the correspondence described in 3.3.4. and also on the coherence established by the Météo-France services (e.g., if both Guadeloupe and Martinique are at level "orange", then Dominica is also coloured orange).

Table 4.A.3 Cross tabulation of lead time and impact of hazard with risk level

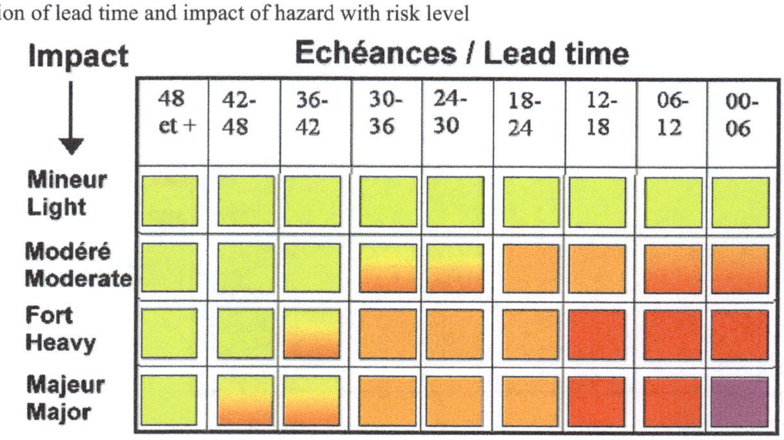

Pictograms of the threatening hazard are placed next to the island or islands concerned whenever the colour is not green. If an island is threatened by several hazards, the pictograms corresponding to each of these phenomena are placed on the map. When a cyclone event occurs, its number or name must be mentioned. Furthermore, official warning coordinated by the NHC on the Lesser Antilles in real time must be displayed on the basis of the four warning levels TRA, HUA, TRW, HUW (See example in Fig. 4.A.4).

The vigilance map is produced twice a day, every day around 06:00 and 17:00 local time, even if there is no threat of danger and all «départements» are green. Whenever a vigilance level is activated somewhere or an official NHC cyclone alert is announced, a new map is produced straight away. It is updated whenever the NHC colours or warnings change or when a new bulletin is issued.

Dissemination and Communication

All available media are currently used to disseminate information in real time with the shortest possible notification times to ensure an optimum response. From the orange vigilance level, TV and radio teams can send journalists and transmission equipment directly to the *«départemental»* weather centres to be at the source of the information.

For disseminating the information, Météo-France relies on:

- *Telephone-based systems* including: i) answering machine/kiosk where all bulletins are recorded (audio accessible with an abridged number from landlines and mobile phones), and ii) SMS messages (via two different routes, from Météo-France via a local provider and from the regional authority via a national provider).
- *Digital media* including: i) map and status reports sent by e-mail to a mailing list in security and subscribers (including the media), ii) posting on the internet, iii) of the map: accessible from the home page via a flashing icon and text using the current vigilance colour, and iv) follow-up reports, in text and audio form.
- *Paper based media*, including fax transmission to civil protection and a few subscribers of the map adapted for black and white format and the follow-up reports.

There must be contact between Météo-France and the «départemental» and local civil protection services before anything is disseminated, regardless of the hazard type and risk level.

From Météo-France, the dissemination system is based on automated fax servers and e-mail or ftp management systems, each with backup. Transmission priority is given to civil protection of course, but also to the media who are responsible for passing the information to subscribers with specific needs. Once civil protection receives the information from Météo-France, it has automatic fax and e-mail forwarding systems enabling all security players (some 200 addressees) to receive the maps and follow-up bulletins. When a crisis team adopts collective measures, these are disseminated by the same means from the operations centre. Immediately after the telephone call or the receipt of the follow-up bulletin, the pre-programmed SMS messages are also sent by an automatic system to a substantial dissemination list.

Some bodies forward the information from Météo-France, as well as civil protection once they have received it. Some examples are as follows:

- The media, transmits the information in routine weather reports, radio or television information publications or in special flashes; a chief meteorologist may also be invited to appear on the television news for specific events;
- Amateur radio stations rebroadcast the information on their frequencies from one municipality to another; and
- Local authorities circulate the information even to isolated areas, by car with a loudspeaker.

This Regional System Requires Coordination With RSMC/NOAA/NHC Alerts in Miami (USA)

For the distribution of coordination roles and the provision of tropical cyclone forecasts, the Miami Regional Specialized Meteorological Center (RSMC-Miami) for tropical cyclones has been designated by WMO as the Regional Centre responsible for tropical North Atlantic, Caribbean and the Gulf of Mexico. Whenever a tropical cyclone threatens land, RSMC-Miami broadcasts guidance which includes watches and warnings and ensures that they have been received in each country or island concerned. It must be stressed that issuance of official warnings in each country/territory is the responsibility of its government and agencies. The system is based on the following principle and distinguishes two different types of cyclone including tropical storm, and hurricane, and two lead-times for the arrival of storm or hurricane conditions, including:

- between 36 and 48 hours: watch; and
- less than 36 hours: warning.

There are four types of NHC warning messages including, i) Tropical storm watch (TRA), ii) Hurricane watch (HUA), iii) Tropical storm warning (TRW) and iv) Hurricane warning (HUW). Table 4.A.4 shows the correspondence between the French Antilles vigilance system and the RSMC Miami warning system.

The assessment of the danger by the forecasting team must of course remain flexible for human expertise but nevertheless requires a structure with thresholds set on the basis of risk assessment.

The threshold values are indicative figures to be adapted in accordance with:

- The recent past, in which vulnerability may have been increased (ground already waterlogged, high river levels, coast already subject to a heavy swell, etc.);
- Particular locations (the colour applies to the «département» but the follow-up report should specify and differentiate the areas most affected);
- Specific parameters (energy of the swell).

Table 4.A.4 Tabulation of the RSMC / NOAA / NHC alerts in Miami (USA)

	Tropical depression	Weak tropical storm	Strong tropical storm	Hurricane
48 h -				
36-48 h		Storm watch	Storm watch	Hurricane watch
24-36 h		Storm warning	Storm warning	Hurricane warning
18-24 h		Storm warning	Storm warning	Hurricane warning
6-18 h		Storm warning	Storm warning	Hurricane warning
0-6 h		Storm warning	Storm warning	Hurricane warning

Tropical Depression = first stage of the cyclone (maximum average wind speeds less than 60 km/h rounded).
Tropical storm = maximum average winds speeds between 60 and 120 km/h rounded.
Hurricane = maximum average wind speeds exceeding 120 km/h rounded.

List of Acronyms for the Appendix

BMV	Weather Watch Bulletin
BRAM	Regional Weather Warning Bulletin
BRGM	Geology and Mining Research Bureau (public agency for the management of surface and underground resources and risks)
CEDRE	Centre for Documentation, Research and Experimentation on Accidental Water Pollution
CMO	Caribbean Meteorological Organization
COMSUP	Armed Forces High Command
CP	Command post
CPP	DIRAG Short-Range Forecasting Centre (one per *«département»*, responsible for local, immediate and short-range forecasting and crisis management)
CPS	DIRAG Synoptic Forecasting Centre (in Martinique, responsible for real-time coordination and framing of synoptic forecasting)
CVH	Hydrology Watch Team
DIRAG	Antilles-French Guiana Directorate of Météo-France
DIREN	Directorate of the Environment (decentralized service responsible for promoting sustainable development by implementing policies for the environment)
EMZ	Zone Headquarters (responsible for implementing all prevention, protection and rescue measures; it also makes the resources of the zone available to support the *«départements»* and manage national or interregional reinforcements.
ERR	Reconnaissance and Intelligence Team
EWS	Early-Warning System

GTS	Global Telecommunications System (data transmission system between the meteorological services of all countries, set up by WMO)
HUA	Hurricane watch
HUW	Hurricane warning
ICAO	International Civil Aviation Organization
IGN	National Geographical Institute (public establishment responsible for production, maintenance and dissemination of geographical reference information).
INERIS	National Institute of the Industrial Environment and Hazards (institute responsible for assessing and preventing accidental or chronic risks to humans and the environment)
INMARSAT	INternational MARitime SATellite organization (company and telecommunications system based on several geostationary satellites)
NHC	National Hurricane Center (the WMO Cyclone RSMC for region IV in Miami)
NOAA	*National Oceanic and Atmospheric Administration (United States)*
ORSEC	Emergency Device
POLMAR	Marine pollution Device
PPI	Action Plan (for technological risks)
PPR	Hazard Prevention Plan (document produced by the Government which controls land use at the level of the municipalities, in accordance with the hazards they are exposed to).
PSC	Municipal Safeguard Plan (emergency municipal plan which surveys the risks in the municipality and the available resources, the organization for issuing warnings, information, protection and support of the population in relation to these risks).
PSFP	Emergency Plan for Heavy Rainfall
PSSC	Specialized Cyclone Emergency Plan
RETEX	Feedback (a posteriori assessment of an event)
RSMC	Regional Specialized Meteorological Center
SCHAPI	Central Hydrology Service for Flood Forecasting (national agency bringing together meteorologists and hydrologists for 24-7 structure and operational support to the Flood Forecasting Services)
SHOM	Navy Hydrographic and Oceanographic Service
SIDPC	Interministerial Defence and Civil Protection Service ("départemental" service assisting the Prefect in risk and crisis management and in preparing safeguard and assistance measures for the population)
SMS	Short Message Service
SWB	Special Weather Bulletin
TRA	Tropical storm watch
TRW	Tropical storm warning
WMO	World Meteorological Organization

Chapter 5
The Warning Management of the Deutscher Wetterdienst

Gerhard Steinhorst and Rainer Vogelgesang

Abstract The Deutscher Wetterdienst (DWD) operates a three-tiered warning management system (time dimension) that differentiates between five risk levels. The warning system is co-ordinated with disaster management authorities which, in Germany, is the responsibility of the Länder (States) so that its organisation is consequently very diverse. Warnings are provided to users through various channels. The general public is informed over the Internet. There are closed online user groups for fire brigades and the winter maintenance service. The DWD is committed to the single-voice principle, according to which only warnings issued by Germany's National Meteorological Service should be binding for disaster mitigation purposes. The warning system is continuously updated and optimised. Using Bavaria as an example, this document illustrates the structural and organisational provisions underlying the "Special Severe Weather Emergency Plan".

5.1 Overview of the German Weather Warning System and the Role of the Deutscher Wetterdienst

5.1.1 Background to the Establishment of the Weather Warning System of the Deutscher Wetterdienst

The Deutscher Wetterdienst (DWD) is a federal authority under the Federal Ministry of Transport, Building and Urban Affairs. It was established on 11 November 1952 in response to the Law on the Deutscher Wetterdienst. In its most recent amendment (10 September 1998), the DWD Law calls for "*the issuance of official warnings of weather phenomena that could become a danger for public safety and order, especially concerning the impending danger of floods*". The present warning system was developed between 2000 and 2003.

5.2 Governance and Institutional Arrangements (National to Local Levels)

5.2.1 Policy, Institutional and Legal Framework to Support Emergency Planning and Response

Germany is a federal nation comprised of 16 Länder (States). Disaster management is the task of each individual Land (State) and its organisation differs from one Land to another. Civil protection is a central pillar in the basic concept of the Federal Republic of Germany's national safety architecture. It is part of the government's duty to protect and assure safety. The Federal Government and the Länder have implemented specific rules in laws to guarantee the protection of the population as well as the security of supplies in an emergency. The Civil Defence Act, various fire and disaster management laws and the emergency services laws provide the umbrella under which the Federal Government, the Länder and the municipalities operate a joint assistance system for the protection of citizens.

The DWD has signed an administrative agreement with each Land concerning co-operation in the field of disaster management. A second agreement concluded by the DWD governs cooperation with the Federal Office for Civil Protec-

Gerhard Steinhorst, Deutscher Wetterdienst
Rainer Vogelgesang, Deutscher Wetterdienst

M. Golnaraghi ed., *Institutional Partnerships in Multi-Hazard Early Warning Systems*,
DOI 10.1007/978-3-642-25373-7_5, © Springer-Verlag Berlin Heidelberg 2012

tion and Disaster Assistance (BBK). As the Higher Federal Authority, under the Federal Ministry of the Interior (BMI), the BBK performs duties in the area of civil protection and disaster assistance. The BBK operates a Joint Information and Situation Centre (GMLZ) on behalf of the Federal Government and the Länder. The aim of the GMLZ is to improve co-operation between the Federal Government and the Länder and between different federal ministries, as well as with national, international and supra-national organisations and between Germany and other countries, regarding the mitigation of major hazard and damage situations. It is, additionally, intended to provide continuous, up-to-date information about such situations to a variety of relevant units. The Joint Information and Situation Centre (GMLZ) uses both the German Emergency Preparedness Information System (deNIS) and a continually expanding network of internal and external experts. Essential meteorological information for disaster relief is made available to deNIS by the DWD.

A further cooperation agreement has been signed between the DWD and the German Federal Agency for Technical Relief (THW). The THW is a federal agency and the sole governmental disaster relief organisation. Its chief duty is to provide technical assistance in the wake of a disaster (e.g. by fighting floods or erecting emergency accommodation). As a result of this agreement, the DWD supplies meteorological information to support the THW's missions and facilitate planning.

Flood forecasting is the responsibility of the Länder where specific flood response centres are operated. Their forecasts rely on the meteorological information and forecasts supplied by the DWD.

5.2.2 National to Local Emergency Planning and Related Linkages to Early Warning Systems

A general definition of official warnings is given by the DWD Law. Specific warning criteria have been defined by the DWD and the Länder in administrative agreements. DWD services are incorporated in the emergency plans of fire brigades. The DWD also participates in the planning and execution of LÜKEX crisis management exercises that involve multiple German federal states.

5.2.3 Organizational Structure of DWD's Warning Management

The warning system established by the DWD for the issuance of warnings is operated around the clock. As illustrated in Fig. 5.1, it comprises three tiers – "Early Warnings", "Pre-Warnings" and "Warnings" of severe or extreme weather. The multi-level structure of the warning system ensures that the information issued becomes increasingly detailed and specific as a severe weather event approaches (Fig. 5.2). More specifically:

- Early Warning (level 1) is concerned with medium-term weather hazards for the period of the upcoming 2 to 7 days. All findings are compiled by the DWD into a "Weekly Weather Hazard Forecast", which is made available as a preventive disaster control measure and to inform the public;
- Pre-Warning (level 2) are produced for a time window of 12 to 48 hours. The results are summarised in "Warning Situation Reports" that describe the expected warning situation. Pre-Warnings of impending severe weather are issued on a district level in the case of large-scale, non-convective, weather events. Large-scale Pre-Warnings of severe weather, on a Land level, are issued for convective weather events.
- Warnings (level 3) of the warning management system encompasses current, district-related Warnings that are issued no more than 12 hours before the arrival of the warning event.

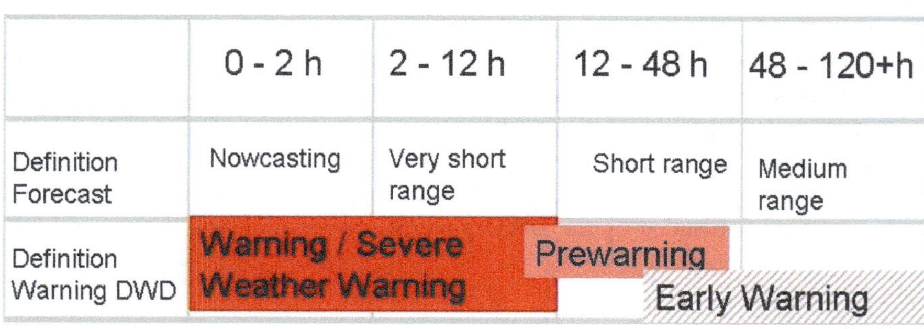

Fig. 5.1 The DWD's three-tiered warning system

Figure 5.3 illustrates the warning process along with the users of the warnings.

Early Warnings are issued for the whole of Germany by the Weather Forecasting and Advisory Centre (VBZ) housed at the Deutscher Wetterdienst's (DWD's) headquarters in Offenbach. Pre-Warnings and (severe) Weather Warnings are issued for their respective areas of responsibility by the six Regional Centres (RZ), located in Hamburg, Potsdam, Essen, Leipzig, Stuttgart and Munich, as well as by the regional component of the VBZ (Fig. 5.4). Each Regional Centre's area of responsibility comprises one or more Länder or, in other words, it is based on administrative units. The Warnings refer to individual districts or non-district cities (approximately 400 administrative units in total (Fig. 5.5). As discussed later, DWD's warnings are disseminated through various communication channels.

Public authorities and disaster relief units at the Länder and federal levels, fire brigades, technical relief units and the police are among the major users of the DWD's warnings (Fig. 5.6). For fire brigades, the Deutscher Wetterdienst (DWD) has developed a specialised, Internet-based, information system called FeWIS (Weather Information System for Fire Brigades) that delivers compact, specifically tailored, warnings. Information is, likewise, supplied to a database at the Joint Information and Situation Centre (GMLZ) of the Federal Office for Civil Protection and Disaster Assistance (BBK). In the event of a weather-related disaster, the situation centres are advised by federal ministries or relief units such as the German Federal Agency for Technical Relief (THW) or the German Red Cross. If necessary, a DWD expert takes part in the disaster relief team meetings and provides the team members with on-the-spot consultation.

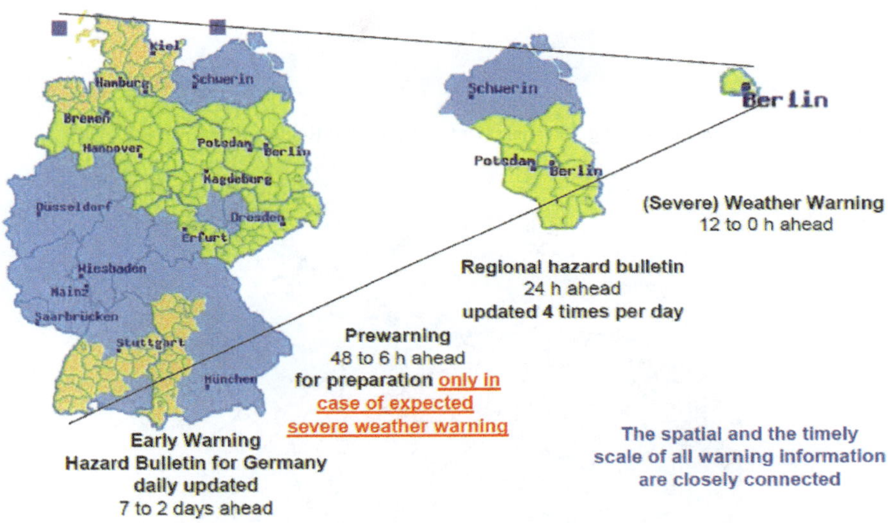

Fig. 5.2 The warning information becomes increasingly specific as a severe weather event approaches

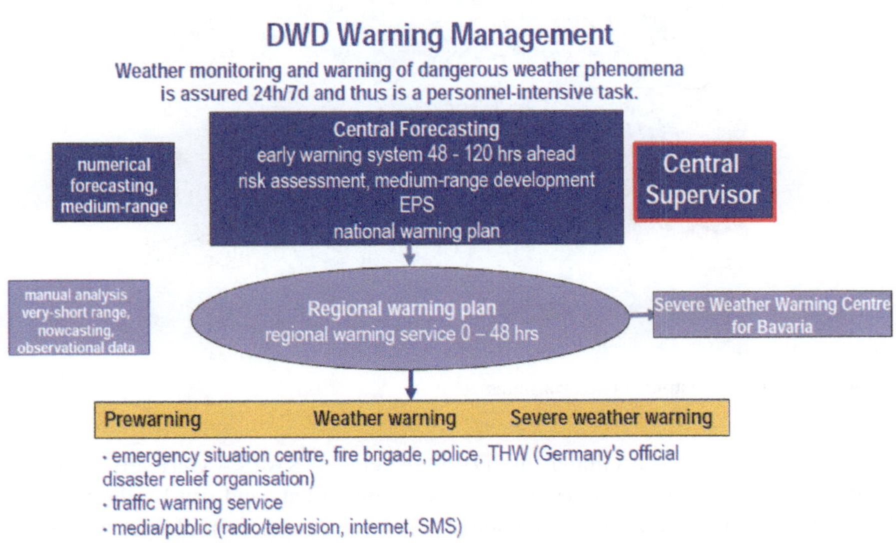

Fig. 5.3 Warning management of the DWD – the flow of warning information

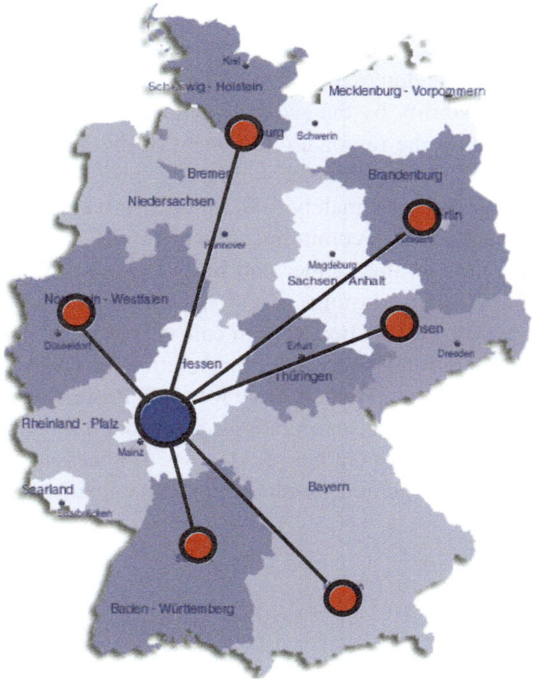

Fig. 5.4 The locations of DWD's Weather Forecasting Advisory Centre (VBZ) and Regional Centres (RZ)

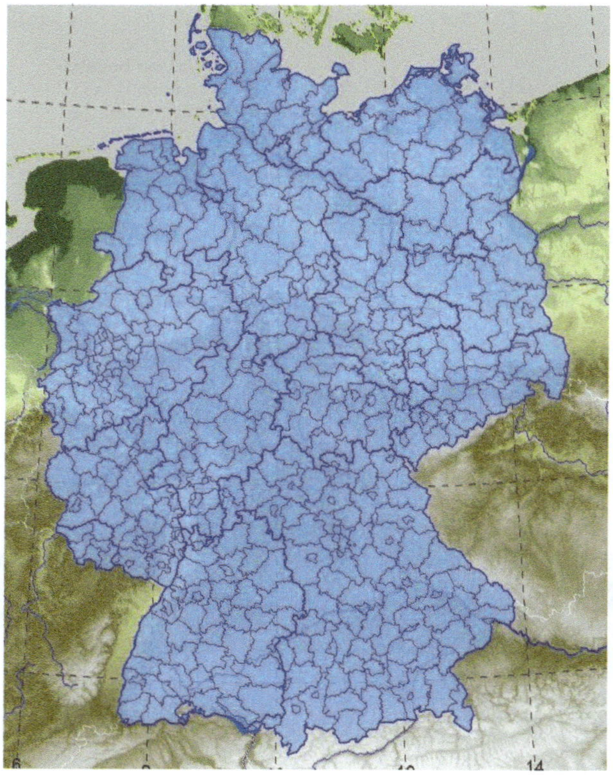

Fig. 5.5 The areas for which DWD issues district-related warnings

Working with the road authorities in the Länder, the DWD developed SWIS – Germany's Road Weather Information System – to optimize the deployment of personnel, vehicles, de-icing agents and grit by the winter maintenance service. This system provides invaluable support for winter maintenance service management. Reports are sent by fax or e-mail and there is also a closed user group for SWIS, which benefits from supplementary, up-to-date information on the Internet.

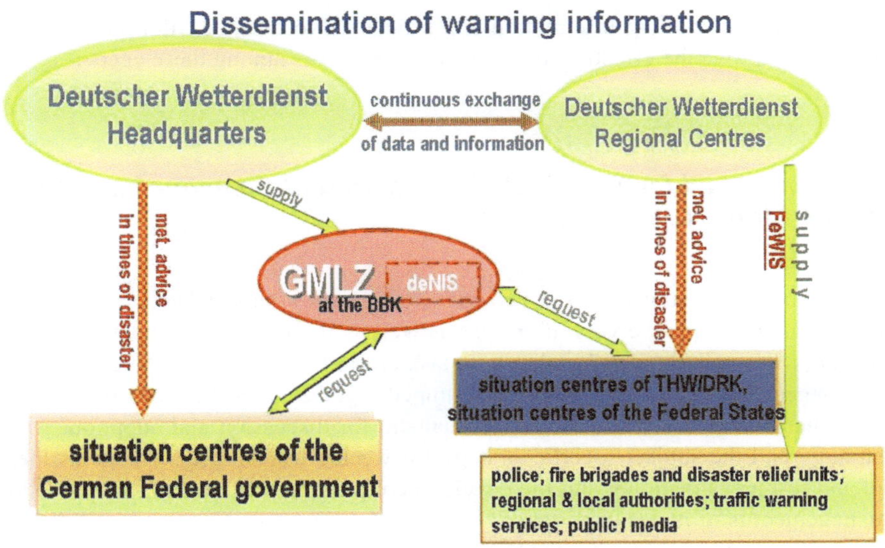

Fig. 5.6 The dissemination of DWD warnings to federal and regional institutions

The flood response centres of the Länder are other key customers that the DWD supplies with information, from the Early Warning stage onwards, as part of its tasks in the field of preventive flood control. In an emergency, meteorological data, model outputs, specialized data products and warning bulletins from the DWD form the basis for flood calculations by the regional flood control centres, as well as for the resulting disaster mitigation measures. Last but not least, the DWD attaches particular importance to informing the public about imminent weather hazards.

5.2.4 Institutional Capacities and Concept of Operations (Coordination and Operational Collaboration)

In accordance with Sect. 4 of the DWD Law, "*the Deutscher Wetterdienst supports the Länder in carrying out their responsibilities with regard to disaster control, especially in cases of extreme weather events*". If a disaster occurs, a meteorologist from the DWD can, at the request of the Länder government concerned, participate in the meetings of the disaster relief team. Moreover, a telephone warning service is established at the DWD's Weather Forecasting and Advisory Centre or the Regional Centres to provide the disaster relief teams of the Länder and districts with the necessary information. More specifically, the DWD cooperates with the following organizations and institutions:

- At Länder level, with the situation centres at the ministries of the interior of the Länder or their ministries for the environment, such as the situation centre for disaster control in case of a nuclear emergency;
- At regional level, with the regional police departments that are responsible for several districts;
- At district level, with the control or operation centres at the district administrations and fire brigade units;
- At local level, with the local authorities, to some extent;
- And with the operation centres of the German Federal Agency for Technical Relief (THW) and the German Red Cross (DRK).

5.2.5 Utilization of a Multi-Hazard Approach (Financial and Budgetary Aspects)

The DWD issues the warnings of weather hazards and all related information autonomously and independently of any other authority or disaster management organisation. Close co-operation with these bodies is assured by the DWD's 6 Regional Centres from which disaster management institutions are provided with advice. Providing information about the weather and issuing weather warnings are the sole responsibility of the DWD which, against the backdrop of the "single voice" principle, attaches particular importance to its role as the sole issuer of official warnings. The DWD maintains a close relationship with the users of its Warnings and, in particular, with the various disaster management authorities, in order to be in a position to optimize warning information and products. No detailed information is, however, available on how this interaction helps to improve the effectiveness and use of resources among these agencies.

5.2.6 Key Factors Considered for Sustainability of Different Components of Early Warning Systems With a Multi-Hazard Approach

The DWD is a federal authority and subject to the Federal Government's budget law. As such, it is obliged to plan its revenues and expenditures annually for the year after next as well as the three subsequent years. The planning information is organised according to so-called "titles", rather than task areas. In other words there is no budget that is explicitly planned for warning management. The DWD's plans are lumped together with the plans of all other federal authorities, including the ministries, then submitted to the German parliament for discussion and adoption. All planning is based on the DWD's overall strategy and the subject-specific strategy for weather forecasting. The goals identified therein set the priorities for financial and resources planning and the development plan. In this context, the provision of information and services to disaster management is of major importance.

Training programmes are organised each year for DWD staff members responsible for weather forecasting. The topics covered by the courses include recent developments in numerical weather prediction as well as methods for forecasting extreme weather phenomena. The DWD trains its weather forecasting personnel within the framework of a broadly based meteorological education and training programme, one component of which is the improvement of warning management competences. There is no budget that is explicitly planned by the DWD for warning management. However, the provision of information and services to disaster management is of major importance in the context of the DWD's overall strategy and the subject-specific strategy of weather forecasting. Within the framework of the DWD's strategic orientation in the field of weather forecasting and warning management, the work processes have been reorganised and human resources considerably reduced by centralization and semi-automation. In consequence, the warning services are provided in a most efficient manner.

The DWD does not carry out any assessment of the cost-effectiveness of its warning activities. It is to be assumed, however, that the economic value of the DWD's warning management greatly exceeds its costs, as this is regularly demonstrated to be the case in cost-effectiveness studies on other National Meteorological Services. The services provided for warning management are included in other tasks of weather forecasting and the overall services provided by the DWD are recorded in the business accounts. The DWD operates its warning management by means of its own internal capacities with, in this context, no contribution from other authorities on the federal or the regional level.

5.3 Utilisation of Risk Information in Emergency Planning and Warnings

5.3.1 Organizational Responsibilities and Arrangements for the Development of Risk Information

The governments of the Länder are largely autonomous in disaster management. Since disaster management is the responsibility of the Länder, risk information is compiled by the regional government authorities. With disaster control being the responsibility of the Länder, it is very difficult to introduce national standards, such as those developed by the Federal Office for Civil Protection and Disaster Assistance (BBK) for the whole of Germany. There are some university institutes and ministerial authorities at Länder level that address specific agricultural issues or social impacts of disaster management, however no detailed information is currently available on which agencies are involved.

The Deutscher Wetterdienst (DWD) has elaborated or contributed to the elaboration of the following hazard-related products and services in support of disaster prevention and preparedness:

- Collaboration on German DIN Standards on constructional loads such as wind-induced loads, snow and ice loads, loads on overhead lines;
- Climate maps (e.g. mean annual number of hot days, 10-year statistics on the date of first snow cover (≥ 5 cm));
- 10- and 30-year station data statistics (e.g. number of days with a daily wind peak over 10 Bft, number of days with daily precipitation totals ≥ 50 mm);
- Expert opinions for project planning purposes (upon individual request), for example expert reports on wind speeds.

The DWD also offers the WESTE-KAT service that supplies disaster management organisations with climate data for disaster prevention or for the assessment of a past disaster situations. In addition, the DWD's Hydrometeorology Department, which helps to solve hydrological problems in the German regions through its collaboration on the relevant bodies, should be mentioned in this connection.

5.3.2 Hazard Assessment, Quantification and Mapping (National to Local)

The major natural hazards in Germany are coastal and inland flooding, landslides, avalanches and snow load-caused collapses of roofs (e.g. of sports halls). Particular significance must also be given to windstorms with extremely high wind speeds, thunderstorms with hail and lightning discharges, and continuous precipitation events with extreme amounts of precipitation followed by flooding. The river flood warning lies within the responsibility of the flood response centres of the Länder.

The DWD does not retain any systematic records of weather-related damage or the costs incurred.

However, it is a member of the European Severe Storms Laboratory (ESSL) and has access to ESSL's database of convective severe weather events in Europe. There are plans to extend this database to include other severe weather phenomena. This information will facilitate statements about the frequency of occurrence of specific extreme weather events. The ESSL database, however, contains no systematic data about damage attributable to severe weather.

5.3.3 Assessment of Vulnerabilities and Exposure (National to Local)

The Deutscher Wetterdienst (DWD) has, for many years, carried out extreme-value statistical analyses aimed at the practical objectives of hydrology and water management. One result of these analyses is the KOSTRA Atlas "Heavy precipitation totals in Germany", published in 1997 on the basis of the 1951–1980 time series. The Atlas contains a statistical analysis of heavy precipitation data, showing also the frequency of exceedance and probability of recurrence. The KOSTRA Atlas can be taken as a basis for technical flood protection in the sense of disaster preparedness. Faced with the climate changes over the past 20 years, the decision was taken to continue the KOSTRA analyses and, at the same time, extend and update the database. The result of this is the KOSTRA-DWD 2000 Atlas entitled "Heavy precipitation totals in Germany (1951 to 2000)". As mentioned earlier, the DWD's WESTE-KAT service also supplies disaster management organisations with climate data that can be used to assess exposure to hydrometeorological hazards.

5.3.4 Storage and Accessibility of Disaster and National Hazard Risk Information

The DWD's KOSTRA Atlas can be purchased by anyone from one of DWD's partners.

5.3.5 Development and Utilization of Hazard/Risk Information to Support Emergency Planning and Warnings

On its website, the DWD offers access to a warning system named WESTE-KAT. WESTE-KAT is provided only for disaster management organisations and supplies these agencies with climate data for preventive disaster management purposes or for the evaluation of a past disaster situations. The DWD has developed this system in collaboration with the disaster management organisations.

All warnings of extremely severe weather are issued by the DWD accompanied by a supplementary text that describes the effects of the expected extreme conditions. Supplementary information may also be provided for all other weather and severe weather warnings. This information could, for example, relate to the possible consequences of the severe weather, or define "do's" and "don'ts" for persons affected by the event. It is published on the basis of an internal DWD regulation and is not stipulated by the Federal Government.

5.4 Hazard Monitoring, Forecasting, and Mandates for Warning Development

5.4.1 Organizational Responsibilities for Monitoring, Forecasting and Development of Hazard Warnings

The DWD has been given the statutory task of "*issuing official warnings of weather phenomena that could become a danger for public safety and order, especially concerning the impending danger of floods*". However, warnings of severe weather events are also issued by private meteorological providers – something that is not forbidden in Germany. Nevertheless, the DWD is of the opinion that only official warnings issued by Germany's National Meteorological Service should be binding for disaster mitigation purposes (single-voice principle). Effective protection against weather-related damage can only be achieved if public institutions and the general public receive warnings from one source. The German Federal Government supports the single-voice principle. In almost all other countries, official warnings are disseminated by the national weather service and a clear set of rules has also been formulated by the WMO on this subject (CBS, RA VI).

Table 5.1 shows the various weather phenomena for which the DWD issues warnings along with their classification by warning Types. The warnings do not refer exclusively to severe weather but also to selected weather phenomena of a less hazardous nature.

The DWD also issues warnings to safeguard marine shipping, the sea coast and lakes. In addition, it warns against extreme heat and UV radiation and publishes a forest fire danger index. Type III river flood warnings are the responsibility of the flood response centres in the Länder rather than the DWD. However, these centres are supplied with all essential meteorological data by the DWD.

Pursuant to the law on the prevention of radiation (StrVG), the DWD is also responsible for monitoring the radioactivity of air and precipitation. The DWD makes its measurements and forecasts available to the Federal Office for Radiation Protection (BfS) and the Federal Ministry for the Environment, Nature Conservation and Nuclear Safety (BMU). Public authorities at the federal and regional levels as well as the ministries in the Länder use this information as the basis for their recommendations on further measures to be taken for protection.

Table 5.1 Weather phenomena for which the DWD issues warnings (N.B.Type depends on the federal state)

Weather phenomena	Hazard type	Warning responsibility
Violent gusts	I	DWD
Hurricane-force gusts	I	DWD
Extreme hurricane-force gusts	I	DWD
Severe thunderstorm	I	DWD
Intense heavy rainfall	I	DWD
Strong continuous rain	I	DWD
Extremely strong continuous rain	I	DWD
Heavy snowfall	I	DWD
Extremely heavy snowfall	I	DWD
Heavy snowdrifts	I	DWD
Black ice	I	DWD
Heavy thaw	I	DWD
Wind gusts	I	DWD
Storm gusts	I	DWD
Violent storm gusts	I	DWD
Thunderstorm	I	DWD
Heavy rainfall	I	DWD
Continuous rain	I	DWD
Snowfall	I	DWD
Snowdrifts	I	DWD
Slippery roads	I	DWD
Fog	I	DWD
Frost	I	DWD
Severe frost	I	DWD
Flood	III	Flood warning centres of the Länder
Avalanches	III	LfU
Forest fire	III, I*	Forestry administrations of the Länder
Radioactivity	III	BfS, BMU

5.4.2 Organizational Collaboration and Coordination for Development of Hazard Warnings

The DWD's warning management has been described earlier and the public authorities that use the DWD's warnings for the performance of their own duties have also been identified. The cooperation between the DWD and disaster relief units in Bavaria is described in the following sections as an illustrative example. It should be kept in mind, however, that cooperation and coordination mechanisms and processes differ from one Land to another.

In 2003, the Bavarian Ministry of the Interior (BStMI) and the Deutscher Wetterdienst (DWD) agreed on the establishment of a "Bavarian Severe Weather Warning Centre (UWZB)" at the DWD Regional Centre in Munich. Analyses had shown that 90% of all disasters recorded in Bavaria since 1993 were weather-related. Following a successful pilot project, the Bavarian Severe Weather Warning Centre (UWZB) officially commenced operation at the Munich Regional Centre in mid-2004.

The Bavarian Severe Weather Warning Centre provides services linked to the protection of life and property to the Bavarian population in general and the Bavarian disaster control and safety authorities in particular. It is part of the DWD's Regional Centre in Munich and operates 24 hours a day in standby mode. In the event of severe weather, it is activated and manned with an appropriate lead time on the basis of an on-call duty system. Thanks to the UWZB's help, it has been possible to set up an integrated severe weather warning and alarm system, characterised by high temporal and spatial detail, for the Bavarian authorities and population, which has been regularly improved since 2004.

The Bavarian Severe Weather Warning Centre's (UWZB's) responsibilities are as follows:

- To issue warnings/severe weather warnings and "now cast" bulletins, which are regionalised down to districts and sent directly by fax, text message and e-mail to the relevant authorities (situation centres, district governments, police headquarters, district administrations, municipalities) as well as to fire brigades, the German Federal Armed Forces and relief units such as the THW, the Bavarian Red Cross, water and alpine rescue units, etc. These warnings are also available to anyone on the DWD's website. In addition, the regional warning situation report from the same morning is provided to the disaster control authorities to facilitate their daily planning.
- To alert and provide individual consultation to the disaster control authorities according to the "Special Severe Weather Emergency Plan" which has been in force in Bavaria since 2005 (Fig. 5.7). This plan decrees that the severe weather warnings issued by the UWZB should be assessed in light of the local risk situation by authorised persons, appointed for this purpose in the towns and municipalities. These persons should initiate timely and adequate measures as appropriate (e.g. for the duration of large public festivals, by closing underpasses, subways and underground car parks in case of heavy rainfall, or by reducing the water level in reservoirs to counter the threat of flooding). The success of the alarm system hinges on the speed at which the UWZB's severe weather warnings are communicated to the relevant authorities. The towns and municipalities receive these warnings both from the UWZB, which sends a text message as an advance signal, and from the responsible rural district office, via a suitable alarm system, fax or e-mail. Provisions are made for alarms to be dealt with by an alternative unit during times when the authority concerned is not manned (at nights and weekends).
- To act as a point of contact for the relevant disaster control and safety authorities in Bavaria and establish a dialogue with them. In this connection, the Bavarian Severe Weather Warning Centre (UWZB) makes use of the warning telephone service installed at the Munich Regional Centre for public authorities as well as FeWIS, the Weather Information System for Fire Brigades, which is available in Bavaria down to the level of the lower disaster control agencies or, in other words, the rural district offices and non-district cities. It also utilises special Intranet sites belonging to the rural district offices by means of which towns and municipalities with no access to FeWIS can obtain warnings or weather and radar information published by the DWD for their particular region. The combination of the warning telephone system and the Intranet pages allows the UWZB to provide optimal consultation to users.
- To warn the Bavarian population of imminent severe weather by issuing radio bulletins (embodied in Bavarian law) in the form either of emergency information or of official emergency warnings in the event that the severe weather develops rapidly and takes on an extreme form. As soon as a severe weather warning is issued, the UWZB transmits a radio bulletin, tailored to the needs of the media, to the radio, television and teletext services. Bayerischer Rundfunk (BR – the public-service radio station in Bavaria) takes priority over private radio stations in the Free State of Bavaria's severe weather warning management system. It receives the severe weather radio bulletins in a special XML format so that they can be called up in a suitable form (at the push of a button) on BR's Intranet by all relevant editorial teams (weather, traffic, news, etc.) the minute they have been compiled by the meteorologist. This direct transmission to BR is part of an agreement between the Bavarian Ministry of the Interior (BStMI), the Munich Regional Centre and Bayerischer Rundfunk (BR).
- To co-operate with flood forecasting services, the Bavarian avalanche warning service and road maintenance authorities. These institutions receive detailed daily information about weather conditions – even if the weather situation is not severe – or weather and severe weather warnings as and when appropriate. The exchange of information and contacts is more intense in severe weather situations.
- To provide information about severe weather events to the media, depending on the situation.

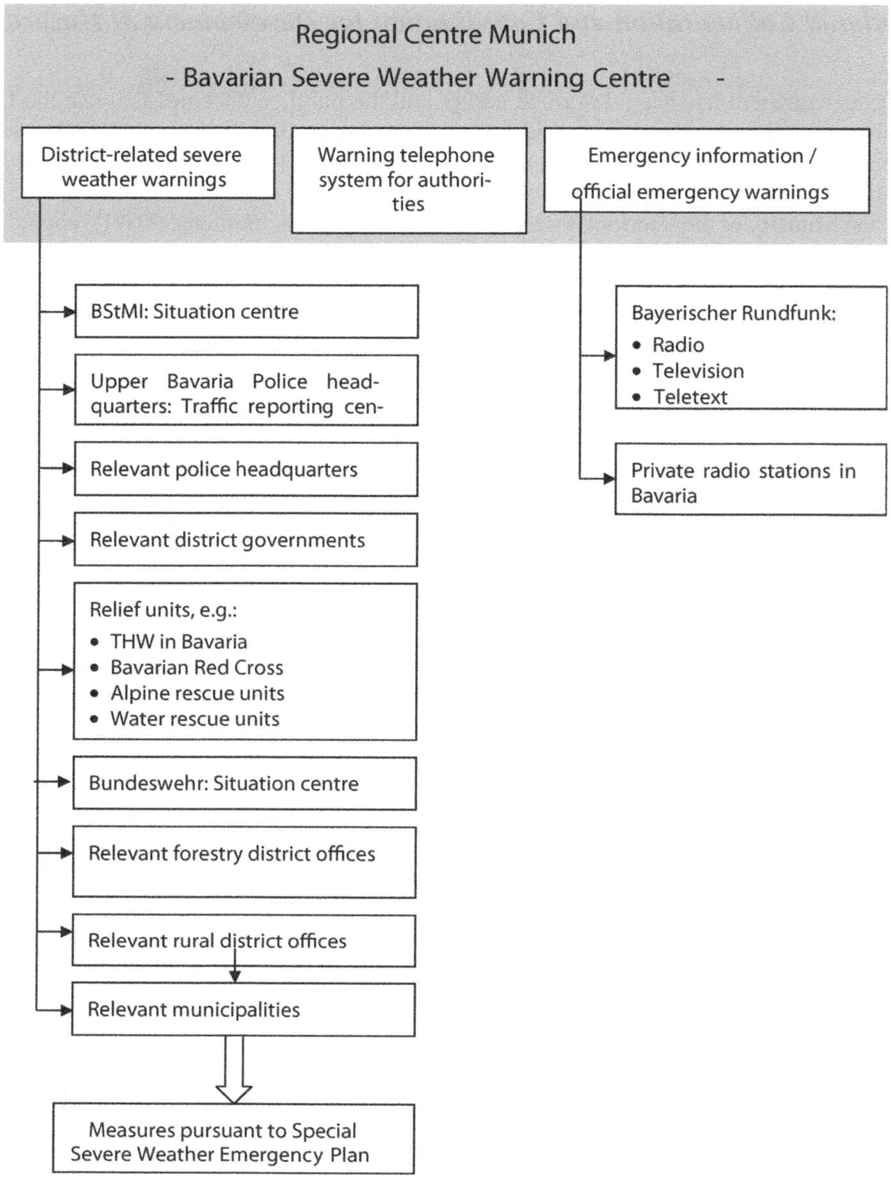

Fig. 5.7 Bavarian Severe Weather Warning Centre procedure for alerting disaster relief units and transmitting emergency information to radio stations

The close co-operation in severe weather situations has resulted in the successful implementation of several vital measures, including:

- Timely evacuation of large camp sites (e.g. in 2004);
- Intensive support for flood forecast centres (e.g. timely intervention at Lake Forggensee in 2005 to enable the flood water of the River Lech to be controlled in an orderly manner. The Swabian district government stressed that without this intervention, several bridges over the Lech would have been washed away);
- Timely information to the organisers of large public festivals (e.g. Forchheim 2007, 2008).

The Bavarian Severe Weather Warning Centre (UWZB) has demonstrated its effectiveness as an institution for mitigating severe weather situations, as testified by letters from the Bavarian Minister of the Interior in 2005 and 2007. In another letter dated 22 February 2007, the Minister paid tribute to the co-operative spirit displayed by the UWZB, public authorities and radio stations in their response to Storm "Kyrill". Bayerischer Rundfunk is extremely satisfied with the direct information channel described above.

As a final note, there is no authority or institution in Germany that coordinates the issues of disaster management on an inter-authority and national level. Consequently, the DWD has several contact persons for disaster management including, for example, one who is responsible for coordinating the cooperation with disaster management organisations at the federal level and another responsible for operational cooperation on the Länder level.

5.5 Development of Understandable, Authoritative, Recognisable and Timely Warnings

5.5.1 Warning Message Development Cycle

As outlined previously, the DWD operates a three-tiered warning management system consisting of Early Warnings, Pre-Warnings and district-related Warnings, based on the principle of increasing spatial and temporal refinement. The following sections provide further details on DWD's weather warnings.

- *Early Warnings* – Early knowledge of significant weather events (significant warning events and (extreme) severe weather events) expected within a supra-regional radius (250 to 700 km) is communicated in the "Weekly Weather Hazard Forecast" as a probability forecast (possible, probable, very probable) and taken into account in the short and medium-range local area forecasts (from two days ahead).
- *Pre-Warnings* – Knowledge of significant weather events (all warning events) expected within a regional radius (50 to 250 km) is taken into account in the short-range local area forecasts (up to two days ahead) and communicated in the regional warning situation report (Regional Centre area), with an advance warning time of 24 hours or less.
- *District-related Warnings* – A specific weather or (extreme) severe weather warning always refers to an individual district, possibly with a further breakdown according to altitude. The lead time is the time from when the warning is issued until the time when it actually takes effect. It is, invariably, a compromise between providing concrete information to the user/customer at the earliest possible stage and achieving the finest possible spatial and temporal differentiation, in combination with the required accuracy. This also applies, analogously, to the duration (period of validity) of a warning.

A lead time (i.e. the time period preceding the expected warning event within which a warning should be issued) is defined for each such event. The warning event starts at the end of the lead time and lasts until the end of the validity period specified in the warning. A maximum period of validity is likewise fixed for each warning event. If a particular warning event is expected to last longer, the anticipated duration of the event can be mentioned in the warning text. In this case, the warning must be renewed and re-issued.

Supplementary texts of a general nature are included in severe weather warnings. They describe the potential damage that can be caused by the severe weather and list a series of "do's" and "don'ts". These supplementary texts must be selected from a catalogue or formulated on the basis of a catalogue of standard texts. They are designed with the needs of the general public in mind.

The DWD's warning system differentiates between four warning levels, determined by the degree of danger associated with the expected weather conditions: – (simple) weather, significant weather, severe weather and particularly extreme and severe weather. A set of correspondingly intense weather phenomena (i.e. conforming to defined threshold values) is specified for each warning level. Pre-Warnings are, moreover, issued for selected severe weather warnings.

All warnings are made available to the public over the Internet on the DWD's warning pages (Fig. 5.8). They are shown in an overview diagram for Germany as a whole and in more detailed diagrams for the individual Länder. Those districts for which warnings have been issued appear in different colours according to the warning level. If several weather warnings with different levels exist for the same district, this district is represented in the colour corresponding to the level with the highest degree of danger. A hazard symbol (warning triangle) is used to indicate that a Pre-Warning has been issued for a particular Land (State).

By clicking on the appropriate symbol, the map can be made to show only warnings relating to a single weather phenomenon. In addition to the static overview of the current warning situation, the DWD also releases a film showing how this situation is expected to develop in the next 24 hours.

Weather information tailored to the needs of special users is accessible online to closed user groups. SWIS and FeWIS are typical of such groups that profit from Deutscher Wetterdienst (DWD) warnings. In the Road Weather Information System (SWIS) user group, specific information is provided for winter maintenance services and road maintenance authorities. FeWIS stands for "Weather Information System for Fire Brigades" and is particularly targeted at the needs of professional fire services. It delivers a warning overview, a warning situation report, Pre-Warnings and district-related Warnings (see Fig. 5.9). Each FeWIS member can specify the threshold values above which they would like to receive information or, in other words, they can change the default thresholds set by the DWD.

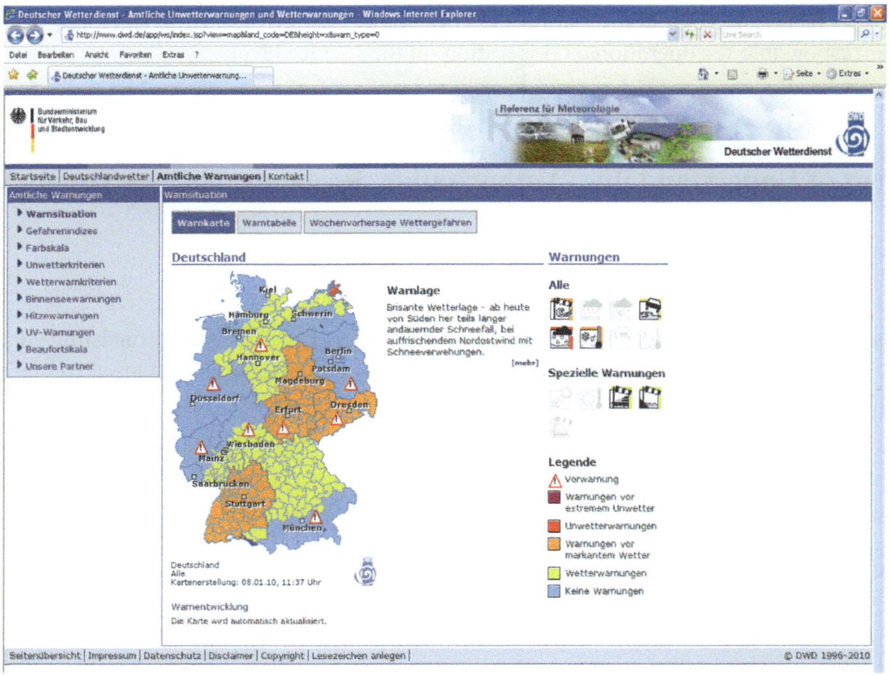

Fig. 5.8 The DWD's on-line warning map

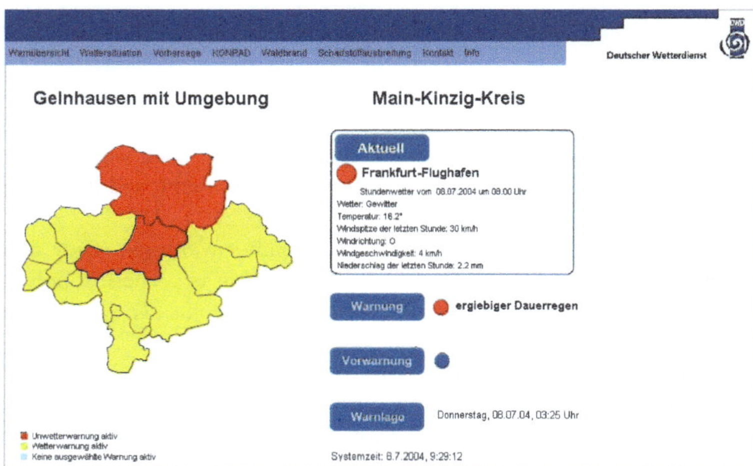

Fig. 5.9 District-related warnings in FeWIS

Table 5.2 Warning levels and colours used for districts on the on-line warning map

Warning level	Expected weather conditions	Colour
Official warning of particularly extreme and severe weather	Extremely dangerous weather conditions	Dark Red
Official severe weather warning	Very dangerous weather conditions	Red
Official warning of significant weather	Dangerous weather conditions	Ochre
Official weather warning	Possibility of weather-related risks	Yellow
No warning	No weather-related risks	Blue

FeWIS also provides many other types of information – images and loops taken by the weather radar (Fig. 5.10), satellite loops, supplementary information about temperatures, precipitation, wind peaks, frost penetration depths, snow depths, forecast texts, etc. – as well as the forest fire danger index (Fig. 5.11).

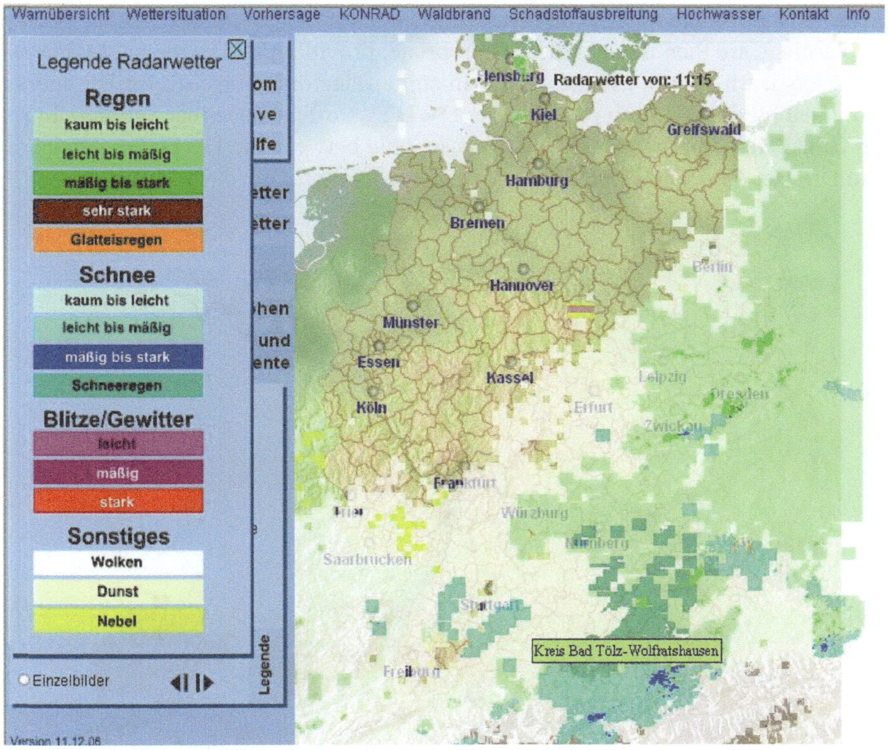

Fig. 5.10 Weather radar image in FeWIS

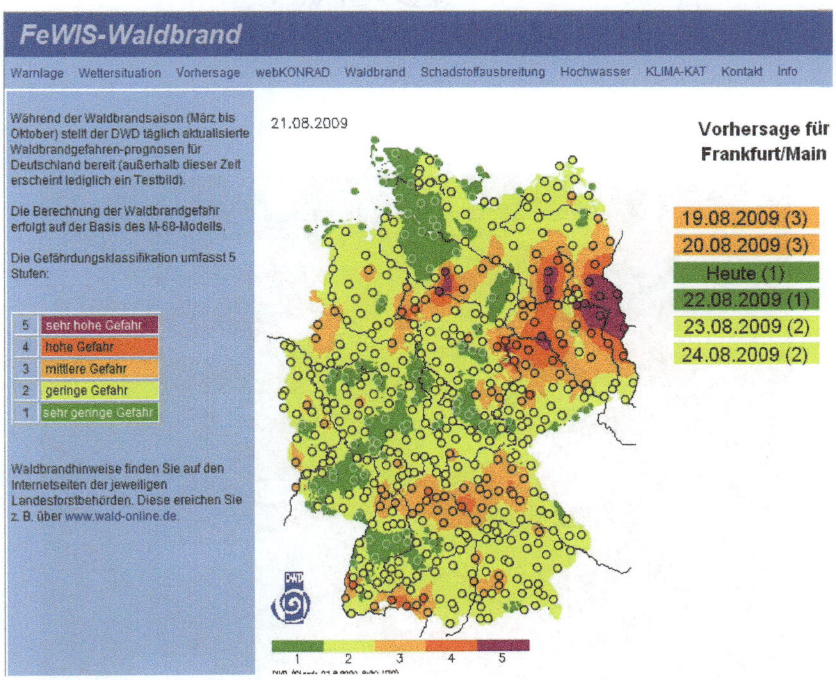

Fig. 5.11 Forest fire danger index in FeWIS

Direct access to the webKONRAD system (training essential), which evaluates radar information and is hence a valuable source of data about the intensity and direction of movement of thunderstorms in convective weather conditions (Fig. 5.12), is also available through FeWIS. In the event of high pollution levels, the DWD can produce a forecast of pollutant dispersion for FeWIS users (Fig. 5.13).

Specific warning criteria or threshold values can be agreed with users within the framework of written agreements with the Regional Centres. Warnings are then issued where appropriate, providing it is in the public interest to prevent the hazards concerned. FeWIS users learn how to use the system in an introductory course. Refresher courses or presentations of new features can be arranged by the DWD on request. The DWD also offers a seminar entitled "Introduction to Meteorology for Disaster Relief Units". In addition, regular meetings are held between the DWD, the Länder and the Federal Office for Civil Protection and Disaster Assistance (BBK) to improve cooperation.

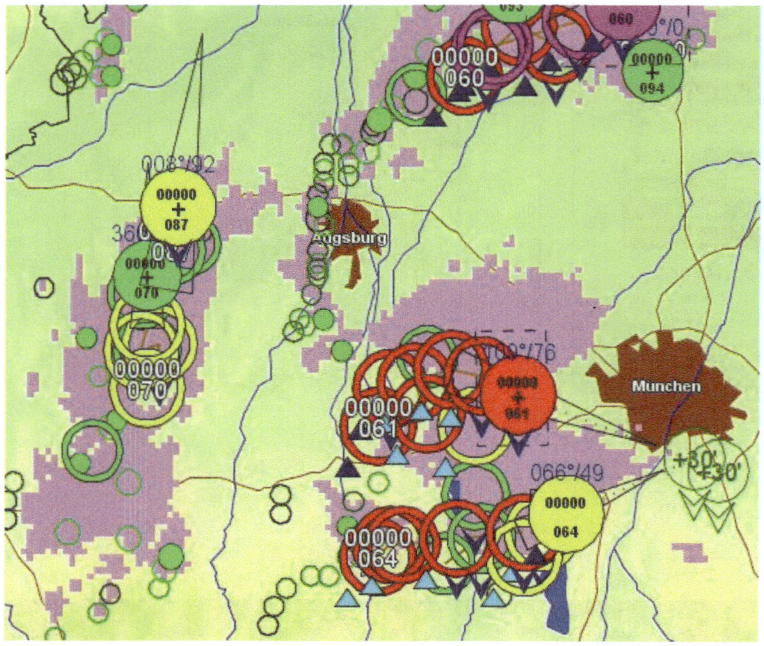

Fig. 5.12 Intensity and direction of movement of thunderstorms in FeWIS

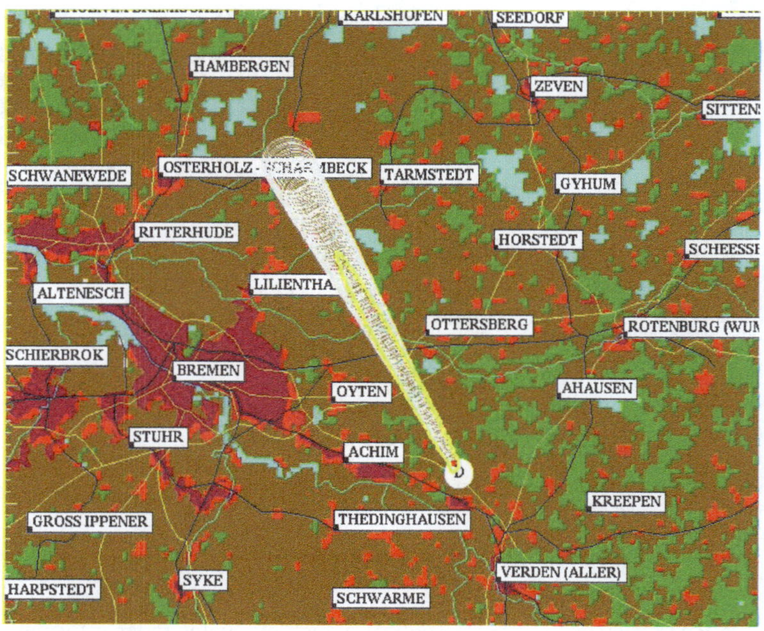

Fig. 5.13 Dispersion model of pollutant emissions

5.5.2 Warning Message Improvement Cycle

Various changes to the warning criteria for slippery roads and thunderstorms have been implemented since the warning system was first introduced in 2003, in response to demands from users (utility companies, disaster relief units). Criteria for warning against vibrations of overhead lines were, for example, formulated by the DWD as a reaction to the catastrophic effects of snowfall on electricity supplies in northern Germany. In addition, the FeWIS system is regularly optimized to take account of users' wishes. The verification of warning quality has, moreover, led to the definition of mandatory lead times for warnings and validity periods. The results of a survey launched in 2007 on public warning are shown in Fig. 5.14.

The lead times used are a compromise between warning of a severe weather as early as possible and delivering a statement which is as reliable as possible regarding the place and time of occurrence of the severe weather. The lead times for all warning events are defined in tabular form, with Table 5.3 illustrating some examples.

Over the years, the DWD has also broadened the scope of its warning activities:

- To protect the population against the risks of excessive exposure to UV, the Deutscher Wetterdienst has published UV Index forecasts since 1995. From 1 April to 31 August, district-specific figures appear on the DWD's website.
- A heat/health warning system was developed by the DWD as a reaction to the 2003 heat wave in Western Europe. As soon as the 'perceived temperature' exceeds a defined threshold, the DWD issues a heat warning. These warnings are regionalized down to district level. They are transmitted directly to the local authority partners for redistribution to institutions such as nursing and old people's homes and hospitals. Furthermore, the general public has free-of-charge access to all valid heat warnings on the DWD website.

The DWD has, in addition, implemented a quality management system in accordance with DIN EN ISO 9001:2000 and was awarded certification in 2004. A strategic process called "Weather Forecasting and Warning Services" is defined within this system. The management reviews of this process take place on a regular basis and also include improvements to the DWD's warning management.

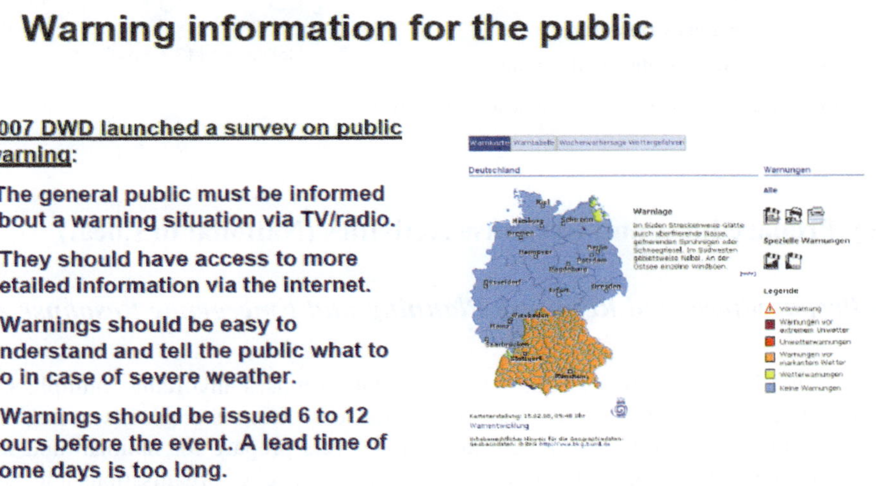

Fig. 5.14 The results of a survey on public warning (2007)

Table 5.3 Lead times of selected warning events

Warning event	Warning	Prewarning	Early warning
Violent storm gust	6–1 hr	48 hrs ahead	120–48 hrs ahead
Continuous rain	12–2 hrs	48 hrs ahead	96–48 hrs

5.6 Warning Dissemination Mechanisms

The Deutscher Wetterdienst's (DWD's) warnings are disseminated via the Internet, fax-on-demand services, text messages, e-mail and, in specific cases, by telephone. They are, moreover, placed on an ftp server from which they can be downloaded by customers. The specialized FeWIS and SWIS weather information systems for fire brigades and road and winter maintenance services should also mentioned here as examples of dissemination mechanisms for closed online user groups. For information to the general public, the DWD uses the Internet (http://www.dwd.de), as illustrated earlier in Fig. 5.8, and relies on the voluntary support of radio and television companies. Broadcast stations are, however, generally under no obligation to disseminate the DWD's warnings on television and radio since, except in Bavaria, there are no legal regulations applicable to this area. Nevertheless, public service stations, in particular, have increasingly included DWD warnings of critical weather situations in their news bulletins and several broadcasters also display these warnings on their websites. Figure 5.15 illustrates the distribution of warning information to the general public and via FeWIS.

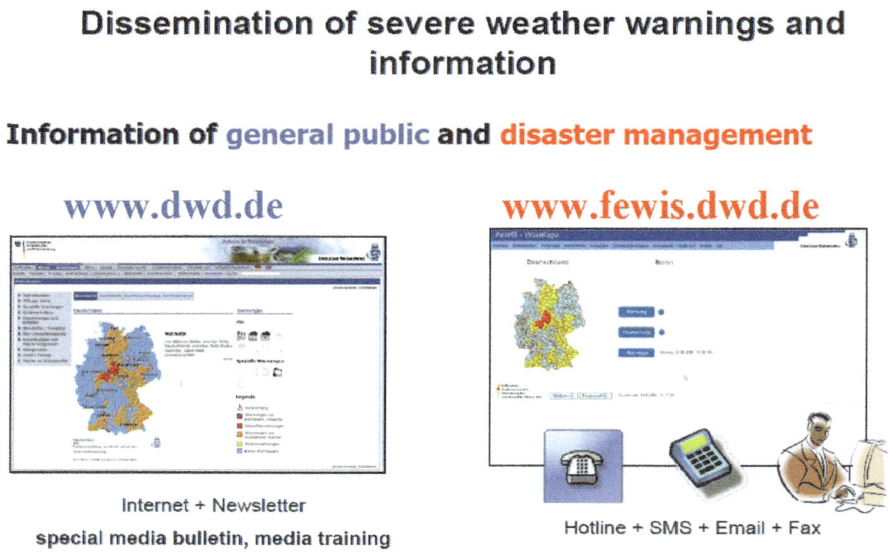

Fig. 5.15 Dissemination of warning information to the public and via FeWIS

5.7 Emergency Preparedness and Response Activities (National to Local)

5.7.1 Disaster Preparedness and Response Planning and Emergency Response Activation

The DWD participates in LÜKEX exercises, crisis management exercises involving multiple German federal states, which take place approximately once every two years. The purpose of these exercises is to help emergency task forces and crisis management teams at Federal Government and Land levels prepare for national and longer-lasting, extreme crisis situations, together with private operators of critical infrastructure, relief organisations and associations. LÜKEX is a so-called "strategic staff framework exercise" that enables the emergency preparedness of management and coordination teams to be tested. However, no measures are actually implemented on the ground during these exercises and no emergency and service vehicles or units are moved. The DWD provides the necessary "artificial weather" in the form of weather charts, reports and dispersion models while meteorologists at the Weather Forecasting and Advisory Centre (VBZ) in Offenbach and the various Regional Centres offer advice on the current weather situation.

5.7.2 Community Response Capacities

If a disaster occurs, a meteorologist from the DWD can attend the meetings of the disaster relief team, at the request of the Länder government concerned.

5.7.3 *Public Awareness and Education*

The DWD raises public awareness of its on line warning pages through press releases, an annual report and brochures. In addition, the DWD's website provides an explanation of the criteria for weather warnings. The warnings describe the risks and explain the general rules of conduct.

5.8 Development of Warnings and Related Products and Services for the Disaster Risk Management User Community

The following illustrate some examples of developments undertaken by the Deutscher Wetterdienst (DWD) as part of continuing efforts to provide the best possible services to the disaster risk community:

- The users of FeWIS, the weather information system especially for fire brigades, are initially trained in the use of the system. On request, the DWD also holds refresher seminars or update presentations. FeWIS is continuously improved in response to user requirements, resulting from the feedback received, for example, during training seminars;
- Since its introduction in 2003, the DWD's warning system has been modified, for example in the warning criteria for 'black ice' and 'thunderstorm', in response to user requirements (energy suppliers, disaster management). As a consequence of the catastrophic effects of heavy snowfalls on the power network in northern Germany, the DWD has introduced warning criteria for conductor vibration;
- Since 1995, the DWD has been providing UV index forecasts to protect the population against the risks of excessive UV exposure. These forecasts, which are compiled district-wise, are published on the DWD's website from the 1st of April until 31st of August;
- Following the 2003 heat wave in Western Europe, the DWD has developed a heat warning system. As soon as the 'perceived temperature' exceeds a certain threshold, the DWD issues a heat warning.

5.9 Improvement of Overall Operational Framework of Early Warning Systems

Cooperation between the stakeholders usually takes place at the Länder level bilaterally between the DWD (represented for example by the head of one of the Regional Centres) and the representatives of the Länder. This cooperation includes participation in the meetings of disaster management organisations and the holding of user conferences and training programmes for closed user groups such as for the FeWIS user group (Fig. 5.16). Storm "Kyrill", which raged throughout Germany in January 2007, is a good example here. In March 2007, the DWD invited representatives of several disaster relief units to its Headquarters with the aim of identifying possible weaknesses and jointly developing suggestions for future improvements (feedback event: "Lessons Learned from Kyrill").

Fig. 5.16 Co-operation between DWD and disaster management institutions

5.10 Examples of Previous Events Where the Operational Early Warning System Has Led to Improvements in Emergency Preparedness and Prevention

Extremely heavy precipitation occurred in the Elbe catchment area during the first half of August 2002. This precipitation lasted more than 48 hours. From 12 August 2002, unusually extreme flooding was recorded in Saxony as a result of the heavy rainfall which affected the entire region. Warnings that heavy precipitation was likely the following weekend were issued by the Deutscher Wetterdienst (DWD) on 8 and 9 August 2002. Nevertheless, co-operation between the DWD and the flood response centre was hindered by misunderstandings regarding the impact of such enormous amounts of precipitation. Measures to improve collaboration between the DWD and the Länder concerned have since been implemented.

Storm "Kyrill", which raged in January 2007, was predicted by the DWD five days before it actually spread to Germany. There is little doubt that the DWD's accurate warnings helped save human lives and curtail the damage to property. Thanks to the availability of these early warnings, the North Rhine-Westphalian Ministry of the Interior's Situation Centre was able to increase its staffing levels and an emergency task force was set up at the Federal Office for Civil Protection and Disaster Assistance (BBK). Personal contact with the meteorologist at the Regional Centre was a very important decision-aid when it came to civil protection measures, for example. The opportunity to talk to the meteorologist directly was also instrumental in improving the quality of the experts' assessments and raising their awareness of the situation. The high regional content of the warnings enabled the relevant disaster relief units to be alerted specifically. As a reaction to the temporary overloading of the DWD's servers, a new system is now in place that guarantees the supply of information to FeWIS users and facilitates access to DWD warnings at any time.

The strong wind field of the hurricane-force low pressure area "Emma" traversed Germany with heavy thunderstorms on 1 March 2008. Seven days previously, the DWD had warned of this storm event in its "Weekly Weather Hazard Forecast", based on the results of a numerical weather prediction and statistical follow-up processes. Both the general public and the disaster relief units were able to be warned in good time.

5.11 Overall Lessons Learned and Future Steps for Improving National Meteorological and Hydrological Services Contribution in Early Warning Systems Particularly Focusing on Institutional Coordination and Cooperation

A measure of particular importance was the establishment of a contact officer for fire brigades and relief units to coordinate cooperation with the Länder and pursue the improvement of the information systems such as FeWIS. The contact officer is also responsible for optimizing the communication with fire brigades and relief units and for improving training seminars for the users.

Another instrument which has proved its value is the "cross-border exchange of warnings" (i.e. the direct exchange of warnings through predefined channels between the Regional Centres of neighbouring countries). The cross-border exchange of warnings is the subject of various bilateral agreements concluded between the Deutscher Wetterdienst and the National Meteorological Services of Austria, France and Switzerland. As a result of the agreements, there is also a regular exchange of information between the Meteorological Services.

The warning system developed between 2000 and 2003 has proved to be a success. Thanks to a continuous improvement process, it is regularly adapted to changing requirements and extended by the addition of new products. In the next few years, the DWD plans to increase the level of automation in its warning compilation procedure and centralize organisation as a way to further optimize its warning bulletins. By guaranteeing warnings of a consistently high quality, the DWD will establish the single-voice principle and become accepted as the sole binding source of information about dangerous weather phenomena, not just for disaster relief units but also for the general public.

Taken overall, it can be stated that the successful co-operation between the DWD and the disaster management institutions is a key factor for the effective performance and further development of the warning management in Germany.

Acknowledgements
The authors would thank the support from the following agencies who have provided references and input, including:

Deutscher Wetterdienst (DWD)
VBZ (Weather Forecasting and Advisory Centre)
WVPK (Planning and Co-ordination Division)
KU 4 (Hydrometeorology Department)
D 63067 Offenbach/Main
Frankfurter Str. 135

Bundesamt für Bevölkerungsschutz und Katastrophenhilfe (BBK)
(Federal Office of Civil Protection and Disaster Assistance)
E-mail: bbk-leitung@bbk.bund.de
Provinzialstraße 93
D 53127 Bonn

Bundesministerium des Innern (BMI)
(Federal Ministry of the Interior)
Alt-Moabit 101 D
D 10559 Berlin

Bayerisches Staatsministerium des Innern (BStMI)
(Bavarian Ministry of the Interior) Sachgebiet ID4
Odeonsplatz 3
D 80539 München

List of Acronyms

BBK	Bundesamt für Bevölkerungsschutz und Katastrophenhilfe (Federal Office for Civil Protection and Disaster Assistance)
BfS	Bundesamt für Strahlenschutz (Federal Office for Radiation Protection)
BMU	Bundesministerium für Umwelt, Naturschutz und Reaktorsicherheit (Federal Ministry for the Environment, Nature Conservation and Nuclear Safety)
BR	Bayerischer Rundfunk (public-service radio station in Bavaria)
BStMI	Bayerisches Staatsministerium des Innern (Bavarian Ministry of the Interior)
CBS	WMO Commission for Basic Systems
deNIS	deutsches Notfallvorsorge-Informationssystem (German Emergency Preparedness Information System)
DWD	Deutscher Wetterdienst
ESSL	European Severe Storms Laboratory
FeWIS	Feuerwehr-Wetter-Informationssystem (Weather Information System for Fire Brigades)
GMLZ	Gemeinsames Melde- und Lagezentrum (Joint Information and Situation Centre)
KOSTRA	Koordinierte Starkniederschlags-Regionalisierungs Auswertungen (Co-ordinated evaluation of the heavy precipitation totals in Germany)
LfU	Bayerisches Landesamt für Umwelt (Bavarian Environment Agency)
LÜKEX	LänderÜbergreifende Krisenmanagement-Übung/EXercise (Crisis management exercises involving multiple German federal states)
RA VI	WMO Regional Association VI (Europe)
RZ	Regionalzentrale (Regional Centres)
StrVG	Strahlenschutzvorsorgegesetz (law on the prevention of radiation)
SWIS	Straßen-Wetter-Informations-System (Road Weather Information System)
THW	Technisches Hilfswerk (German Federal Agency for Technical Relief)
UV	Ultraviolet
UWZB	Unwetterwarnzentrale Bayern des DWD (DWD's Bavarian Severe Weather Warning Centre)
VBZ	Vorhersage- und Beratungszentrale (Weather Forecasting and Advisory Centre)
WMO	World Meteorological Organization

Chapter 6
Multi-Hazard Early Warning System of the United States National Weather Service

Harold 'Jim' Keeney Jr., Steve Buan and Laura Diamond

Abstract The United States began to develop its national early warning system in the late 1940s as part of the post-war effort to invest in reducing the impact of tropical cyclones, flooding, tornadoes, drought, winter storms, severe hail and wind, tsunamis and other hazards that threatened its citizens. Initial steps included the erection of a network of weather surveillance radars, the creation of regional river forecast centers and the development of a comprehensive, multi-agency emergency broadcast system. Those initiatives were followed by the creation of the National Response Framework, the Incident Command System, the multi-channel Emergency Alert System, and policies at the federal, state and local level that promoted broader hazard awareness, risk reduction and emergency preparedness.

The National Oceanic and Atmospheric Administration's National Weather Service is authorized by statute and Executive Order to forecast and issue alerts and warnings associated with hydrometeorological events as well as tsunamis. This mission is achieved through dedicated, long-term partnerships with other federal agencies (most notably the United States Geological Survey, United States Army Corps of Engineers, and the Federal Emergency Management Agency), which leverage the expertise and infrastructure throughout the Federal Government. Simultaneously, the National Science Foundation and other federal and state-level funding sources promote university-led advancements in early warning methods and technology and training of young engineers, hydrologists, meteorologists, computer scientists, etc.

While the various hazard-specific early warning systems in the United States were not originally designed as a part of a single, multi-hazard system, they share common elements and are linked with the country's varied emergency response systems. The National Response Framework guides the institutional coordination and cooperation associated with planning and response to major emergencies. Specific capabilities have been developed to address threats from low-probability/high-impact events such as tsunamis, tropical cyclones, earthquakes, volcanoes, and even terrorism.

The Federal Emergency Management Agency (FEMA) of the United States Department of Homeland Security in partnership with dozens of state and local emergency management agencies, 27 federal agencies, the private sector, the American Red Cross and other non-governmental organizations, plays a crucial role in assisting citizens to build, sustain, and improve the Nation's capability to prepare for, protect against, respond to, recover from, and mitigate all hazards. FEMA's coordination and cooperation with partner institutions often dictates the response to major emergency events. No other federal agency has a greater role than FEMA in helping the Nation mitigate and prepare for hazards.

Multi-hazard early warning systems are only as good as their weakest subsystem. Like many other countries, the United States has learned that unless its multi-hazard early warning system functions from "end-to-end", avoidable losses and suffering may still occur. Its most spectacular lesson in recent years was Hurricane Katrina – a hazard scenario that was recognized well in advance. The failures in the Nation's early warning system during that event are widely acknowledged to have occurred because of weaknesses in risk awareness, warning/communications, and response, despite strengths in forecasting. Other emergencies, especially major disasters such as Hurricane Andrew (1992), the Great Midwest Flood (1993), the Northridge Earthquake (1994), and the 9/11 Terror Attacks (2001), have inspired improvements in emergency preparedness, risk awareness, and hazard mitigation.

It is not necessary for these lessons to be painfully re-learned by others around the world. The National Weather Service hopes that this document improves global understanding in the good practices of institutional coordination and cooperation as they relate to multi-hazard early warning systems.

Harold 'Jim' Keeney, Jr., Deputy Chief, Meteorological Services, Integrated Services Division, National Weather Service, Central Region, Kansas City, MO
Steve Buan, Service Coordination Hydrologist
Laura Diamond, Hydrologist, North Central River Forecast Center, National Weather Service, Chanhassen, MN

M. Golnaraghi ed., *Institutional Partnerships in Multi-Hazard Early Warning Systems*,
DOI 10.1007/978-3-642-25373-7_6,

6.1 Overview of Early Warning Systems and the Role of National Meteorological and Hydrological Services

6.1.1 Background

Prior to 1951, there was no method that the U.S. Government could use to broadcast warnings to citizens in the event of an emergency. However, radio stations and networks were able to interrupt normal programming and issue a bulletin in the event of an emergency, as happened during the attack on Pearl Harbor on December 7, 1941, as well as the first successful tornado warning near Tinker Air Force Base in Oklahoma City, Oklahoma, in 1948. President Harry S. Truman established CONELRAD in 1951 (Control of Electromagnetic [or Electronic] Radiation) as the first U.S. Government-organized means of emergency broadcasting. CONELRAD was designed primarily for civil defense warnings to the U.S. public and eventually incorporated all U.S. amateur radio broadcasts.

CONELRAD was replaced by the Emergency Broadcast System (EBS) on August 5, 1963, which was later replaced with the Emergency Alert System (EAS) in 1997. The official EAS is designed to enable the U.S. President to speak to the entire U.S. population within 10 minutes of an emergency. This official national EAS has never been intentionally activated. The EAS covers all forms of radio and television broadcasts, including AM, FM, VHF, UHF, cable and satellite television, and satellite radio.

The Emergency Alert System is supplemented by numerous secondary dissemination systems and services. These systems can be publicly or privately owned and operated. The use of the Internet to disseminate warning information is the most prevalent. Community siren systems, alarm systems in buildings, and subscription services that communicate with hand-held devices, are some other examples. Most of these secondary communication systems disseminate the official environmental hazard warnings to their end-users.

6.1.2 History of the U.S. National Meteorological and Hydrological Services

NOAA's National Weather Service (NWS) is responsible for providing national meteorological services and a significant portion of national hydrological services, and for operating the early warning system for hydro-meteorological hazards faced by the United States. The National Weather Service had its beginning in the early history of the United States.

During the early and mid-1800s, weather observation networks began to grow and expand across the United States. Although most basic meteorological instruments had existed for over 100 years, it was the telegraph that was largely responsible for the advancement of operational meteorology during the 19th century. With the advent of the telegraph, weather observations from distant points could be "rapidly" collected, plotted, and analyzed at one location. The ability to observe and display simultaneously observed weather data, through the use of the telegraph, quickly led to initial efforts toward the forecasting of weather.

During the late 1940s, the military gave the U.S. Weather Bureau a new and valuable tool – 25 surplus radars – thus launching the network of weather surveillance radars still in use today. In 1946, the first River Forecast Centers (RFC) were established to provide forecasts of hazardous conditions on the Nation's rivers. The advent of computer technology in the 1950s paved the way for the formulation of complex mathematical weather models, resulting in a significant increase in forecast accuracy. In 1970, the name of the Weather Bureau was changed to the National Weather Service and the agency became a component of the Commerce Department's newly created National Oceanic and Atmospheric Administration (NOAA).

6.2 Governance and Institutional Arrangements (National to Local Levels)

6.2.1 Policy, Institutional and Legal Frameworks to Support Emergency Planning and Response

The U.S. National Response Framework *(NRF)* (http://www.fema.gov/pdf/emergency/nrf/nrf-core.pdf) provides the guiding principles that enable all response partners to prepare for and deliver a unified national response to disasters and emergencies – from the smallest incident to the largest catastrophe. The Framework establishes a comprehensive, national, all-hazards approach to domestic incident response. The *National Response Framework,* which is maintained by the U.S. Department of Homeland Security's (DHS) Federal Emergency Management Agency (FEMA), is a guide for how the United States conducts an all-hazards response. The term "response" as used in this *Framework* includes immediate actions to save lives, protect property and the environment, and meet basic human needs. The *Framework* is built upon scalable, flexible, and adaptable coordinating structures to align key roles and responsibilities across the Nation, linking all levels of government, non-governmental organizations, and the private sector. It outlines the key response principles, participants, roles, and structures that guide the Nation's response operations. It provides structures for implementing national-level policy and operational coordination for domestic incident response (incidents include actual or potential emergencies or all-hazard events that range from accidents and natural disasters to terrorist attacks).

The legal authorities that guide the structure, development, and implementation of the NRF are statutes, regulations, Executive Orders, and Presidential directives and can be found in the *List of Authorities and References* (http://www.fema.gov/pdf/emergency/nrf/nrf-authorities.pdf). This list and the associated summaries are not exhaustive and should not be used as a substitute for the authorities themselves. The statutes and regulations are organized by subject matter, except for some principal emergency authorities that are listed first. The Executive Orders of the President and the Homeland Security Presidential Directives (HSPDs) are listed separately.

The NRF includes an overview of the *Stafford Act Support to States* (http://www.fema.gov/pdf/emergency/nrf/nrf-stafford.pdf), which outlines actions Federal agencies are likely to take to assist State, tribal, and local governments affected by a major disaster or emergency. The DHS National Operations Center continually monitors potential major disasters and emergencies. When advance warning is received, DHS may deploy – and may request that other Federal agencies deploy – liaison officers and personnel to a State emergency operations center to assess the emerging situation. Key operational components that may be activated include the National Response Coordination Center (NRCC), Regional Response Coordination Center (RRCC), Joint Field Office (JFO), and Disaster Recovery Centers (DRCs). A RRCC may be fully or partially activated. Facilities, such as mobilization centers, may be established to accommodate Federal personnel, equipment, and supplies.

Immediately after a major incident, tribal and/or local emergency personnel respond and assess the situation. If necessary, those officials seek additional resources through mutual aid and assistance agreements and the State. State officials also review the situation, mobilize State resources, use interstate mutual aid and assistance processes such as the Emergency Management Assistance Compact to augment State resources, and provide situation assessments to the FEMA regional office. The Governor activates the State emergency operations plan, declares a state of emergency, and may request a joint Preliminary Damage Assessment. The State and Federal officials conduct that assessment in coordination with tribal/local officials, as required, and determine whether the impact of the event warrants a request for a Presidential declaration of a major disaster or emergency. Based on the results of the damage assessment, the Governor may request a Presidential declaration in order to access Federal assistance, as needed.

After a major disaster or emergency declaration, a RRCC coordinates initial regional and field activities until a Joint Field Office is established. Regional teams assess the impact of the event, gauge immediate State needs, and make preliminary arrangements to set up field facilities. (If regional resources are, or may be, overwhelmed or if it appears that the event may result in particularly significant consequences, DHS may deploy a national-level Incident Management Assistance Team). Depending on the scope and impact of the event, the NRCC carries out initial activations and mission assignments and supports the Regional Response Coordination Center. The Governor appoints a State Coordinating Officer to oversee State response and recovery efforts. A Federal Coordinating Officer, appointed by the President in a Stafford Act declaration, coordinates Federal activities in support of the State. The Unified Coordination Group (UCG) coordinates field operations from a Joint Field Office. In coordination with State, tribal, and/or local agencies, Emergency Support Functions assess the situation and identify requirements. Federal agencies provide resources under FEMA mission assignments or their own authorities.

Mutual aid agreements are a fundamental aspect of institutional cooperation that occurs during major emergencies. The Emergency Management Assistance Compact (EMAC) of 1996 is congressionally ratified organization that provides form and structure to interstate mutual aid. Through EMAC, a disaster-affected state can request and receive assistance from other member states quickly and efficiently, resolving two key issues at the onset: liability and reimbursement. As immediate response priorities are met, recovery activities begin.

Federal and State agencies assisting with recovery and mitigation activities convene to discuss needs. The Stafford Act Public Assistance program provides disaster assistance to States, tribes, local governments, and certain private non-profit organizations. FEMA, in conjunction with the State, conducts briefings to inform potential applicants of available assistance and how to apply. Throughout response and recovery operations, FEMA Hazard Mitigation program staff at the Joint Field Office look for opportunities to maximize mitigation efforts in accordance with State hazard mitigation plans.

As the need for full-time interagency coordination at the Joint Field Office decreases, the Unified Coordination Group (UCG) plans for selective release of Federal resources, demobilization, and closeout. Federal agencies work directly with disaster assistance grantees (i.e., State or tribal governments) from their regional or headquarters offices to administer and monitor individual recovery programs, support, and technical services.

The following chart (Fig. 6.1) summarizes the Stafford Act support to States.

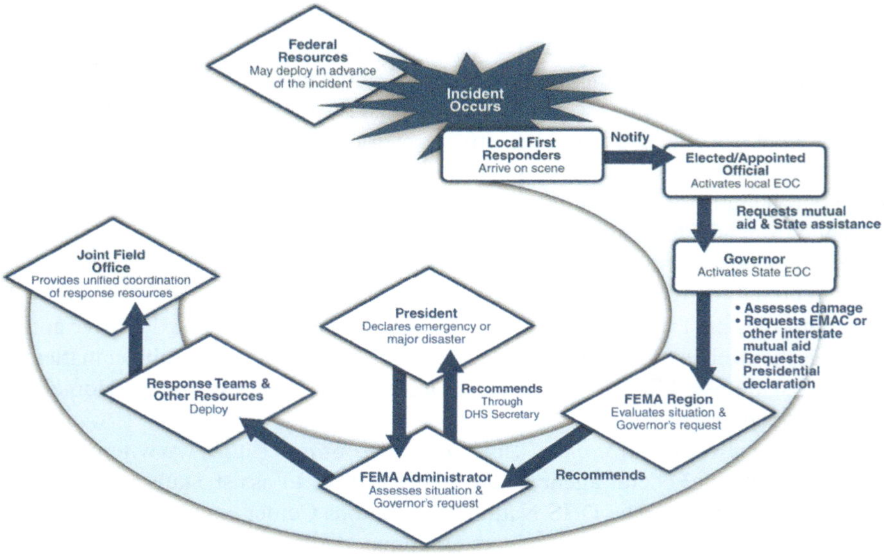

Fig. 6.1 Flow of Incident Support and Response Coordination

Federal departments and agencies routinely manage the response to incidents that do not require the assistance of other Federal agencies. In these types of so-called "non-Stafford Act" situations, the NRF provides an overview regarding actions the Department of Homeland Security (DHS) and Federal agencies are likely to take to support a Federal department or agency that has requested DHS assistance in handling an incident that would generally fall under the requesting department's or agency's jurisdiction (http://www.fema.gov/pdf/emergency/nrf/nrf-fedtofed.pdf).

When a Federal entity with primary responsibility and authority for handling an incident requires assistance, that agency may request DHS coordination of Federal multiagency assistance. Generally, the requesting agency funds the participation of other Federal departments and agencies in accordance with provisions of the Economy Act unless other pertinent authorities exist. Key operational components that may be activated include the National Response Coordination Center (Washington, DC), RRCC (FEMA Region Office), and JFO (at the site of the disaster). To initiate Federal-to-Federal support, the requesting agency submits a request for assistance to the DHS via the National Operations Center (NOC). Requests include a summary of the situation, types and amount of resources needed, financial information, and any other appropriate details. Upon approval of the request, the Secretary of Homeland Security issues an operations order to the NOC. The NOC, through the NRCC, coordinates the activation of the appropriate Emergency Support Functions. The Secretary of Homeland Security designates a Federal Resource Coordinator (FRC) to manage Federal resource support. In circumstances requiring extraordinary coordination, the Secretary may appoint a Principal Federal Official to serve as his or her representative in the field. The requesting agency designates a senior official to work in coordination with the FRC as part of the UCG to identify and define specific support requirements. The requesting agency also provides comptrollers to the NRCC, RRCC, and JFO, as appropriate, to oversee financial management activities. A Regional Response Coordination Center may be fully or partially activated to facilitate the deployment of resources until a Joint Field Office is established. Facilities, such as mobilization centers, may be established to accommodate personnel, equipment, and supplies. Once established, the JFO serves as the focal point for coordinating Federal assistance to the requesting agency and on-site incident command structure.

Within the framework outlined above, and in very broad terms, U.S. national meteorological and hydrological services are reflected in legislation and legal frameworks through the following priorities for service to the Nation: protection of life, protection of property, and promotion of the Nation's welfare and economy. As areas of responsibility grew across the country, some important Acts were passed to further the work within the Department of Commerce (DOC) including:

- The Air Commerce Act of 1926 added a new area of responsibility for providing weather services. The rapid growth of civil aviation created a demand for such services.
- The Flood Control Act of 1938 (33 USC 706). This Act came into being as a result of the need for a more effective warning and flood control system. The Flood Control Act of 1938 authorized allotment of funds from appropriations for flood control for the establishment, operation, and maintenance of an information service on precipitation, flood forecasts, and flood warnings, whenever in the opinion of the Chief of Engineers and the Secretary of Commerce such service is advisable in connection with flood control purposes.

- The Federal Aviation Act, Public Law 85-726 (49 USC 1463), effective August 23, 1958, broadened the scope of responsibility of the Secretary of Commerce concerning air navigation.
- The Environmental Science Services Administration (ESSA) was formed on July 13, 1965. In the words of President Johnson, ESSA would provide "a single national focus to describe, understand, and predict the state of the oceans, the state of the upper and lower atmosphere and the size and shape of the earth." (ESSA was merged with other capabilities in 1970 to create NOAA.)

The National Weather Service (NWS) observes and reports the weather of the United States and its possessions and issues forecasts and warnings of weather and flood conditions that affect the Nation's safety, welfare and economy; develops the National Meteorological Service System; participates in international meteorological and hydrological activities, including exchanges of meteorological data and forecasts; and provides forecasts for domestic and international aviation and for shipping on the high seas. While military services are not part of the mission, the National Weather Service is responsible for providing many basic services to the Department of Defense. These services are coordinated through the Office of the Federal Coordinator for Meteorology. The Department of Defense operates its own weather service agencies to fulfill specialized and unique requirements, and provides a robust capability to support civilian agencies in their response to major emergencies.

6.2.2 National to Local Emergency Planning and Related Linkages to Early Warning Systems

The *National Incident Management System* (NIMS) provides a systematic, proactive approach to guide departments and agencies at all levels of government, non-governmental organizations, and the private sector to work seamlessly to prevent, protect against, respond to, recover from, and mitigate the effects of incidents, regardless of cause, size, location, or complexity, in order to reduce the loss of life and property and harm to the environment. NIMS works hand in hand with the *National Response Framework* that was outlined in Sect. 2.2.1. NIMS provides the template for the management of incidents, while the NRF provides the structure and mechanisms for national-level policy for incident management. (http://www.fema.gov/pdf/emergency/nims/NIMS_core.pdf).

The components of NIMS work together in a flexible, systematic manner to provide the national framework for incident management, from Preparedness, Communications and Information Management, Resource Management, Command and Management, to Ongoing Management and Maintenance.

Preparedness

Effective emergency management and incident response activities begin with a host of preparedness activities conducted on an ongoing basis, in advance of any potential incident. Preparedness involves the integration of assessment planning, procedures, protocols, training, exercises, drills, personnel qualifications, licensure, certification, evaluation and revision.

Communications and Information Management

Emergency management and incident response activities rely on communications and information systems that provide a common operating picture to all command and coordination sites. NIMS describes the requirements necessary for a standardized framework for communications and emphasizes the need for a common operating picture. This component is based on the concepts of interoperability, reliability, scalability, and portability, as well as the resiliency and redundancy of communications and information systems.

Resource Management

Resources (such as personnel, equipment, or supplies) are needed to support critical incident objectives. The flow of resources must be fluid and adaptable to the requirements of the incident. NIMS defines standardized mechanisms and establishes the resource management process to identify requirements, order and acquire, mobilize, track and report, recover and demobilize, reimburse, and inventory resources.

Command and Management

The Command and Management component of NIMS is designed to enable effective and efficient incident management and coordination by providing a flexible, standardized incident management structure. The structure is based on three key organizational constructs: the Incident Command System, Multiagency Coordination Systems, and Public Information.

Ongoing Management and Maintenance

Within the auspices of Ongoing Management and Maintenance, there are two components: the National Integration Center and Supporting Technologies.

National Integration Center (NIC)

Homeland Security Presidential Directive 5 required the Secretary of Homeland Security to establish a mechanism for ensuring the ongoing management and maintenance of NIMS, including regular consultation with other Federal departments and agencies; State, tribal, and local stakeholders; and Non-Governmental Organizations (NGOs) and the private sector. The NIC provides strategic direction, oversight, and coordination of NIMS and supports both routine maintenance and the continuous refinement of NIMS and its components. The NIC oversees the program and coordinates with Federal, State, tribal, and local partners in developing compliance criteria and implementation activities. It provides guidance and support to jurisdictions and emergency management/response personnel and their affiliated organizations as they adopt or, consistent with their status, are encouraged to adopt the system. The NIC also oversees and coordinates the publication of NIMS and its related products. This oversight includes the review and certification of training courses and exercise information.

Supporting Technologies

As NIMS and its related emergency management and incident response systems evolve, emergency management/response personnel will increasingly rely on technology and systems to implement and continuously refine NIMS. The NIC, in partnership with the Department of Homeland Security Science and Technology Directorate, oversees and coordinates the ongoing development of incident management-related technology, including strategic research and development.

As mentioned previously, the Emergency Alert System (EAS) is the alert dissemination mechanism for the multi-hazard early warning system. The EAS is jointly coordinated by the Federal Communications Commission (FCC), Department of Homeland Security, the National Weather Service and the commercial broadcast industry in the United States. The latter two organizations utilize the EAS as a primary means for warning the U.S. public of severe weather (e.g., tornadoes, floods), natural hazards (e.g., tsunamis, earthquakes, wildfires and volcanoes), technological hazards (e.g., chemical releases, oil spills, and nuclear power plant emergencies), and national emergencies (e.g., terrorist attacks). The National Weather Service (NWS) triggers the Emergency Alert System through its NOAA Weather Radio All Hazards system that broadcasts across 1000+ transmitters throughout the United States. The Department of Homeland Security manages the entire EAS system. The EAS regulations and standards are governed by the Public Safety and Homeland Security Bureau of the Federal Communications Commission (FCC). Each State and several territories have their own EAS plan which administers the system in that jurisdiction. Coordination of the State EAS plans typically takes place through a State EAS committee that includes the state Emergency Management Agency, the commercial broadcast industry, and the NWS.

6.2.3 Organizational Structure for Implementing the Plans

That National Incident Management System and the National Response Framework are designed to improve collective capabilities and efficiencies. During incidents requiring coordinated Federal support, the Framework provides the guidelines and procedures to integrate capabilities and resources into a cohesive, coordinated, and seamless national framework for incident management. The NWS role within the National Response Framework is to provide decision support services to national, regional and local emergency management stakeholders within any activation/command structure that is implemented to respond to the impending hazard. The National Incident Management System (NIMS) provides the command/support structure definition.

6.2.4 Institutional Capacities and Concept of Operations (Coordination and Operational Collaboration)

The development of early warning systems in the United States has been largely hazard-specific as opposed to multi-hazard, although significant integration of synergistic elements has occurred over time. No single concept of operations exists for all forms of hydro-meteorological early warning generation, since each hazard requires a unique combination of institutions and systems. For example, tropical cyclone prediction is quite different from flash flood prediction, even though they may share common capacities in satellite and radar analysis, as well as early warning dissemination.

The National Weather Service (NWS) specifies its operations and services through a series of directives that are constantly updated and posted to its website: http://www.nws.noaa.gov/directives. These directives control the meteorological and hydrological services produced by the NWS' various hazard-specific operations, which include 13 River Forecast Centers, 122 Weather Forecast Offices, and nine national specialized centers that support activities such as tsunami warning, marine weather forecasts, aviation weather forecasts, and tropical cyclone prediction. These hazard-specific units comply with these national directives and establish Standard Operating Procedures that meet national requirements and that also reflect their unique circumstances and capabilities. Importantly, NWS' Office of Climate, Water, and Weather Services has the specific mandate of supporting the development and maintenance of operation plans. Its Operations and Requirements Division implements and integrates NWS climate and hydrometeorological forecast and warning services nationwide. General guidance on early warning system conceptualization and concept of operations planning is being developed by the NWS and the Cooperative Program for Operational Meteorology, Education and Training (COMET), and is available at: http://www.meted. ucar.edu/hazwarnsys.php.

6.2.5 Utilization of a Multi-Hazard Approach

The Emergency Alert System was built on a structure conceived in the 1950s when over-the-air broadcasting was the best-available technology for widely disseminating emergency alerts. It leverages over-the-air broadcasts by television and radio stations, as well as the NOAA Weather Radio All-Hazards system. While the NOAA/NWS weather radio system has been upgraded to an all-hazard warning capability, the entire EAS has not leveraged digital technologies yet. Because of this it does not have *complete* interoperability with recent emerging technologies such as the Internet and cell phones. The U.S. President and Congress are driving the agencies responsible for the EAS to work with the private sector to develop digital capabilities that can make the U.S. all-hazards early warning system interoperable with emerging technologies. This is a work in progress.

The Department of Homeland Security (DHS) is developing the "Integrated Public Alert and Warning System" (IPAWS) as the next-generation infrastructure of alert and warning networks in the United States. IPAWS will expand upon the traditional (audio-only) radio and television Emergency Alert System (EAS) by providing one message over more media to more people before, during, and after a disaster. The vision of IPAWS builds upon and maintains an effective, reliable, integrated, flexible and comprehensive system that enables the American people to receive alert and warning information through as many means as possible. IPAWS ensures the President can alert and warn the public under all conditions. IPAWS will provide Federal, State, territorial, tribal, and local warning authorities the capabilities to alert and warn their respective communities of all hazards impacting public safety and well-being via multiple communications pathways. To be more specific, IPAWS will:

- Diversify and modernize the EAS;
- Create an interoperability framework by establishing or adopting standards such as the Common Alerting Protocol;
- Enable alert and warning to those with disabilities and to those without and understanding of the English language;
- Provide the President, Federal, State, territorial, tribal, and local emergency managers access to more communication pathways to ensure alerts and warnings reach the maximum number of Americans over as many pathways as possible;
- Partner with the National Weather Service to enable seamless integration of message transmission through national networks.

The NWS' Office for Climate, Water, and Weather Services' Integrated Operations Branch develops plans and writes policies for common or integrated operations and service functions including:

- Data acquisition and quality control;
- Meteorological analysis and decision support systems;
- Internal warning and forecast coordination, including WFO, RFC, and NCEP Service Center operational procedures and interactions;
- Communication infrastructure, including dissemination and service back up.

The Branch facilitates and coordinates service changes and monitors the impacts on service effectiveness and operational efficiency.

6.2.6 *Key Factors Considered for Sustainability of Different Components of Early Warning Systems with a Multi-Hazard Approach*

The all-hazards early warning system in the United States, the Emergency Alert System (EAS), is an inherent government function mandated by law. For inherent government functions, U.S. federal agencies have historically focused solely on completing the goal and do not conduct extensive research (e.g. cost benefit analysis) prior to implementation. What is known at the time of the mandate is leveraged by the responsible agencies as best as possible. This was the case with the EAS in the 1990s and is currently the case with the Integrated Public Alert and Warning System (IPAWS).

Due in part to this challenge, the U.S. Government formed the Government Accounting Office (GAO) in 1921. The GAO audits, evaluates, and investigates Federal Government programs for the U.S. Congress. They publish their findings in reports that are available on their web site: www.gao.gov. Internationally, the GAO is unique in that its auditors conduct not only financial audits, but also engage in a wide assortment of performance audits. In March 2007, the GAO issued a report on the EAS to help direct the agencies responsible for the development of IPAWS. They concluded, *"To improve the media's ability to issue emergency alerts, GAO recommends that DHS and FCC develop a plan to verify (1) the dependability and effectiveness of the EAS relay system, and (2) that EAS participants have the training to issue effective EAS alerts. Also, DHS and FCC should establish a forum for stakeholders to address the challenges of implementing an integrated alert system. In response, DHS agreed with the intent of our recommendations. FCC provided technical comments."* The complete report is available at: www.gao.gov/new.items/d07411.pdf.

When a project requires private sector expertise the U.S. Government contracts out through a competitive bid process. While this is not as intensive as cost benefit analysis research, it does ensure a level of cost effectiveness and that deliverables for the project are met. An example of this contract solicitation process for DHS' IPAWS project is available at: https://www.fbo.gov/index?s=opportunity&mode=form&tab=core&id=18eca683f41a39d2922bace118e66178&_cview=1.

Funding mechanisms for the all-hazards early warning system in the United States are driven by the annual Congressional Appropriations process that provides federal agencies with their operating budget (http://www.senate.gov/reference/resources/pdf/97-684.pdf). This is how the agencies responsible for the EAS and IPAWS development (FEMA, FCC and NWS) derive their funding for the development, maintenance, sustainability and improvement of the early warning system and emergency preparedness activities. Funding is then transferred to state and local government entities through various grant programs. Private sector entities that are supporting development or maintenance of the U.S. early warning system are reimbursed through the individual contracts awarded through the aforementioned competitive bid process.

6.3 Utilization of Risk Information in Emergency Planning and Warnings

6.3.1 *Organizational Responsibilities and Arrangements for the Development of Risk Information*

When threatened by natural, technological, or human caused emergencies or disasters, citizens expect elected or appointed leaders to take immediate action to deal with the problems. They expect the government to marshal its resources, channel the efforts of voluntary organizations and private enterprises in the community, and solicit assistance from outside the jurisdiction if necessary. The elected leaders in each jurisdiction are legally responsible for ensuring that necessary and appropriate actions are taken to protect people and property from the consequences of emergencies and disasters. This is accomplished in all States and most local jurisdictions by developing and maintaining viable all-hazard, all-threat emergency plans by statute or ordinance. Each jurisdiction's plans must reflect what that community will do to protect itself from *its* unique hazards and threats with the unique resources *it* has or can obtain. Congress also recognizes State and local emergency management responsibilities in the Robert T. Stafford Disaster Relief and Emergency Assistance Act, as amended:

- "It is the intent of Congress, by this Act, to provide an orderly and continuing means of assistance by the Federal government to State and local governments in carrying out their responsibilities to alleviate the suffering and damage which result from ... disasters" (Sec. 101(b), emphasis added).
- "The purpose of this title is ... to vest responsibility for emergency preparedness jointly in the Federal government and the States and their political subdivisions" (Sec. 601).

The President of the United States has identified emergency planning as a national security priority, which must be conducted in an atmosphere of trust and mutual understanding. If the planning process is accomplished properly, it provides a methodical way to think through the entire life cycle of a potential crisis. It directs how a community envisions and shares a desired outcome, selects effective ways to achieve it, and communicates expected results.

With the United States' system of emergency management, the local government must act first to respond to the public's emergency needs. Depending on the nature and size of the emergency, State, Territorial, regional compact organization, and Federal assistance may be provided to the local or Tribal jurisdiction. The focus of the local Emergency Operations Plan is on the emergency measures that are essential for protecting the public. At the minimum, these measures include warning, response, emergency public information, evacuation, and shelter.

One of the main components of the Emergency Operations Plan is the identification of the hazards that could affect the local jurisdiction. The hazard/risk assessments must provide sufficient information to enable the jurisdiction to identify and prioritize appropriate response, preparedness, planning, and mitigation actions to reduce losses from identified hazards.

The hazard/risk assessment planning information is specific to each state/local level and includes:

- the probable location of hazards or the areas that may be affected by them;
- the severity and magnitude of hazard events;
- the probability of future hazard events, and
- historical occurrences of hazard events in the region.

Where possible, the locations or geographic areas that may be affected by the hazards are identified and mapped by the local jurisdictions. The areas that may be affected by weather-related natural hazards are far more difficult to identify. Flooding, for example, may affect a specific area while severe winter weather, drought, heat waves and even earthquakes (though not a weather-related phenomenon) are usually widespread events that affect an entire state or region. Some locations, because of specific characteristics of the local infrastructure, environment, socio-economic conditions or a combination of these elements, may be more susceptible to specific natural hazards.

The hazard/risk assessment identifies disasters that can precipitate "cascading" hazards, i.e., those additional hazards resulting from a natural disaster. Cascading hazards can include interruption of power and communications; interruption of the water supply; business and economic interruptions; computer failure and/or loss of essential records; transportation interruptions; and health and/or environmental hazards. Any of these cascading hazards, alone or in combination with the direct adverse effects of a disaster, can potentially impact emergency response operations in affected communities. For example, severe cold weather can result in "Ice Dams" on rivers, causing severe flooding. Cascading hazard models depict either potential threat (threat model) sequences or document historic ones (consequence model).

Once a State or community identifies its risks/vulnerabilities, it can take measures to mitigate and prepare its social, political, environmental, and economic environment along with its citizens so it can become a more resilient community. The preparedness measures for natural weather-related risks are normally implemented through the local offices of emergency management/preparedness. These offices work closely with their local National Weather Service Forecast Office to create an integrated system to communicate risk and warnings. This system takes into account new and emerging technologies, social networks, and the changing socioeconomic and demographic characteristics of the general population.

The Federal Emergency Management Agency (FEMA) is the lead governmental organization responsible for hazard and vulnerability assessment. That risk assessment process provides the foundation for the rest of the mitigation planning process, and includes four basic activities: (1) identify hazards, (2) profile hazard events, (3) inventory assets, and (4) estimate losses. This process measures the potential loss of life, personal injury, economic injury, and property damage resulting from natural hazards by assessing the vulnerability of people, buildings, and infrastructure to natural hazards. While there are countless programs, centers, and tools available at various levels of government, academia, and the private sector, the sources listed on the FEMA website provide a good sample of those commonly used for conducting a multi-hazard risk assessment in the United States. These are listed in Table 6.1.

Table 6.1 Example of tools for conducting a multi-hazard risk assessment in the United States (source: FEMA)

Type	Activities and tools
Multi-Hazard	HAZUS-MH Environmental Planning and Historic Preservation Program Natural Hazards Gateway Coastal Services Center National Geophysical Data Center National Water and Climate Center National Climatic Data Center Natural Hazards Center Global Change Master Directory
Hurricane	National Hurricane Program National Hurricane Center Information Historical Hurricane Tracks
Flood	Flood Hazard Mapping Floodplain Management Community Rating System Emergency Management Institute Community Rating System Resource Center National Dam Safety Program National Inventory of Dams (NID) Association of State Dam Safety Officials Nationwide River Gage Site for Flooding Information National Flood Risk Management Program
Wildfire	U.S. Fire Administration Wildland Fire Assessment System National Interagency Fire Center National Wildfire Hazard Mitigation Programs Database
Earthquake	National Earthquake Hazards Reduction Program National Earthquake Information Center
Tribal	Bureau of Indian Affairs HUD Office of Native American Programs EPA American Indian Tribal Portal EPA Tribal Programs Indian Health Service Area Offices and Facilities

More information about these programs is available at: http://www.fema.gov/plan/mitplanning/risk.shtm

6.3.2 Hazard Assessment, Quantification and Mapping (National to Local)

Natural Hazards

Natural hazards historically are responsible for the greatest loss of life and property throughout the United States. Most of the major disaster declarations over the past 8 years are due to natural hazards – severe storms, floods, fires, tornadoes and ice storms have been especially destructive. The U.S. Natural Hazard Statistics provide statistical information on fatalities, injuries and damages caused by weather related hazards, which can be accessed at: http://www.nws.noaa.gov/ om/hazstats.shtml

The following cumulative statistics are compiled from information contained in *Storm Data*, a report comprising data from NWS forecast offices in the 50 states, Puerto Rico, Guam and the Virgin Islands.

Frequency of Occurrence/Location/Level of Impacts

Natural hazards occur daily and are dependent upon geographic region as well as season. Flooding can occur in the form of flash-flooding from thunderstorms year-round in the United States. Flooding can also occur from heavy snow melt in the northern portions of the United States. Tornadoes are a high-risk hazard to several states and can occur throughout the year, but mostly occur in the summer months, with the highest concentration in the Midwest and Gulf States. Hazardous material spills occur daily and cause significant damage to property, although the risk to lives is very low and depends on the hazardous material agent as well as the amount that was involved during the incident (Fig. 6.2).

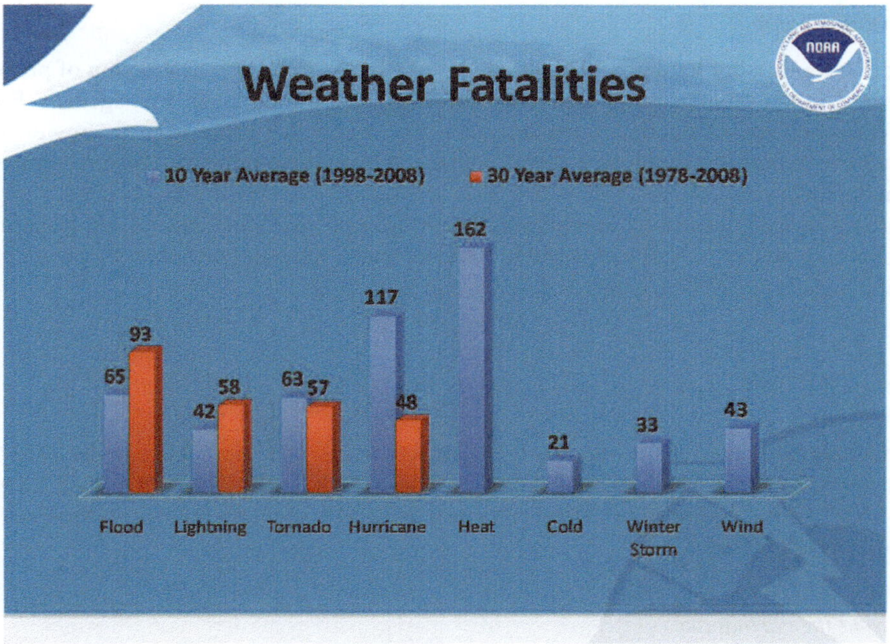

Fig. 6.2 Weather-related fatalities in the United States (1978–2008)

Increase of Risks

Several hazards have increased due to settlement in hazardous areas, especially along coastlines or seismically active areas. In 2003, an estimated 153 million people, or roughly 53 percent of the U.S. population, lived in 673 coastal counties; such counties account for ten of the 15 most populous cities in the United States (Rubin 2006). This movement of people into hazardous areas has caused great economic loss. Based on information from the Spatial Hazard Event and Losses Database for the United States, weather-related events account for almost three-quarters of the total hazard losses – in excess of $470 billion – between 1960–2005, with hurricanes, tropical storms, and floods as the leading causes. Hurricane Katrina alone cost more than $125 billion.

Some Major Disasters (of the Last 50 Years) and Their Associated Improvements:

Hurricane Betsy 1965

Hurricane Betsy had long-lasting and record-breaking impacts. It resulted in 76 deaths and caused more than $1 billion in damage (roughly $8 billion in 2005 dollars). The limited technology of the 1960s made tracking the storm challenging.

Improvements following the hurricane:

A major piece of legislation passed after the hurricane was the National Flood Insurance Act (NFIA) of 1968. This Act authorized the Federal Government to provide flood insurance to property owners on the condition that local governments adopt regulations specifying that the insurance would be provided only if the applicant communities adopted land-use regulation. With the NFIA, the Federal Government's role expanded into land use regulation and policy, which was normally the terrain of the local governments.

The Great Midwest Flood of 1993

In the summer of 1993, continuous rain fell across the Midwestern United States. Most of the Midwest received over 12 inches of precipitation; parts of Iowa, Kansas, Minnesota, Missouri, and Nebraska got more than 24 inches of rain. As rivers overflowed and dams and levees failed, the result was the largest and most damaging flood event ever to occur in the United States. Damages totaled $15 billion and at least 50 people died. Thousands of people evacuated, and some never did return to their homes.

Improvements following the flood:

The Great Mississippi River Basin Flood of 1993 caused loss of life and extensive destruction of property. The winter following the floods, the Clinton administration convened the Interagency Floodplain Management Review Committee to conduct a comprehensive review of floodplain management. The report that came from the review called for shared responsibility for floodplain management among Federal, State, and local governments. It also recommended restrictions on developments in floodplains.

A post event analysis revealed extensive opportunities to infuse new research, science and technology into the Hydrological Services Program of NOAA's National Weather Service, which ultimately resulted in the creation of the Advanced Hydrologic Prediction Service, or AHPS, which is described in Sect. 2.4.2.

The Northridge Earthquake 1994

The Northridge Earthquake of January 17, 1994, was the most destructive U.S. earthquake in the 1979–2001 interval, and it stood as FEMA's most expensive disaster (until 9/11). As of July 31, 2006, FEMA funding in all categories for the Northridge quake totaled $6.97 billion – almost three times as much as FEMA spent on Hurricane Andrew. The quake killed 72 people and caused over 9,000 injuries.

Improvements following the earthquake:

Improvements made following the Northridge earthquake were mitigation related. Nearly $1 billion in new mitigation efforts were designated to strengthen integrity (building codes) of four local hospitals damaged by the quake.

The Indian Ocean Tsunami Catastrophe 2004

The disaster that was the Indian Ocean earthquake and tsunami of 2004 had lessons learned that continue to be worked on to the present. The United Nations lists a total of 229,866 people lost, including 186,983 dead and 42,883 missing. On February 9, 2005, President Bush asked Congress to increase the U.S. commitment to a total of $950 million.

Improvements following the tsunami:

Through international efforts a tsunami warning system was established in the Indian Ocean Region. The U.S. tsunami warning program was extended to include the Atlantic and Caribbean. Numerous improvements were also made to the mitigation and preparedness planning for the tsunami hazard in the United States.

This is a great example of the world uniting to provide international relief and sustained emergency management improvements moving forward. In the United States, just about every Federal agency contributed in some form to these efforts. The NWS leads the tsunami warning program in the United States, under the authority provided to NOAA by the 2006 Tsunami Warning & Education Act (see Appendix 1 for details).

Hurricane Katrina/Rita 2005

In early September 2005, the Gulf States of the United States were hit by two Category 3/4 tropical cyclones that created massive economic losses and human suffering, most spectacularly in the historic but vulnerable city of New Orleans. The United States was reminded that its multi-hazard early warning system is only as good as its weakest link. Unless it works well from "end-to-end" – i.e., all the way to the final stage of communicating and ensuring correct actions are taken by at-risk communities – a worst-case scenario event that was anticipated years in advance could still have horrific impact.

> An EWS is only as good as its weakest link.
>
> *The experience of Hurricane Katrina in September 2005 underlined that to be successful, an EWS must work well from "end-to-end" – i.e., all the way to the final stage of communicating and ensuring correct actions are taken by those at-risk.*

Improvements following Hurricane Katrina/Rita:

The failures in the early warning are widely acknowledged to have occurred because of weaknesses in risk awareness, warning/communications, and response, and despite strengths in forecasting. The NWS continues to improve its capability to forecast hurricanes, and its institutional counterparts at the federal, state and local levels have made wide-ranging improvements to risk awareness and communication.

6.3.3 Assessment of Vulnerabilities and Exposure (National to Local)

Vulnerability Assessments

Vulnerability assessments (hazard assessments) are conducted at the community level during mitigation plan development or through the 5 year updating cycle. Most hazard assessments follow FEMA's Mitigation Risk Assessment Process. This includes the following steps:

1. *Identifying hazards* – This is accomplished at the city, county, or state level and involves going back and looking at what hazards have affected a community.
2. *Profiling hazards* – Profiling hazards involves looking at each potential hazard, defining its characteristics and researching the history of the hazard's impact on the jurisdiction. Data for hazard history can include many different sources, based on the specific hazard. Some of the most popular sources are: NOAA, Climatology offices, joint analysis centers, National Pipeline Mapping System, GIS divisions, local newspaper archives, state departments of transportation, National Climatic Data Center, local fire and police departments, National Severe Storms Laboratory, U.S Nuclear Regulatory Commission, state offices of emergency management, FEMA, and state departments of natural resources.
3. The third step in the hazard assessment process involves *assessing a jurisdiction's vulnerability*, including inventorying critical assets, and then estimating their losses. This is usually accomplished by using GIS to map the infrastructure within hazardous zones. Since flooding is a natural hazard that has defined boundaries, most communities will map the assets they have in the floodplains. Assets are defined by FEMA as: lifeline utilities and other infrastructure, such as communication lines and energy services. Also included are dams, police and fire stations, city halls, substations, schools, day care centers, and warning sirens. The second part of this step is to estimate the losses of the infrastructure if the hazard were to occur. Loss estimation tools such as FEMA's HAZUS software can be used in this step, or it can be accomplished by getting the information directly from the source, such as a building operator or the jurisdiction's assessor's office.

Most of the above work is undertaken in consultation with the National Weather Service by other government entities or by consulting and academic researchers.

Mapped Vulnerabilities

Mapping vulnerabilities is usually the job of the local governments, but can also be accomplished by the State. This is mainly accomplished by city or county emergency management agency or the GIS division. Information regularly mapped includes special populations, such as people with disabilities or immigrant groups that may not have an idea of the hazards that face them, due to either their background or because they are new to the area. This information is gathered by using the U.S. Census Bureau. To validate this information, feedback is gathered from the cities (if information was gathered by the county) or through public meetings so that all impacted parties and stakeholders have the opportunity to provide input.

6.3.4 Storage and Accessibility of Disaster and National Hazard Risk Information

Local hazard and vulnerability archival systems are stored at county emergency management offices, city offices, and at the state emergency management offices who forward mitigation plans to the Federal Government (mitigation plans are updated every 5 years). Hazard and vulnerability data is made available to the public through several public meetings during the mitigation planning process. These data are also provided to the public through city and county websites, where most mitigation plans are available to the public, so that information on jurisdictional hazards can be seen. The quality management framework process in place to keep hazard assessment data current varies by government and jurisdiction. Most hazard assessments are completed during mitigation plan updates every 5 years. However, in the case of natural hazards that affect a community for the first time, some communities will re-assess their hazards immediately following the event.

The National Hazards Center at the University of Colorado-Boulder is a clearinghouse for hazard information and archived hazard assessment data (http://www. colorado.edu/hazards/). The mission of the Natural Hazards Center is "to advance and communicate knowledge on hazards mitigation and disaster preparedness, response, and recovery". Using an all-hazards and interdisciplinary framework, the Center fosters information sharing and integration of activities among researchers, practitioners, and policy makers from around the world; supports and conducts research; and provides educational opportunities for the next generation of hazards scholars and professionals.

6.3.5 Development and Utilization of Hazard/Risk Information to Support Emergency Planning and Warnings

Risk information (hazard assessments) is the basis for all jurisdictional emergency planning. Identifying the community-specific hazards helps with gap assessment and effective response planning by incident type. Risk information is integrated into local emergency plans and, in some cases, comprehensive plans. Most often, comprehensive plans identify future land-use development trends that include infrastructure developments, in relation to the hazardous areas already identified by the community.

Including potential impacts within a warning message helps Emergency Managers/planners deliver a call to action, based on the severity of the event and the affected populations. Full-scale training exercises are regularly held to identify potential impacts of varying scenarios and those impacts are then used to create appropriate warning messages for use in real-time warning generation.

6.4 Hazard Monitoring, Forecasting, and Mandates for Warning Development

6.4.1 Organizational Responsibilities for Monitoring, Forecasting and Development of Hazard Warnings

The U.S. agencies responsible for the primary warning to the public for the indicated hazards are detailed in Table 6.2.

For Type II and Type III hazards, the NWS provides meteorological and hydrologic guidance for decision support. Additional Type II support examples include: guidance for debris flow hazards for recently burned areas through flash flood guidance support; and issuance of avalanche watch/warning and air quality hazard alerts, in cooperation with the USGS and state agencies.

6.4.2 Organizational Collaboration and Coordination for Monitoring, Forecasting and Development of Hazard Warnings

Information is delivered to the local forecast offices in a variety of ways for use in product and service generation. The information from Weather Service Radar, Automated Surface Observing Systems (ASOS), satellite and ground observations as well as data from our national centers allow forecasters to create needed public forecasts and warnings. Observational networks deployed by other government agencies utilize the NOAA satellite network for accurate and timely communication services for meteorological and hydrological observations.

The overall suite of products the NWS generates can be described on the order of minutes for warnings and alerts, hours for watches, days for forecasts, 6 to 14 days for threat assessments, months for guidance products, and seasons and years for climate outlooks. As one moves from warnings to outlooks, forecast uncertainty increases.

The Hydrological Services Program of NOAA's National Weather Service collaborates with the local, national, and international community to provide essential warnings and advisories of potentially hazardous conditions involving water (Fig. 6.3).

Table 6.2 Agencies responsible for the primary warning to the public for hazards in the United States

Weather Phenomena	Hazard Type	Warning Responsibility
Tornado (rotational high winds)	Type I	United States National Weather Service
Flash flood	Type I	
Strong winds	Type I	
Hailstorm	Type I	
Thunderstorm or lightning	Type I	
Heavy snow	Type I	
Freezing rain	Type I	
Dense fog	Type I	
Tropical cyclone	Type I	
Storm surge	Type I	
Coastal flooding	Type I	
Heat wave	Type I	
Cold wave	Type I	
Drought	Type I	
River flooding	Type I	
Marine hazards (storm, sea ice, icebergs, etc.)	Type I	
Sandstorm	Type I	
Hydrometeorological hazards to aviation (i.e., turbulence, icing)	Type I	
Smoke, dust or haze	Type I	
Tsunami	Type I	
Space Weather (Geomagnetic and Solar Storms, Radio Blackouts)	Type I	
Landslide or mudslide	Type II	United States Geological Survey
Earthquakes	Type III	
Volcanic events	Type III	
Airborne hazardous substances (e.g., nuclear, biological, chemical)	Type III	Department of Homeland Security/State Emergency Management Agencies/Nuclear Regulatory Commission
Waterborne hazards (e.g., nuclear, biological, chemical, oil spills)	Type III	
Desert locust swarm	Type III	United States Land Management Agencies (e.g. Bureau of Land Management, United States Forest Service, National Park Service, Bureau of Indian Affairs, State Fire Agencies – coordinated through the National Interagency Fire Center)
Avalanche	Type II	
Forest or wild land fire	Type III	

The National Weather Service River Forecast Centers have developed a spatially and temporally continuous modeling system for all watersheds across the United States. These models assimilate internally generated meteorological observations and forecasts with hydrological observations from other government agency partners to produce a suite of hydrological forecast and warning products. These products range from soil moisture state conditions used to calculate flash flood potential to river level hydrographs predicting flood crest elevations.

The Advanced Hydrologic Prediction Service (AHPS) is provided through modernization of the National Weather Service (NWS) hydrology program. The need for AHPS is addressed in the NWS Strategic Plan for Weather, Water, and Climate Services and provides NOAA's customers additional time to prepare for floods or droughts, more accurate water forecasts and flood warnings, and better information for decision making. AHPS builds upon existing services and systems, including the NWS' 13 River Forecast Centers (RFCs) and 122 Weather Forecast Offices (WFOs); ongoing partnerships with other government and private sector organizations; modernization programs including the Next Generation Weather Radar (NEXRAD), Geostationary Satellite System (GOES), Automated Surface Observation System (ASOS),and Advance Weather Interactive Processing System (AWIPS) and the National Weather Service River Forecast System (NWSRFS) platform. Outputs include:

- More accurate and comprehensive predictions of river height and flood potential;
- Longer term forecasts, from days to months, with probabilistic information to assist in risk based decision making;

- High resolution, visually oriented forecast products at 4,000 locations nation wide;
- Distribution of graphical information on NWS Web sites;
- Hydrographs, providing near term river height and flow forecasts;
- Probabilistic forecasts, providing graphical forecasts from days to months into the future;
- Flood inundation maps, providing a graphical display of forecasted flood areas;
- A color coded flood severity index indicated in Fig. 6.4.

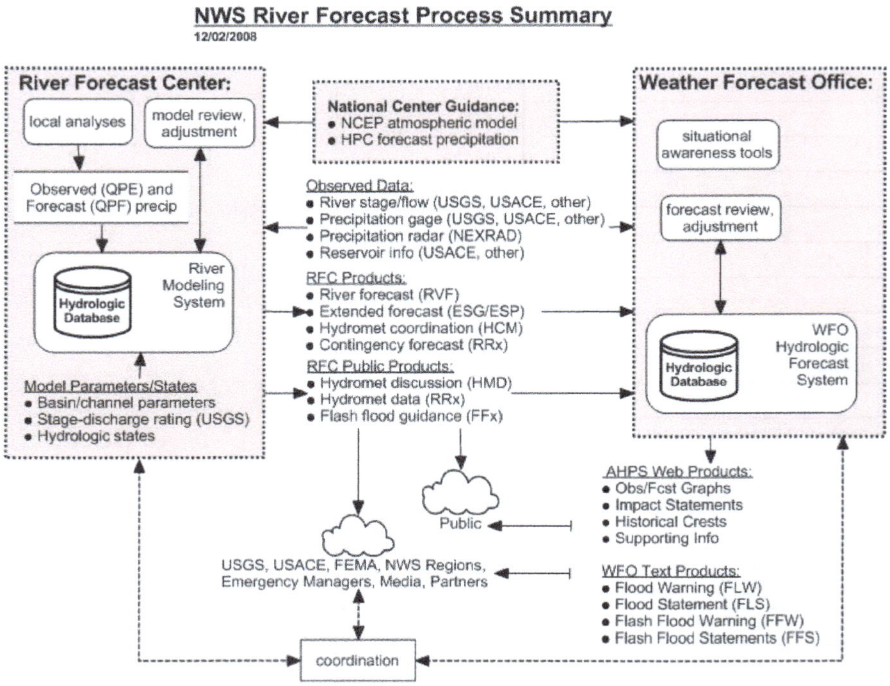

Fig. 6.3 NWS River Forecast Process Summary (NWS Service Assessment Report 2009)

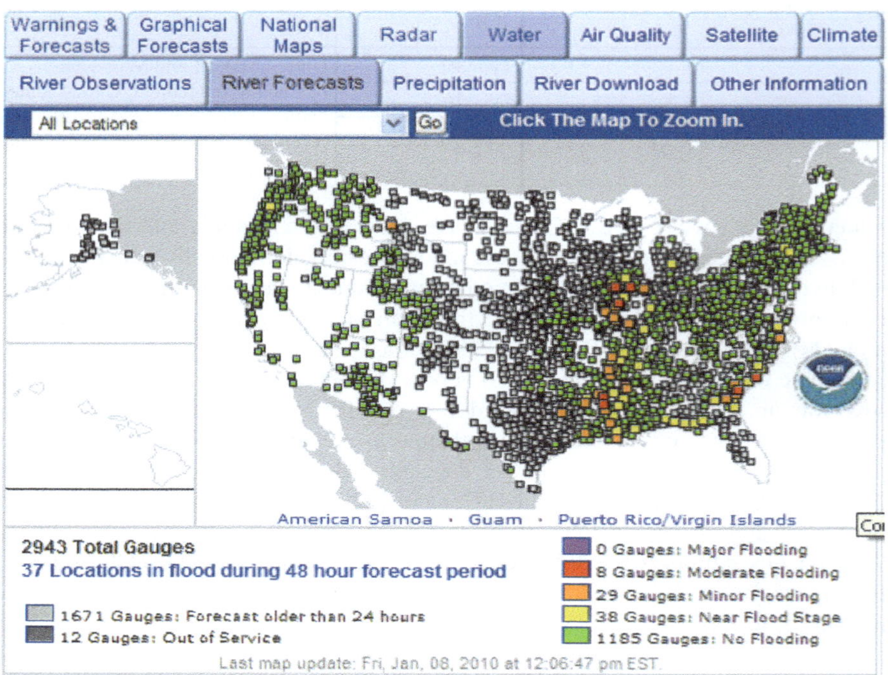

Fig. 6.4 Advanced Hydrologic Prediction Service (AHPS) forecast points

In a 2006 survey of the impact of weather forecasts, the National Center for Atmospheric Research investigated four interdependent aspects of early warning information: sources, perceptions, uses and values of Americans (Lazo et al. 2009). Close to nine out of 10 adult Americans obtain weather forecasts an average of more than three times each day, and households place an average value of 10.5 cents on every weather forecast obtained. This equates to an annual value of $31.5 billion. In comparison, the cost of providing forecasts by government agencies and private companies is $5.1 billion. U.S. adults obtain an estimated 300 billion forecasts each year, were generally satisfied with weather forecasts, and had fairly high confidence in forecasts with a lead time of one to two days. Gaining a better understanding of people's attitudes and behaviors toward forecasts also provides valuable information to emergency managers, and to improving early warning system effectiveness.

6.5 Development of Timely, Authoritative, Recognizable, and Understandable Warnings

6.5.1 Warning Message Development Cycle

The NWS uses the "*Ready-Set-Go*" concept for its watches and warnings. Outlooks are used to prepare for an event: as the event becomes increasingly evident a watch is issued; and once severe weather is imminent, a warning is issued.

The first step in the overall warning process is the get "*Ready*" for an event. The NWS uses outlook products for a variety of parameters to help set the stage for a possible event. Some examples of outlook products include those for severe weather, large scale precipitation events, tropical development, fire weather and winter weather. The outlook phase covers the time period up until approximately 72 hours before event onset.

The second stage in the process is to get "*Set*" for an event. This is accomplished by way of watch products. When the NWS forecasters believe there is a good probability for an event a "Watch" is issued. Watches are usually issued days to hours before an event starts. Examples of watch products include tornadoes, severe thunderstorm, tsunamis, fire weather, winter storms, flood, and flash flood events.

The third and final step in the warning concept is the "*Go*" analogy. Warning forecasters in the WFOs watch the WSR-88D for severe weather signatures. Once a storm is identified as possibly becoming severe, a "Warning" is issued. On the graphic below (Fig. 6.5), we see a line of thunderstorms with three individual cells capable of producing severe weather. The warning forecaster will produce polygons around the areas that are in immediate danger. A text product of the warning is produced outlining the type of warning, the areas that are threatened, the length of the warning, and preparedness actions. This warning is disseminated to the customers in a variety of ways.

National Weather Service warnings follow a known, well-publicized format used for both human reading and ingestion by computer system.

Fig. 6.5 Illustrations of NWS Severe Warning "Go"

6.5.2 Warning Message Improvement Cycle

The National Weather Service (NWS) uses a variety of opportunities to further improve its warning services. During and after the warning event users of the information are given the opportunity to comment on the efficiency of the products by way of Internet surveys. This information is received at the local offices and is reviewed and incorporated into operations, if applicable.

Another way improvement is obtained is by the use of verification statistics after all events. Three main statistics are generated: probability of detection (POD), false alarm ratio (FAR), and critical success index (CSI). For episodes where these numbers are not within satisfactory values, post-event reviews are performed at the offices involved. This involves a complete review of the event including staffing, radar interpretation, and product preparation. Areas found to be in need of improvement are identified. Office personnel are made aware of the findings with the hope for improvements during the next event.

A third way improvements are identified is by way of post-storm service assessments. The NWS conducts service assessments to evaluate its performance after catastrophic weather events. For these events, assessment teams composed of experts in and outside of the organization study what happened before, during and after the event. Interviews with the survivors, local emergency management and media outlets are conducted with the hope of getting a complete picture of the service provided. The team will complete a report recommending changes in NWS procedures, products and services with the hope of improving overall performance in future event. For examples of past service assessments please visit the following web site: http://www.nws.noaa.gov/os/assessments/index.shtml

6.6 Warning Dissemination Mechanisms

All-hazards warning dissemination in the United States is described in the *National Response Framework.* The National Response Framework is comprised of the core document, the Emergency Support Function (ESF), Support and Incident Annexes, and the Partner Guides (Fig. 6.6). The core document describes the doctrine that guides our national response, roles and responsibilities, response actions, response organizations, and planning requirements to achieve an effective national response to any incident that occurs, as described in Sect. 2.1.1.

The following documents provide more detailed information to assist practitioners in implementing the Framework:

- *Emergency Support Function Annexes* group Federal resources and capabilities into functional areas that are most frequently needed in a national response (e.g., Transportation, Fire fighting, Mass Care).
- *Support Annexes* describe essential supporting aspects common to all incidents (e.g., Financial Management, Volunteer and Donations Management, Private Sector Coordination).
- *Incident Annexes* address the unique aspects of how we respond to seven broad incident categories (e.g., Biological, Nuclear/Radiological, Cyber, Mass Evacuation).
- *Partner Guides* provide ready references describing key roles and actions for local, tribal, State, Federal, and private-sector response partners.

Fig. 6.6 Documents for implementing the National Response Framework

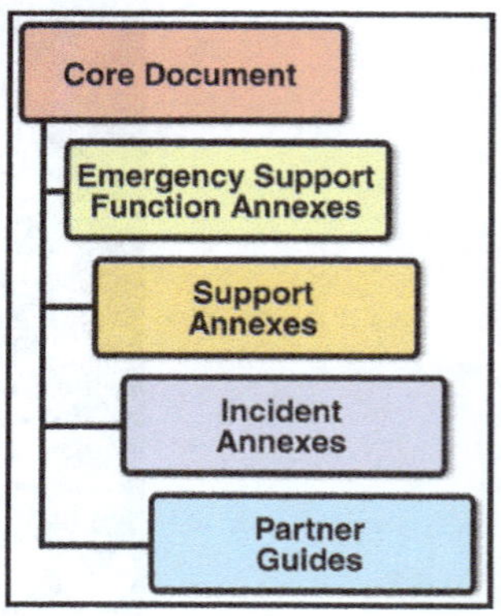

Emergency Support Function (ESF) #2, the *Communications Annex*, supports the restoration of the communications infrastructure, facilitates the recovery of systems and applications from cyber attacks, and coordinates federal communications support to response efforts during incidents requiring a coordinated federal response. ESF #2 also provides communications support to Federal, State, tribal, and local governments and first responders when their systems have been impacted by a disaster or incident. ESF #2 identifies and describes the role of the National Weather Service as follows: *"supports the Emergency Alert System and provides, in coordination with FEMA, public dissemination of critical pre-event and post-event information over the all-hazards NOAA Weather Radio system, the NOAA Weather Wire Service, and the Emergency Managers Weather Information Network."*

The National Weather Service coordinates and maintains its own policy, a *National Directives System*, for the operational implementation of its systems and services. Series 10–17 covers NWS policy for the dissemination of warnings, specialized forecasts, and other products. The NWS produces weather, water, and climate forecasts, warnings and other data. This suite of information forms a national data base and infrastructure that can be used by United States customers and the global community. The NWS delivers critical life- and property-saving products directly to customers and makes available its entire information suite through various government operated systems and through our partners in the media and private sector. The NWS also relays non-weather-related emergency information following the "all hazards" concept – the dissemination of non-weather-related emergency messages from external authorities, in addition to NWS information. These emergency messages include both natural events, such as earthquakes and volcanic activity, and technological events, whether accidental or intentional, such as chemical or biological releases, oil spills, or nuclear incidents.

There are a variety of formal and informal mechanisms for improving dissemination of warnings utilized in the United States. The Government Accountability Office audit, evaluation and investigation process mentioned previously in response 2.2.6, is an example of a formal mechanism. Informal mechanisms utilized by the NWS would include assimilating stakeholder feedback or social science research into future service improvements. Stakeholder inputs are acquired in a variety of formats but primarily through meetings and workshops, and web-based feedback. In some instances, the NWS has utilized surveys and focus groups to gauge potential service improvements. The NWS dedicates a position in every one of its 122 Weather Forecast Offices, 13 River Forecast Centers, 6 Regional Headquarters and National Headquarters, to serve as the lead for customer feedback and service enhancements. In NWS Weather Forecast Offices and Headquarters, that position is known as the Warning Coordination Meteorologist (WCM) while in the River Forecast Centers it is called the Service Coordination Hydrologist (SCH).

Warning Coordination Meteorologists (WCMs) and Service Coordination Hydrologists (SCHs) lead the effort to identify and meet customer requirements for weather and hydrologic service and information in a number of ways. They conduct area-wide evaluations of WFO products and services by ensuring they are generated according to established national and regional directives. They lead the preparedness program which involves ensuring all people within the county warning area are aware of the weather hazards which threaten their safety and property and are prepared to take action if and when these threats become reality. Teaching is paramount in this endeavor, which involves conducting education and training courses with media, emergency managers, schools and clubs in the area. As needed, they also lead or serve as members on Flood or Disaster Survey teams. In addition, they assist in planning the execution of drills to test federal, state and local operational plans and procedures. More information on the position may be found at:
http://www.weather.gov/om/wcm_jobaid/job_description.shtml

To properly identify the strength of tornadoes for example, WCMs will perform storm damage surveys after most events. The damage areas are surveyed on foot and information is gathered from structural damage that is observed as well as interviews with storm victims. This information is then reviewed and a final strength value is given to the tornadic storm. The values for tornadoes are identified by one of five values in the Enhanced Fujita Scale. For specific values please visit: http://www.spc.noaa.gov/faq/tornado/ef-scale.html

The following are some of the early warning systems used in the United States today. These systems include technologies supplied from private and public organizations and include a wide range of tools that help provide warnings to citizens and first responders regarding events that could threaten their locality.

1. Public T.V. and Radio Media:

   The longstanding collaboration between the National Weather Service (NWS) and the media has served the American public very well, especially during severe weather events. Today, this public-private collaboration continues to evolve, both responding to and driving the revolution in information technology and making use of the products and services resulting from the recent modernization of the National Weather Service.

   Because of outstanding past performances in many hazardous weather situations, today's media enjoy a high level of trust and confidence from the public and have become an indispensable element of a region's hazardous weather warning team. Electronic media (traditionally radio and television, but today also includes cable) provide opportunities for nearly instantaneous communication to the public, a vital feature in a rapidly evolving severe weather situation such as a tornado or a flash flood.

Recently, to help increase coordination with T.V. and radio media sources, the NWS helped create Integrated Warning Teams (IWT) in certain geographic areas. This group brings partners from the National Weather Service, emergency management, first responders, news media, academia, local health departments, private sector communications companies and other associated players. The main goal of the IWT program is to initiate dialogue between IWT members to start coordination and improve warnings that increase the proper public response to severe weather events.

2. "All Hazard" NOAA Weather Radios:

   The NWS has also developed a nationwide network of radio stations broadcasting continuous weather information directly from the nearest National Weather Service office to local citizens. The "All Hazards" National Weather Radio broadcasts official Weather Service warnings, watches, forecasts and other hazard information 24 hours a day, 7 days a week.

3. Emergency Alert System (EAS):

   The Emergency Alert System is a national warning system in the United States put into place in 1997 and is jointly coordinated by the Federal Communications Commission (FCC), Federal Emergency Management Agency (FEMA), and the National Weather Service (NWS). The official EAS is designed to enable the President of the United States to speak to the United States within 10 minutes. The EAS regulations and standards are governed by the Public Safety and Homeland Security Bureau of the FCC. Each State and several territories have their own EAS plan with 30+ radio/TV stations located throughout the United States designated as National Primary Stations and the Primary Entry Point (PEP) System to distribute Presidential messages to other broadcast stations and cable systems.

4. "Outdoor" Warning Systems:

   Many communities have installed "outdoor warning sirens" (also referred to as an air raid siren, civil defense sirens, tornado siren, tsunami siren, or also rarely referred to as the doomsday siren). This is a mechanical or electronic device (modern-day sirens are electrically-powered whether they are electronic or mechanical) for generating sound to provide warning of approaching danger and sometimes to indicate when the danger has passed. Initially designed to warn of air raids in World War II, they were adapted to warn of nuclear attack and of natural phenomena such as tornadoes. Some newer sirens have the ability to broadcast voice messages over large areas, depending on winds and noise. These electronic sirens differ from electromechanical sirens in that they rely on a series of large loud speakers to produce sound.

5. Internet Warning Systems/Data:

   These systems include Email, Short Message Service (SMS), Multimedia Messaging Services (MMS) and Really Simple Syndication (RSS) service feeds, which are a family of web formats used to publish frequently updated digital weather information to any subscriber. The National Weather Service also supplies web based "Polygon" data that can be integrated into real-time GIS mapping software allowing "outdoor" warning sirens within the "polygon" area to be automatically activated. This system is the basis for many local emergency alert systems such as the Faith Based Notification System where warnings are sent via SMS text messages to pagers, cell phones and email to administrative members of the church who then disseminate the warning to church members.

6. Emergency Managers Weather Information Network (EMWIN):

   EMWIN is a suite of data access methods which generates a live stream of weather and other critical emergency information. Each data stream method has unique advantages and includes digital weather information received through radio, Internet push pull technologies utilizing email and satellite broadcast.

7. NOAA Weather Wire Service:

   NWWS is a satellite data collection and dissemination system operated by the National Weather Service. Its purpose is to provide local, State and Federal Government, commercial users, media, and private citizens with timely delivery of meteorological, hydrological, climatological, and geophysical information. The vast majority of NWWS products are weather and hydrologic forecasts and warnings issued around the clock from 122 NWS offices nationwide. An important element of the NWWS mission is providing rapid delivery of critical NWS-issued severe weather warnings and watches. All products in the NWWS data stream are prioritized, with weather and hydrologic warnings receiving the highest priority (watches are next in priority). This allows special handling and delivery of warning products ahead of other less critical weather forecast products. NWWS delivers severe weather and storm warnings to users in 10 seconds or less from the time they are issued, making it the fastest delivery system available for these very time sensitive products.

8. The National Warning System (NAWAS):

   NAWAS is the primary national system for emergency communications among Federal, state, and local emergency operations centers. FEMA operates two national warning centers: the FEMA Operations Center (FOC) at Mount Weather, Virginia, and the FEMA Alternate Operations Center (FAOC), in Thomasville, Georgia. NAWAS is a dedicated, 24-hour, specialized telephone line with over 1,660 terminals that can be activated simultaneously but are more typically used in a hierarchical manner based on the region of concern. The system can, with the push of a button, activate terminals for the whole Nation, specific FEMA regions, individual states or local circuits. These warning points are typically located at the state emergency operations center, state police dispatch centers and county/city emergency operations centers, law enforcement dispatch centers, or fire dispatch centers.

9. Reverse 911 Services:

   Most residents in the United States can call 9-1-1 to get police, fire, or emergency medical assistance. In some communities a Reverse 911 Community Notification System is used to call residents and businesses with emergency information. In the event of an emergency, an operator in the 911 Center can identify the affected neighborhood or region of the County and records a message that describes the situation and recommends the protective actions residents should take. The Reverse 911 system will automatically call out to all listed telephone numbers within that geographic area and deliver the recorded message. If phone lines are busy, the system will attempt to redial those numbers a predetermined number of times to make contact. If an answering machine picks up the call, the system is programmed to track numbers not answered and will try and redial those numbers again.

10. Subscription-based notification system:

    These are standardized, secure and certified communication platforms through which local public safety organizations can communicate important neighborhood-level information to residents. Citizens sign up to receive emergency messages through social applications similar to Twitter and Facebook. Citizens can also sign up through private warning subscription providers creating competition to the normal "official" warning process.

11. Route Warnings:

    These warnings are conducted by personnel utilizing emergency vehicles' public address systems to warn citizens in designated neighborhoods. These types of warnings are used in specific types of disasters such as flash flooding, hazardous materials incidents, and severe weather.

12. Electronic Message Boards:

    Many communities utilize electronic message boards along freeways to alert drivers to problems on the road ahead. These messages include severe weather events impacting traffic at that moment (such as impassable roadways due to flooding, ice, etc. and slow traffic due to heavy snow, tornadic activity, heavy rains, etc.). The system also uses message boards to notify drivers of emergency alerts such as evacuations and child abductions (AMBER Alerts).

13. Integrated Public Alert System (IPAWS):

    This system is the Nation's next-generation infrastructure of alert and warning networks expanding upon the traditional audio-only radio and television Emergency Alert System (EAS) by providing one message over more media to more people before, during, and after a disaster. The vision of IPAWS builds and maintains an effective, reliable, integrated, flexible and comprehensive system that enables the American people to receive alert and warning information through as many means as possible. IPAWS ensures the President can alert and warn the public under all conditions. IPAWS will provide Federal, State, territorial, tribal, and local warning authorities the capabilities to alert and warn their respective communities of all hazards impacting public safety and well-being via multiple communications pathways. The Commercial Mobile Alert System (CMAS) is one of the many projects within IPAWS intended to send alerts and warnings to the public utilizing authorized text-based alerts to subscribers of wireless carriers who chose to participate in the program.

14. National Weather Service Automated Warning Systems:

    The NWS has implemented several automated warning systems that are utilized by public and private organizations as well as citizens. These systems include the Automated Flood Warning Systems (AFWS), NWSChat service and the iNWS service.

- The AFWS system connects numerous local flood warning systems, and integrates and shares information from 1700 sensors in 12 states.
- NWSChat has opened up several lines of communication (instant messaging type services) that were not pre-existing, providing for a much faster relay of useful information to vetted public and private entities responsible for public warnings. This system has its own "chat room," in which users can communicate with each other. Additionally, there is a "bot" running which will automatically post to the chat: severe weather warnings, severe weather updates, non-precipitation warnings, local storm reports, SPC watches, some NHC products are automatically posted. (This prevents the NWS from having to retype this information after disseminating these products). The system has recently been expanded by creating "public" rooms for each NWS Forecast Office. These rooms are occupied by any user who chooses to participate and a similar robot that posts the same information as it does to the "real" chat.
- The iNWS System has several services available:
  - iNWS Alerts allow users to configure and receive text message alerts and e-mail message alerts when the National Weather Service issues a watch, warning, or advisory that affects them. Users can configure their alert preferences geographically and by weather phenomena.
  - AHPS Mobile Service allows users to access hydrographs, current and forecasted stages, and river impacts on-the-go from their web-enabled mobile device. Also, users with an iNWS account can login to AHPS mobile and create "River Watch Points" which allows emergency managers to select certain points of interest to view more quickly, in addition to receiving e-mail message alerts whenever the river is observed or forecasted to change stage.
  - iNWS Mobile Service is an application that will install and run on Java enabled mobile devices. It allows you to browse weather service watches, warning and advisories, radar and satellite imagery, observation, and point forecasts all from a simple map interface.
  - iCWSU Service provides an iNWS Mobile Web interface to weather data for the aviation community. Available mobile products include hub forecasts and discussions, TAFs and METARs, storm summaries, convective and tropical outlooks, aviation hazard graphics and imagery. Currently, iCWSU is designed for the San Francisco, California area. Work is being done to expand coverage to the rest of the United States.
  - iNWS Mobile Web Service allows the user to search for weather information by ZIP code or by city and state, just like on the web, including current watches, warning and advisories, current conditions, radar and satellite feeds, and National Weather Service forecasts.

6.7 Emergency Preparedness and Response Activities (National to Local)

6.7.1 Disaster Preparedness and Response Planning and Emergency Response Activation

Emergency preparedness and response plans are targeted to individual needs of vulnerable communities, authorities and emergency respondents in two phases:

1. Historical data for vulnerable communities and the response capabilities of emergency respondents is used in planning for hazardous mitigation for vulnerable areas and individuals. For example, historical data available in the hydrological section of the NOAA weather forecast Internet page provides data showing historical flood levels and current river flow levels. This data allows advance and "current time status" evaluations of potential flooding issues and aids in emergency alerts, advance preparation and response. Historical data also allows a more potent response to areas considered "vulnerable" due to repeated crisis issues for that area and/or individuals.
2. Evaluation of need and potential need are considered when developing advance planning for specific areas of vulnerable populations or geographic locations. Emergency Preparedness plans will include vulnerable populations (economic, medical, physical limitations, age and handicap, non-English speaking and concentrated areas of need such as assisted living centers) in advance. Priority response preparation would include table-top exercises, review meetings with emergency responders and volunteers. An example of advance preparation for medical needs would include planning for dialysis patients and dialysis clinics if there would be wide spread power outages in the area.

The geographic layout of specific communities (counties) will give a conceptual forecast of population concentrations and locations of "in-place" emergency responders (fire/police) and potential areas of concern. Advance planning for Faith-Based Organizations (FBOs) or community shelters should also consider the best routes of transportation and access for large areas of need when planning a shelter system. This allows a coordinated response and the highest level of service in crisis situations.

The FBO (Safe Shelter Partnership) approach allows Emergency Preparedness Managers to work within specific areas and communities by training the congregational members in emergency response and then training the FBO to educate the citizens in the community in which the FBO resides. Emergency Preparedness Managers are actively involved in training volunteers (CERT, RACES), disseminating materials at public events and through public service meetings. The FBO approach is described further in Sect. 2.7.2.

Emergency preparedness/response plans are kept at local emergency management agencies. Most emergency management agencies will also host their plans on their websites; however, these are usually plans that are censored so that secure information such as terrorism-related annexes and other secure information are not given to the public.

Emergency plans are exercised yearly by all emergency response agencies/emergency management agencies to test the plan. Evacuation plans are either worked into an emergency operations plan or are part of a stand-alone evacuation plan. Evacuation plans are most common in coastal communities that have the history of experiencing hurricanes and need to get the populations out during mandatory evacuation orders. The plans identify the established routes and are usually tested via table-top exercises where all agencies having responsibility come to talk through evacuation scenarios. Lessons learned from previous incidents are always looked at when updating emergency operations plans.

The role of NOAA within disaster response planning

In the Midwest portion of the country the major emergency preparedness concern is usually weather-related. The National Weather Service plays a key role in helping Emergency Preparedness Managers in many ways. Some examples:

- The National Weather Service and Emergency Preparedness Managers have jointly developed advanced communication processes and procedures (radio and web-based).
- Emergency Managers and NWS jointly attend emergency response-related meetings and conferences and have established a consensus regarding levels of severe weather and how to alert the public.
- There are joint exercises between emergency preparedness managers and the National Weather Service, teleconference calls and updates, as well as public service alerts.
- The National Weather Service web-based Weather Forecast page is very useful in planning for severe weather and alerting public support departments of city and county governments to potential issues.

6.7.2 Community Response Capacities

Local Emergency Preparedness Managers usually belong to either a metro council (if urban) and/or a regional council. The NWS is often represented locally in these organizational meetings. In some cases, the councils have paid for community-based surveys that show expectations, levels of understanding of alerts and warnings, and what the public considers the best source for receiving severe weather information.

As described in Sect. 2.2, the National Response Framework recognizes the frontline roles that individuals, households, communities, State and tribal governments play. Communities and local jurisdictions are not only the first line of defense when incidents occur, they are also responsible for preparedness and planning efforts.

Citizen Corps

For example, a program called Citizen Corps brings together local leaders from governmental and non-governmental organizations, and the private sector, to prepare for and respond to incidents. Citizen Corps councils are typically sponsored by elected or appointed officials and/or emergency managers. These Councils provide leadership and support for programs that educate, train, and engage community volunteers to support emergency management and responders.

Tribal Communities

The Federal Government recognizes the special status and rights of Native Americans, including their entitlement to self-governance. As such, tribal governments are responsible for coordinating resources to address actual or potential incidents. When local resources are not adequate, tribal leaders seek assistance from the Government. For certain types of Federal assistance, tribal governments work with the State, but as sovereign entities they can elect to deal directly with the Federal

Government for other types of assistance. In order to obtain Federal assistance via the Stafford Act, a State Governor must request a Presidential declaration on behalf of a tribe.

The tribal leader (i.e. Chief) is responsible for the public safety and welfare of the people of that tribe. As authorized by tribal government, the tribal leader:

- Is responsible for coordinating tribal resources needed to prevent, protect against, respond to, and recover from incidents of all types. This also includes preparedness and mitigation activities.
- May have powers to amend or suspend certain tribal laws or ordinances associated with response.
- Communicates with the tribal community, and helps people, businesses, and organizations cope with the consequences of any type of incident.
- Negotiates mutual aid and assistance agreements with other tribes or jurisdictions.
- Can request Federal assistance under the Stafford Act through the Governor of the State when it becomes clear that the tribe's capabilities will be insufficient or have been exceeded.
- Can elect to deal directly with the Federal Government in certain circumstances.

State and tribal officials typically take the lead to communicate public information regarding incidents occurring in their jurisdictions. It is essential that immediately following the onset of an incident, the State or tribal government, in collaboration with local officials, ensures that:

- Communication lines with the press are open, questions receive prompt responses, and false rumors are refuted before they spread.
- Information about where to receive help is communicated directly to victims and victims' families.

Faith-Based Organizations

The Faith-Based Organization (Safe Shelter Partnership) is a key component of community-focused preparedness and is driven by Emergency Preparedness Managers. Local FBOs are trained in many areas of community assistance and preparedness:

- The local congregation is trained to be self-sufficient in their homes during times of crisis. They can expand their "windows of resiliency" from 3 to 7 to 14 days if necessary.
- The FBO is trained to manage its own facility (Safe Shelter Partnership) as a shelter during times of need. They receive training from the Emergency Preparedness Manager, the Red Cross, Salvation Army, and local Health Departments. They develop advanced relationships with local hospitals, Family Service Division, United Way, Medical Reserve Corps, NOAA, and local volunteer assistance groups such as CERT and RACES. The FBO (Safe Shelter Partnership) is then trained to educate the community in their area on how to be self sufficient in their homes.
- The FBO (Safe Shelter Partnership) is also trained on how to respond to community issues including debris management and damage assessment.
- The FBO (Safe Shelter Partnership) is considered an extension of the Emergency Operations Center and is trained in various forms of communications including HAM radio, NOAA All Hazards radio, web based "call down" systems, and emergency band radio.

The training involved includes the merging of government, non-government and volunteer agencies to develop the Safe Shelter partnership program with faith-based organizations. Training centers on the concept of keeping these organizations involved at the local level. Training is on-going and is given to the FBO group at least quarterly. An individual FBO is a part of an overall plan that could involve up to 14 different FBOs in one portion of the county. The FBOs are trained to be interchangeable and but also to be self-sufficient. Exercises are planned on an annual basis.

NOAA is a key component of the FBO (Safe Shelter Partnership). In the NOAA training for the FBO, the link between the EOC, FBO and the National Weather Service is explained. The FBO is then better suited to train its congregation and community (or sponsor training by Emergency Preparedness Management). The EOC is the key component for communicating the NOAA information to the FBO, but the FBO is required to have an All Hazards Radio(s) in their facility.

Community focused organizations that educate and train the public are usually volunteer organizations such as the Red Cross and Salvation Army. Other cities and counties have Community Emergency Response Teams (CERT) that train and educate their staff in various subjects regarding emergency management so that in times of disaster, they are able to assist emergency management and response agencies with light search and rescue, perimeter control, debris management, and volunteer donations, etc. Community and volunteer education organizations such as the Red Cross and Salvation Army train their volunteers on a weekly and monthly basis. Other cities and counties have Community Emergency Response Teams that train and educate their staff in various subjects regarding emergency management so that in times of disaster, they are able to assist emergency management and response agencies with light search and rescue, perimeter control, debris management, and volunteer donations, etc. CERT training is usually held every month with a different subject of study each time.

Lessons learned from previous incidents disasters are always looked at when thinking about future capacity building strategies. However, most emergency management departments and response agencies have low budgets and are unable to purchase a lot of equipment to build capacities to fill gaps. Some larger cities do have access to Homeland Security grants and are able to purchase equipment to build first responder capacities.

StormReady

To help communities across the country gain an upper hand in preparation for disasters, a grass roots effort entitled "*StormReady*" began at a local Weather Forecast Office in Tulsa, Oklahoma back in 1999. The program helps arm America's communities with the communication and safety skills needed to save lives and property before and during weather events. StormReady helps community leaders and emergency managers strengthen local safety programs. Communities are better prepared to save lives from the onslaught of severe weather through advanced planning, education and awareness. Since its national implementation in 2000, there have been 1588 StormReady sites recognized in 49 U.S. states, Puerto Rico and Guam. The program was expanded to include areas threatened by Tsunamis and the "TsunamiReady" campaign now has 74 sites recognized in 10 U.S. states, Puerto Rico and Guam.

The guidelines for successful participation in the StormReady program are based on population and these guidelines can be found in Appendix 2. There are three steps to becoming StormReady or TsunamiReady: written application, verification visit(s), and local board action. Applicants with jurisdiction over a community and unincorporated areas of the surrounding county need only submit one application with the combined populations. The application requests a basic accounting of emergency technology and a brief narrative describing preparedness and planning activities. The narrative aids in assessing such things as the hazardous weather plan, exercises and public safety programs. After reviewing the application, the StormReady Chairperson will assign a team to visit the applicant and formally discuss the application. A completed application can be found in Appendix 3. For more detailed information on the programs please go to: http://www. stormready.noaa.gov/

6.7.3 *Public Awareness and Education*

An example of public awareness and outreach can be demonstrated within the Mid-America region. The Metropolitan Emergency Managers Committee (MEMC), in cooperation with NOAA, is very active in educating the public on the forms of weather warnings and the information that is usable and available. The education is focused on understanding the function and use of outdoor weather sirens and their effectiveness and limitations. The training also addresses the need to use TV and radio stations' weather alerts and updates and the need to purchase and use the NOAA All Hazards Weather radio. The MEMC has printed small pocket-sized brochures that explain the various types of alerts and classifications (severe weather watch/severe weather warning) and this information is dispensed in various ways, including "Home Shows", athletic events and public meetings. The local TV stations have developed a Severe Weather DVD and it has been dispensed to school children and at professional baseball games. The local TV stations also permit Public Service announcements that are intended to educate the audience on weather safety and procedures.

Local communities are educated on how warnings will be disseminated through pamphlets, media, and public education campaigns. These have been the most reliable sources for disseminating information. Though several studies and research projects have been completed and the research is ongoing, we still have no real idea how a certain portion of the population will respond when they hear warning messages. What is known is that the general population usually responds to the more credible sources that issue the message.

The National Weather Service developed a public awareness campaign know as "Turn Around, Don't Drown", to educate people domestically and around the world of the dangers of flooding (Fig. 6.7). In fact, this national program is a trademarked slogan for flood safety. The program even provides roadway signs approved by the Federal Highway Administration for local agencies to post and distribute in areas prone to flooding.

On-going public information/education campaigns are built into community education programs and the courses are then usually run through the school system. This enables emergency management to advertise through the school system. Public awareness campaigns are also advertised through county or city websites, community education sites, and some (if funding is available) can be advertised through mass media, such as radio and local television stations.

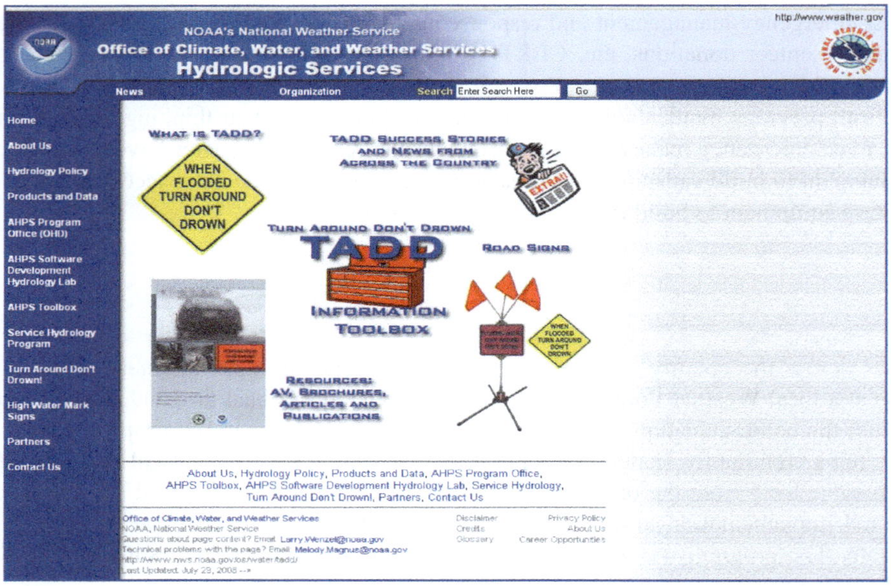

Fig. 6.7 NWS web resources for its "Turn Around, Don't Drown" campaign

The NWS has found that 80 percent of flood-related deaths in South Texas occur as a result of people driving through low-water crossings, walking along the banks of flooded areas, or playing in floodwaters. People underestimate the force and power of water. Six inches of fast-moving flood water can knock over an adult, and it takes only two feet of rushing water to float most vehicles. More than half of all flood fatalities result from automobiles being swept downstream. The Turn Around, Don't Drown (TADD) campaign was launched on May 22, 2003, with a news conference at NWS Southern Region Headquarters located in Fort Worth, Texas. The partners included the NWS, Federal Alliance for Safe Homes (FLASH), and the Texas Division of Emergency Management. Informational material, posters, and bumper stickers were provided, along with a demonstration of a new TADD web page. In an effort to reach as many people as possible, Guerrero and his TADD partners held conference calls with representatives from NWS regions across the country to coordinate a nationwide campaign.

Testimonials across the country have confirmed the success of the campaign and the need to bring its message to a broader audience. In May 2005, through a grant provided by the Allstate Foundation, FLASH, NWS, and Southwestern Insurance Service (SWS), the foundation expanded the Turn Around, Don't Drown campaign in Texas. They targeted the major cities collectively known as flashflood alley – Dallas, Houston, San Antonio and Austin. The cities had alarming flood-related fatalities. Outdoor billboards driving home the flood safety message have been erected in them. The city of San Antonio has placed bumper stickers displaying the slogan on all police, fire, and city vehicles. The message has also been spread through the local media via public service announcements (PSA), distribution of bumper stickers by the Texas Floodplain Management Association, animated presentations, and informative FLASH flood safety flash cards. In 2005 NOAA designated a week in March as Flood Safety Awareness Week.

According to records from the National Climatic Data Center, from January 2007 to July 2007, Texans have experienced more than 900 flooding events, more than doubling the ten-year average of 450 events per year. Over 2,100 flash flood warnings have been issued for the year; again more than double the 10-year average of 903. The number of flooding incidents has increased in the state of Texas, but flood-related fatalities have slightly declined across the U.S.

More information on FEMA best practices in mitigation is available at: http://www.fema.gov/mitigationbp/.

More information about the "Turn Around, Don't Drown" campaign is available at:
http://www.nws.noaa.gov/os/water/tadd/.

6.8 Development of Warnings and Related Products and Services for the Disaster Risk Management User Community

The U.S. National Weather Service continually looks for ways to better the services provided to the general public and other partners. Stakeholder inputs are acquired in a variety of formats but primarily through meetings and workshops, and web-based feedback. Every year a session is held at the American Meteorological Society (AMS) meeting to provide those in attendance with information on upcoming products and services and in turn the NWS solicits feedback. In some instances, the NWS has utilized surveys and focus groups to gauge potential service improvements. As mentioned earlier, the NWS dedicates a position in every one of its 122 Weather Forecast Offices, 13 River Forecast Centers, 6 Regional Headquarters and National Headquarters, to serve as the lead for customer feedback and service enhancements. In the weather forecast offices and headquarters the positions are referred to as the Warning Coordination Meteorologist (WCM), and in the River Forecast Centers as the Service Coordination Hydrologist (SCH). WCMs and SCHs are the "face" of the NWS to both internal and external stakeholders and are the primary NWS service representatives. They are responsible for working with the Emergency Management community to develop community- and hazard-specific messaging, based on the identified hazard risks within a local area.

As was mentioned earlier in the document, another way the NWS identifies potential service improvements is by means of post-storm service assessments. These assessments are performed after events that cause a large number of casualties and/or damage. The team will study what happened before, during and after the event. A final report recommending changes in NWS procedures, products and services is provided to upper management with the hope of improved products and services for the next event.

6.9 Improvement of Overall Operational Framework of the Early Warning System

Prior to the onset of significant weather or water events, the NWS places emphasis on the concept of Decision Support Services (DSS), and a focus not only on prediction but also on interpretation of impacts. Historically, the suite of outlooks and watches has provided hours to days of lead time leading to an event, when warnings and advisories would then be issued. However, much of this information focused on the expected phenomena and not as much on the impacts to the community.

Regional Operations Centers, along with Weather Forecast Offices and River Forecast Centers, conduct DSS briefings to key partners in the emergency management and media communities to complement the existing suite of information. These DSS briefings focus on impacts to help decision makers understand what the forecast phenomena means to their area of responsibility. Briefings are typically conducted remotely via teleconference and a remote desktop display program which allows the briefer to share a focused slide deck of information. For more significant and widespread events, DSS personnel deploy on-site to work side-by-side with these decision makers. Recent examples of on-site deployments include the Red River of the North Floods of 2009, the 2008 Iowa tornadoes which struck the Little Sioux Boy Scout camp and Parkersburg, and the 2007 Greensburg, Kansas tornado.

Local NWS offices typically review operations after each event to glean best practices and lessons learned. These reviews are informal, at times anecdotal, with a more immediate focus of fine tuning the office operation. Local emergency management and media feedback help drive this process. For more significant events, the NWS conducts formal service assessments to analyze the agency's response. These formal assessments directly involve members of the emergency management community, as well as NWS peers from other locations across the country. The formal service assessments provide best practices and lessons learned from all offices involved, and they also provide the agency a road map to implement specific improvements nationwide.

6.10 Examples of Previous Events Where the Operational Early Warning System Has Led to Improvements in Emergency Preparedness and Prevention

2004 Indian Ocean Tsunami

The Indian Ocean Tsunami of December 2004 has raised awareness worldwide of the potentially devastating impacts from episodic events, such as tsunamis. Population density coupled with increasing frequency and duration of storms, sea level rise, and other coastal hazards cause disasters to be more severe and recovery to be slower and less sustainable than in the past.

The Coastal Community Resilience (CCR) Initiative, part of the US Indian Ocean Tsunami Warning System (IOTWS) program, was created to focus upon the community-level capacity building that is essential to the success of any end-to-end warning system. One of the main lessons from the tsunami recovery effort was that single sector planning can't solve the complexity of problems posed by natural hazards. The CCR framework facilitates multi-sectoral planning by initiating dialogue and increasing awareness of hazard risks; characterizing the resilience status and trends of communities; assessing organizational capacity; providing input for local and national planning efforts; and identifying strengths, weaknesses, and gaps in community resilience capacity.

For more information about the IOTWS program, visit www.us-iotws.gov. For more information and case studies about the CCR Initiative, visit www.coastalcommunityresilience.org.

2005 Hurricane Katrina

Hurricane Katrina made landfall in southeast Louisiana on August 29, 2005. Katrina was a then large Category 3 hurricane with winds of 125 mph and a central pressure of 920 millibars (mb). This makes Katrina the third most intense United States land-falling hurricane on record based on central pressure. The storm surge ravaged coastal Mississippi, and several levee breaches occurred in and around New Orleans, Louisiana. The levee breaches and overtopping resulted in floodwaters of 15 to 20 feet covering about 80 percent of the city. An estimated 1,500 direct fatalities occurred with 275,000 homes damaged or destroyed. Estimates of total damage losses in the U.S. total $81.2 billion, and economic losses could be greater than $125 billion. These impacts make Katrina the costliest hurricane in U.S. history and one of the five deadliest hurricanes to ever strike the U.S.

Strongly worded statements emphasizing the horrific impacts of Katrina became a significant moment for the NWS. The language helped reinforce the message from emergency management officials for residents in the area to heed evacuation orders. The unprecedented, explicit detail was used by emergency managers, local officials, and the media to prepare the public for impact and aftermath. From this episode a recommendation was made to ensure templates for these types of statements are made available to all offices involved with future major hurricane landfalls.

Critical communications activities were severely impacted during the storm. Offices were dependent on cell phones and HF amateur radio equipment. Satellite phones were not immediately available as a communications alternative. Because of this, it was recommended that offices should have alternate communications systems in place prior to any significant event that will impact the area.

2008 Midwestern Floods

Heavy rains across the mid-western United States in early June of 2008 caused record breaking floods in several states. The Advanced Hydrologic Prediction Service provided many days advance warning of impending flood conditions based on meteorological and hydrological forecasts of rain and runoff in to rivers. The AHPS information was given broad coverage by the media and used extensively by the federal, state and local response communities to plan their activities in mitigating damage and disruptions to citizens in the affected communities.

2008 Super-Tuesday Tornado Outbreak

During a 12-hour period in the evening and early morning of February 5–6, 2008, 87 tornadoes occurred in nine states killing 57 people in four states, injuring 350 others, and causing $400 million in property damage. This despite the fact that the NWS anticipated the outbreak days in advance, and issued warnings 17 minutes, on average, in advance of all the deadly tornadoes.

2009 Spring Snowmelt Flood Red River of the North

Record breaking fall rains saturated a wide area of the Red River of the North watershed. Heavy winter snows followed setting the stage for record spring snowmelt flooding. The Advanced Hydrologic Prediction Service provided 2 months advance notice of the potential for severe to record flooding in communities along the Red River. Local officials were able to plan their response well in advance of the flooding. This long range planning facilitated a coordinated response by the various government and non-governmental agencies that resulted in a successful flood fight, minimizing property damage by building levees along the river and pro-actively evacuating at-risk populations before the flooding became severe. An annual flood damage report by the USACE estimates the damages prevented by emergency measures in the Red River Basin during 2009 to be nearly $3 billion.

6.11 Overall Lessons Learned and Future Steps for Improving National Meteorological and Hydrological Service Contribution in Early Warning System Particularly Focusing on Institutional Coordination and Cooperation

In its investigation of the 2008 Super Tuesday Tornado outbreak and why so many fatalities occurred despite the advanced warnings, the National Weather Service found that most of the fatalities in this event occurred at night, and over 50 percent of the people interviewed said that they thought that tornado outbreaks only occurred during spring or summer. This caused many of them to minimize the threat of this early February outbreak because of their perception that it was too early in the year and outside the "traditional" tornado season. The investigators issued the following recommendations to, specifically, improve institutional coordination. The status of implementation to date is also noted:

- Require regions to develop severe weather coordination procedures between neighboring offices. *Status:* All weather forecast offices have been made aware of the importance of the coordination process between offices and improvement has been noted.
- Increase its education and outreach on winter season tornadoes and the increasing risk of winter season tornado related fatalities in the south-eastern United States. *Status:* All states in the assessment area have begun to hold Tornado Awareness Campaigns in the fall season as well as the spring season.
- Encourage emergency management agencies to hold Severe Weather Awareness campaigns in the fall in addition to the spring, which focus on the risk of tornadoes and tornado related fatalities that often occur throughout the winter. *Status:* The EM agencies in the assessment area are partnering with local weather forecast offices to get the word out regarding the significance of severe weather possibilities in the fall and winter seasons.
- Expand participation with academic and other partners in social science to study the complete warning process from issuance to response, and systematically incorporate those findings into NWS products and services. *Status:* The National Weather Service has increased its participation in the grassroots movement entitled Weather and Society Integrated Studies (WAS-IS). The group is working to improve the weather enterprise by integrating social science into meteorological research and practice in comprehensive and sustained ways.
- Consider adding a societal impacts program to operational branches of NWSH and the Regions, in order to organize and focus these efforts. *Status:* Although not a dedicated position, NWS Headquarters has identified the need for societal impacts and tasked this as a focal point position to an onboard employee.

The hydrologic science, observation, prediction and management are spread across many federal government agencies making collaboration and coordination difficult during natural disasters such as floods and droughts (Fig. 6.8). The expansion of telecommunication services and Internet resources has made coordination somewhat easier, but has also led to an explosion in the amount of data, modeling, and other information available for scientists and decision-makers to use in a collaborative approach to dealing with natural disasters.

Managers and decision-makers in all sectors of water resources require new and more integrated information and services to adapt to uncertainty, climate and land-use change and increasing demand on limited resources. To meet this challenge, the National Oceanic and Atmospheric Administration (NOAA) is developing the IWRSS consortium – Integrated Water Resources Science and Services – an innovative partnership of federal agencies with complementary operational missions in water science, observation, prediction and management. Consisting initially of NOAA, the U.S. Army Corps of Engineers, and the U.S. Geological Survey, the IWRSS consortium envisions a highly collaborative and integrative framework for providing a seamless suite of water resources information across scales ranging from small hill-slopes to large watersheds, from droughts to floods, and from historical analyses to long-range predictions.

The consortium will unify water science, observation, prediction, and management missions and uniquely assemble key capabilities necessary to achieve these objectives. The IWRSS consortium is intended to be an evolutionary approach for working towards an integrative water resources information system that knits together water resources information, products and services across geographic and organizational scales.

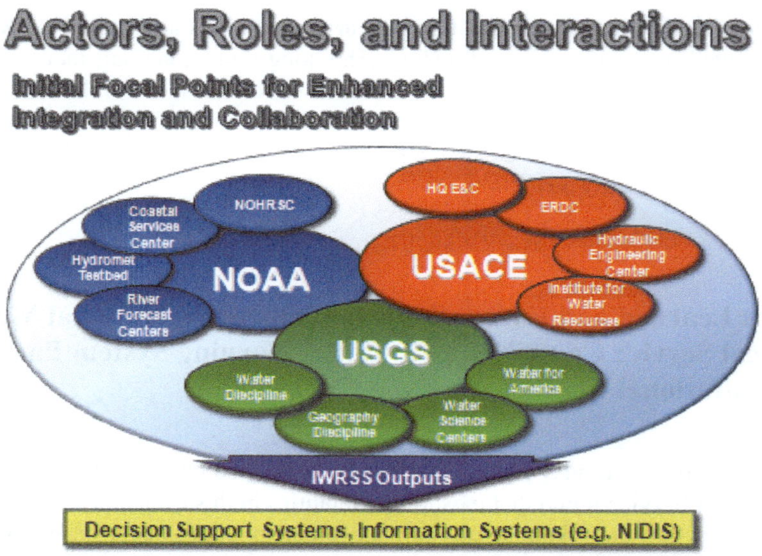

Fig. 6.8 Hydrologic Services Collaborative Partners

High impact events nationwide have yielded opportunities for institutional coordination and cooperation. After events occur, the NWS has worked closely with partners in FEMA to assist them in their mission of response and recovery. FEMA usually asks for information related to the event to assist in their disaster declaration requests. This need has fostered dialogue in the past few years which has led to the development of strong working relationships between the NWS and the various FEMA regions. Better understanding of the different needs of each agency has fostered a quicker process which ultimately rewards the people in the affected area.

Acknowledgements

The following NWS employees, listed in alphabetical order, are acknowledged for their extraordinary efforts to provide information for this report.

Michael Hudson
Chief Operations Officer
Central Region Headquarters NWS
Kansas City, MO

Chris Maier
National Warning Coordination Meteorologist
National Headquarters NWS
Silver Spring, MD

Frank Musonda, Marine Meteorologist/International Program Analyst
International Activities Office
Silver Spring, MD

Firoz Verjee, DRR/Early Warning Systems Specialist
International Activities Office
Silver Spring, MD

The following non-NWS government officials are also acknowledged for their contributions.

Daniel Bovitz, Deputy Director – Emergency Preparedness
Hennepin County, MN

Michael C. Curry, Director – Office of Emergency Preparedness
City of Independence, MO

Joseph Mark Widner, Manager – Emergency Preparedness
City of Independence, MO

All enquiries should be directed to:

Courtney Dragon, Acting Director, NWS International Activities
1325 East-West Highway, W/IA 11436
Silver Spring, MD 20910
United States of America
Email: dan.thompson@noaa.gov
Phone: +1 301 713 0645 x101; Fax: +1 301 587 4524

With copies to the coordinator of NWS activities related to disaster risk reduction:

Lynn Maximuk, Director, NWS Central Region
7220 NW 101st Terrace
Kansas City, MO 64153-2371
United States of America
Email: lynn.maximuk@noaa.gov
Phone: +1 816 268 3130 ; Fax: +1 816 891 8362

References

This document has been compiled from original research and extrapolations from the following public, non-proprietary U.S. Government sources:

COMET MetEd MultiHazard Warning Systems: http://www.meted.ucar.edu/hazwarnsys.php
Department of Homeland Security: www.dhs.gov
FEMA's Progress in All-Hazards Mitigation:
http://www.dhs.gov/xoig/assets/mgmtrpts/OIG_10-03_Oct09.pdf
Federal Emergency Management Agency: www.fema.gov
NRF Resource Center: http://www.fema.gov/emergency/nrf/
National Response Framework: http://www.fema.gov/pdf/emergency/nrf/nrf-core.pdf
List of Authorities and References: http://www.fema.gov/pdf/emergency/nrf/nrf-authorities.pdf
Stafford Act Support to States: http://www.fema.gov/pdf/emergency/nrf/nrf-stafford.pdf
Overview of Federal-to-Federal Support in Non-Stafford Act Situations:
http://www.fema.gov/pdf/emergency/nrf/nrf-fedtofed.pdf
Emergency Support Function, Support and Incident Annexes, and Partner Guides:
http://www.fema.gov/emergency/nrf/
National Incident Management System:
http://www.fema.gov/pdf/emergency/nims/NIMS_core.pdf
Hazard Mitigation Planning Risk Assessment: http://www.fema.gov/plan/mitplanning/risk.shtm
Government Accountability Office: www.gao.gov
GAO Report to Congressional Committees on EMERGENCY PREPAREDNESS, March 2007;
Current Emergency Alert System Has Limitations, and Development of a New Integrated System Will Be Challenging
www.gao.gov/new.items/d07411.pdf
Implementation Support Services for FEMA Integrated Public Alert and Warning System (IPAWS):
https://www.fbo.gov/index?s=opportunity&mode=form&tab=core&id=18eca683f41a39d2922bace118e66178&_cview=1
Lazo, J.K., Morrs, R.E., and Demuth, J.L. (2009). 300 Billion Served – Sources, Perceptions, Uses and Values of Weather Forecasts, Bulletin of the American Meteorological Society.
National Hazard Center http://www.colorado.edu/hazards/
NWS Directives: http://www.nws.noaa.gov/directives
NWS Service Assessment: Central United States Flooding of June 2008 (Draft Report, 2009)
OCWWS Service Assessments: http://www.nws.noaa.gov/os/assessments/index.shtml
OCWWS Natural Hazard Statistics: http://www.nws.noaa.gov/om/hazstats.shtml
Rubin, Clair B. (2006). Emergency Management: The American Experience 1900–2005. Public Entity Risk Institute, Fairfax, Virginia.
Streeter, Sandy (2007). CRS Report for Congress The Congressional Appropriations Process: An Introduction.
http://www.senate.gov/reference/resources/pdf/97-684.pdf.
Service Assessment – Super Tuesday Tornado Outbreak of February 5–6, 2008:
http://www.nws.noaa.gov/os/assessments/pdfs/super_tuesday.pdf

List of Acronyms

AFWS	Flood Warning Systems
AHPS	Advanced Hydrologic Prediction Service
AMS	American Meteorological Society
ASOS	Automated Surface Observation System
AWIPS	Advance Weather Interactive Processing System
CAP	Common Alerting Protocol
CERT	Community Emergency Response Teams
CMAS	Commercial Mobile Alert System
COMET	Cooperative Program for Operational Meteorology, Education and Training
CONELRAD	Control of Electromagnetic [or Electronic] Radiation
COOP	Cooperative Observer Program
CSI	Critical Success Index
DHS	Department of Homeland Security
DOC	Department of Commerce
DOD	Department of Defense
DRC	Disaster Recovery Center
DSS	Decision Support Services
EAS	Emergency Alert System
EBS	Emergency Broadcast System
EMWIN	Emergency Managers Weather Information Network
EOC	Emergency Operations Center
ESF	Emergency Support Function
ESSA	Environmental Science Services Administration
FAOC	FEMA Alternate Operations Center
FAR	False Alarm Ratio
FBO	Faith Based Organization
FCC	Federal Communications Commission
FCO	Federal Coordinating Officer
FEMA	Federal Emergency Management Agency
FOC	FEMA Operations Center
FRC	Federal Resource Coordinator
GAO	Government Accountability Office
GIS	Geographic Information System
GOES	Geostationary Satellite System
HSPD	Homeland Security Presidential Directive
ICS	Incident Command System
IMAT	Incident Management Assessment Team
IPAWS	Integrated Public Alert and Warning System
IWRSS	Integrated Water Resources Science and Services
IWT	Integrated Warning Teams
JFO	Joint Field Office
MACS	Multi Agency Coordination Systems
MCD	Mesoscale Discussion
MEMC	Metropolitan Emergency Managers Committee
METAR	Meteorological Aviation Report
MHEWS	Multi-Hazard Early Warning System
MMS	Multimedia Messaging Services
NAWAS	National Warning System
NEXRAD	Next Generation Weather Radar
NGO	Non-Government Organization
NHC	National Hurricane Center
NIC	National Integration Center
NIMS	National Incident Management System
NOAA	National Oceanic and Atmospheric Administration

NOC	National Operations Center
NRCC	National Response Coordination Center
NRF	National Response Framework
NWS	National Weather Service
NWSRFS	National Weather Service River Forecast System
NWWS	NOAA Weather Wire Service
PDA	Preliminary Damage Assessment
POD	Probability of Detection
RACES	Radio Amateur Civil Emergency Services
RFC	River Forecast Center
RRCC	Regional Response Coordination Center
RSS	Really Simple Syndication
SCH	Service Coordination Hydrologist
SCO	State Coordinating Officer
SMS	Short Message Service
SPC	Storm Prediction Center
TAF	Terminal Aerodrome Forecast
UCG	Unified Coordination Group
U.S.	United States
USACE	United States Army Corps of Engineers
USGS	United States Geological Survey
WCM	Warning Coordination Meteorologist
WFO	Weather Forecast Office

Appendix 1 Partial List of Authorities Relevant to MHEWS

Relevant Statutes and Regulations

1. The Homeland Security Act of 2002, Pub. L. 107–296, 116 Stat. 2135 (2002) (codified predominantly at 6 U.S.C. §§ 101–557),1 as amended with respect to the organization and mission of the Federal Emergency Management Agency in the Department of Homeland Security Appropriations Act of 2007, Pub. L. 109–295, 120 Stat. 1355 (2006), established a Department of Homeland Security (DHS) as an executive department of the United States. The Homeland Security Act consolidated component agencies, including the Federal Emergency Management Agency (FEMA), into the Department. The Secretary of Homeland Security is the head of the Department and has direction, authority, and control over it. All functions of all officers, employees, and organizational units of the Department are vested in the Secretary.

The primary missions of the Department are to:

- Prevent terrorist attacks within the United States;
- Reduce the vulnerability of the United States to terrorism;
- Minimize the damage, and assist in the recovery, from terrorist attacks that do occur within the United States;
- Carry out all functions of entities transferred to the Department, including by acting as a focal point regarding natural and manmade crises and emergency planning;
- Ensure that the functions of the agencies and subdivisions within the Department that are not related directly to securing the homeland are not diminished or neglected except by specific explicit act of Congress;
- Ensure that the overall economic security of the United States is not diminished by efforts, activities, and programs aimed at securing the homeland;
- Ensure that the civil rights and civil liberties of persons are not diminished by efforts, activities, and programs aimed at securing the homeland; and
- Monitor connections between illegal drug trafficking and terrorism, coordinate efforts to sever such connections, and otherwise contribute to the efforts to interdict illegal drug trafficking.

2. The Robert T. Stafford Disaster Relief and Emergency Assistance Act, Pub. L. 93–288, 88 Stat. 143 (1974), codified in 42 U.S.C. §§ 5121–5206 (2007), was also amended in the Department of Homeland Security Appropriations Act of 2007, Pub. L. 109–295, 120 Stat. 1355 (2006), particularly Title VI, the Post-Katrina Emergency Management Reform Act of 2006 (discussed below). The Stafford Act describes the programs and processes by which the Federal Government provides disaster and emergency assistance to State and local governments, tribal nations, eligible private nonprofit organizations, and individuals affected by a declared major disaster or emergency. The Stafford Act covers all hazards, including natural disasters and terrorist events.
3. The Post-Katrina Emergency Management Reform Act (PKEMRA), which is Title VI of the Department of Homeland Security Appropriations Act, 2007, Pub. L. 109–295, 120 Stat. 1355 (2006), clarified and modified the Homeland Security Act with respect to the organizational structure, authorities, and responsibilities of FEMA and the FEMA Administrator.
4. Flood Control and Coastal Emergencies Act, 33 U.S.C. § 701n (2007) (commonly referred to as Pub. L. 84–99), authorizes an emergency fund for preparation for emergency response to, among other things, natural disasters, flood fighting and rescue operations, repair or restoration of flood control and hurricane protection structures, temporary restoration of essential public facilities and services, and provision of emergency supplies of water.
5. The Communications Act of 1934, 47 U.S.C. §§ 151–615b (2007), authorizes the Federal Communications Commission to grant special temporary authority on an expedited basis to operate radio frequency devices, allowing a Federal agency to broadcast public service announcements in the immediate aftermath of an emergency or major disaster.
6. Weather Service Organic Act of 1890, 15 U.S.C. § 313, weather and meteorological reporting/forecasting responsibilities assigned to the Secretary of Commerce; as well as monitoring and recording climatic conditions.
7. Tsunami Warning and Education Act of 2006, 33 U.S.C. §§ 3201 et seq., National Weather Service authority to operate a Tsunami Forecasting and Warning Program. The Program is charged with providing tsunami detection, forecasting and adequate warnings. This Program includes:

    a. operational tsunami detection technology;
    b. tsunami forecasting capability;

c. management of data quality systems;
d. cooperative efforts with the USGS and NSF;
e. capability for disseminating warnings to at-risk States and tsunami communities, and
f. as well as integration of tsunami detection technologies with other environmental observing technologies.

Authority to establish Tsunami Warning Centers which shall continuously monitor pertinent data, evaluate earthquakes as well as deep-ocean buoy data and tidal monitoring stations; disseminate forecasts and tsunami warning bulletins to Federal, State and Local government officials and the public; coordinate with the tsunami hazard mitigation program; and make relevant data available to researchers.

Authority to integrate the tsunami forecast system with other U.S. and global ocean and coastal observations systems, the global earth observing system, global seismic networks, and the Advanced National Seismic System. Authority to conduct a community-based Tsunami Hazard Mitigation Program to improve tsunami preparedness in the U.S. and its territories; maintain a Tsunami Research Program to develop detection, forecast, communications and mitigation science technology, including advanced data collection analysis and assessment techniques; and provide technical assistance and training as part of a Global Tsunami Warning and Mitigation Network; and operate an International Tsunami Information Center in cooperation with the IOC.

Relevant Executive Orders

1. Executive Order 12148, 44 Fed. Reg. 43239 (1979), as amended, by Executive Order 13286, 68 Fed. Reg. 10619 (2003), designates DHS as the primary agency for coordination of Federal disaster relief, emergency assistance, and emergency preparedness. The order also delegates the President's relief and assistance functions under the Stafford Act to the Secretary of Homeland Security, with the exception of the declaration of a major disaster or emergency.
2. Executive Order 12656, 53 Fed. Reg. 47491 (1988), Assignment of Emergency Preparedness Responsibilities, as amended by Executive Order 13286, 68 Fed. Reg. 10619 (2003), assigns lead and support responsibilities for national security preparedness to Federal agencies.

Appendix 2 Stormready Population-Based Guidelines

Guidelines	Population			
	< 2,500	2,500–14,999	15,000–40,000	> 40,000
Guideline 1: Communication				
Established 24 hr Warning Point (WP)	X*	X*	X	X
Establish Emergency Operations Center		X*	X	X
Guideline 2: NWS Information Reception				
Number of ways for EOC/WP to receive NWS warning, etc.	3	4	4	4
Guideline 3: Hydrometeorological Monitoring				
Number of ways to monitor hydrometeorological data	1	2	3	4
Guideline 4: Local Warning Dissemination				
Number of ways for EOC/WP to disseminate warnings	1	2	3	4
NWR – SAME receivers in public facilities	X	X	X	X
Guideline 5: Community Preparedness				
Number of annual weather safety talks	1	2	3	4
Train spotters and dispatchers biennially	X	X	X	X
Host/co-host annual NWS spotter training				X
Guideline 6: Administrative				
Formal hazardous weather operations plan	X	X	X	X
Biennial visits by emergency manager to NWS	X	X	X	X
Annual visits by NWS official to community	X	X	X	X

* Smaller communities (e.g., in Alaska and the U.S. territories) with less than 2,500 residents and no county agency to act as a 24-hour warning point, may form an agreement to have alerts relayed from a neighboring community that has a 24-hour warning point. This partnered agreement shall be documented in the emergency response plans for both communities. The smaller community must designate responsible officials who are able to receive warnings 24/7 from their surrogate 24-hour warning point. NWS recommends the community designate several primary and backup points of contact as the responsible officials. These responsible officials must have the authority and ability to activate local warning systems in a timely manner. It is also recommended that the responsible officials in the smaller community have a 24/7 redundant means to receive alerts, such as NOAA Weather Radio All Hazards

Appendix 3 Completed Stormready Application From Johnson County, Kansas

Print Form

Department of Commerce
National Oceanic & Atmospheric Administration
National Weather Service

StormReady
NATIONAL WEATHER SERVICE

Application Form
OMB Control # 0648-0419
Expires 06/30/2009

Community Information			
County/City/Town	Johnson County	Population	526,000
Primary Point of Contact		**Secondary Point of Contact**	
Name	Nick Crossley	Name	Adam Crowe
Office	Emergency Management	Office	Emergency Management
Title	Director	Title	Assistant Director -- Community Preparedn
Mailing Address	111 S. Cherry St., Ste 100	Mailing Address	111 S. Cherry St., Ste 100
City	Olathe	City	Olathe
State, ZIP	KS 66061	State, ZIP	KS 66061
Phone	913-715-1007	Phone	913-715-1003
e-mail	ncrossley@jocogov.org	e-mail	adam.crowe@jocogov.org

Guideline 1: Communications	
Location of 24-Hour Warning Point	**Location of Emergency Operations Center**
Johnson County Communications Center Sunset Drive Olathe, KS 66061	Johnson County Administration Building 111 S. Cherry St., Ste 100 Olathe, KS 66061

Verification Team General Notes:

Renewal Comments:

Date: Initials:

Note: Please do not write in shaded areas.

Public reporting burden for this collection of information is estimated to average two hours per response, including the time for reviewing instructions, searching existing data sources, gathering and maintaining the data needed, and completing and reviewing the collection of information. Send comments regarding this burden estimate or any other suggestions for reducing this burden to Donna Franklin, National Weather Service, 1325 East West Highway, Room 14456, Silver Spring, MD, 20910.

Statement on confidentiality. Notwithstanding any other provisions of the law, no person is required to respond to, nor shall any person be subjected to a penalty for failure to comply with, a collection of information subject to the requirements of the Paperwork Reduction Act, unless that collection of information displays a currently valid OMB Control Number.

Page 1

Department of Commerce
National Oceanic &
Atmospheric Administration
National Weather Service

Application Form
OMB Control # 0648-0419
Expires 06/30/2009

Guideline 2: NWS Information Reception Equipment			
Warning Point # Required____ # Verif____	Verif	**EOC** # Required____ # Verif____	Verif
☑ NOAA Weather Radio (required if in range)	☐	☐ NOAA Weather Radio (required if in range)	☐
☐ NOAA Weather Wire (subscription)	☐	☐ NOAA Weather Wire (subscription)	☐
☑ EMWIN	☐	☑ EMWIN	☐
☑ Law Enforcement Teletype (LETS)	☐	☐ Law Enforcement Teletype (LETS)	☐
☑ Amateur Radio	☐	☑ Amateur Radio	☐
☑ Pagers* (warning reception)	☐	☑ Pagers* (warning reception)	☐
☑ Television (Local network or Cable TV)	☐	☑ Television (Local network or Cable TV)	☐
☐ Radio Station (AM/FM) - EAS Reception	☐	☐ Radio Station (AM/FM) - EAS Reception	☐
☑ NAWAS	☐	☑ NAWAS	☐
☐ Internet (subscription for alerts)____	☐	☑ Internet (subscription for alerts) Internwarn	☐
☐ Commercial Data Service____	☐	☑ Commercial Data Service WeatherTap	☐
☑ Other* Paging System	☐	☑ Other* Weather Messaging	☐
☐ Other*____	☐	☑ Other* Paging System	☐

List any additional capabilities on a separate sheet

*Capabilities needing explanation:

Verification Team Notes:

Renewal Comments:

Date: Initials:

Note: Please do not write in shaded areas.

Page 2

Department of Commerce
National Oceanic &
Atmospheric Administration
National Weather Service

StormReady
NATIONAL WEATHER SERVICE

Application Form
OMB Control # 0648-0419
Expires 06/30/2009

Guideline 3: Local Weather & Water Monitoring Equipment

Warning Point # Required____ # Verif____	Verif	EOC # Required____ # Verif____	Verif
☐ Anemometer (Wind gauge)	☐	☐ Anemometer (Wind gauge)	☐
☑ Rain Gauge	☐	☑ Rain Gauge	☐
☐ River Gauge	☐	☐ River Gauge	☐
☐ Locally owned Radar	☐	☐ Locally owned Radar	☐
☑ Internet Radar Source WeatherTap	☐	☑ Internet Radar Source WeatherTap	☐
☑ Internet Weather Station JOCO Link Alert	☐	☐ Internet Weather Station ____	☐
☑ TV Radar Source KCTV, WDAF, KCMB	☐	☑ TV Radar Source KCTV, WDAF, KCMB	☐
☐ Other* ____	☐	☑ Other* StormLab	☐
☐ Other* ____	☐	☐ Other* ____	☐

List any additional capabilities on a separate sheet

*Capabilities needing explanation:

Verification Team Notes:

Renewal Comments:

Date: Initials:

Note: Please do not write in shaded areas.

Department of Commerce
National Oceanic &
Atmospheric Administration
National Weather Service

StormReady

Application Form
OMB Control # 0648-0419
Expires 06/30/2009

Guideline 4: Local Warning Dissemination

Warning Point # Required ___ # Verif ___	Verified	EOC # Required ___ # Verif ___	Verified
☑ Outdoor Warning Siren(s)	☐	☑ Outdoor Warning Siren(s)	☐
☑ Cable TV Override	☐	☑ Cable TV Override	☐
☐ Plan for Sirens on Emergency Vehicles	☐	☐ Plan for Sirens on Emergency Vehicles	☐
☑ Telephone Tree to Critical Facilities	☐	☑ Telephone Tree to Critical Facilities	☐
☑ Local Alert Broadcast System*	☐	☑ Local Alert Broadcast System*	☐
☑ Local Pager System* (dissemination)	☐	☑ Local Pager System* (dissemination)	☐
☑ Coordinated Area-Wide Radio Network*	☐	☑ Coordinated Area-Wide Radio Network*	☐
☐ Local Flood Warning System*	☐	☑ Local Flood Warning System*	☐
☐ Other*___	☐	☐ Other*___	☐
☐ Other*___	☐	☐ Other*___	☐

Renewal Comments:

Date: Initials:

Note: Please do not write in shaded areas.

Page 4

Department of Commerce
National Oceanic & Atmospheric Administration
National Weather Service

StormReady

Application Form
OMB Control # 0648-0419
Expires 06/30/2009

Local Government-Owned Buildings in Which Public Traffic is Common

Office	Location or Address	Tone Alert NOAA Weather Radio	Verif	Comments
Warning Point	S. Sunset Dr., Olathe, KS	☒	☐	
EOC	111 S. Cherry St, Ste 100, Olathe	☒	☐	
City Hall	111 S. Cherry St., Olathe, KS	☒	☐	
School Superintendent	N/A	☐	☐	
Sunset Drive Office Buildin	11811 S. Sunset Dr., Olathe, KS	☒	☐	
Health & Human Services	11875 S. Sunset Dr., Olathe, KS	☒	☐	
Northeast Offices	6000 Lamar, Mission, KS	☒	☐	
County Courthouse	100 S. Kansas Ave, Olathe, KS	☒	☐	
		☐	☐	
		☐	☐	
		☐	☐	
		☐	☐	
		☐	☐	
		☐	☐	
		☐	☐	

Verification Team Notes:

Renewal Comments:

Date: Initials:

Note: Please do not write in shaded areas.

Department of Commerce
National Oceanic &
Atmospheric Administration
National Weather Service

StormReady

Application Form
OMB Control # 0648-0419
Expires 06/30/2009

Guideline 5: Community Preparedness

Annual Safety Talks # Required____ # Verif____

	Date	Topic	Location	Speaker
1	01/26/09	Severe Weather Spotting	Olathe Fire Administration Buildin	Andy Bailey, NWS
2	02/04/09	Severe Weather Awareness	Santa Fe Trail Middle School (Ol	Evan Bookbinder, N
3	02/27/09	Severe Weather Awareness	Johnson Co. Administration Buildi	Mike Costello
4	03/09/09	Severe Weather Awareness for Indoo	Johnson Co. Administration Buildi	Adam Crowe
5	04/09/09	Severe Weather Awareness	Shawnee LDS Church	Adam Crowe

List any additional safety talks on a separate sheet

Weather Radio Purchase Program

Has your community/county developed a program to subsidize the purchase of Specific Area Message Encoder (SAME) equipped Weather Radios for its citizens? (Not required) Yes_____ No X_____

If yes, provide details:

Other Community Preparedness Activities

	Date	Activity	Location	Organizer
1	03/01/09	Creation of Twitter Account (@JOCOA	Internet/Twitter	Adam Crowe
2	03/03/09	Presentation -- Emergency Manageme	Blue Valley School	Adam Crowe/BV R
3	03/12/09	Presentation -- Severe Wx Awareness	Johnson County Admin Building	Adam Crowe
4	03/18/09	Presentation -- Severe Weather Aware	Resource Center for Independent	Adam Crowe/Locati
5	04/01/09	Creation of multiple educational & awa	Internet/YouTube	Adam Crowe

List any additional activities on a separate sheet

Renewal Comments:

Date: Initials:

Note: Please do not write in shaded areas.

Department of Commerce
National Oceanic &
Atmospheric Administration
National Weather Service

StormReady

Application Form
OMB Control # 0648-0419
Expires 06/30/2009

Guideline 6: Administrative Tools/Record keeping		Verif	Renewal Year
Formal Hazardous Weather Operations Plan ➢ Procedure for reporting storm damage to the local National Weather Service Office in real-time ➢ EOC Activation Procedures ➢ Spotter Activation Criteria ➢ Local Warning System(s) Activation Criteria		☐ ☐ ☐ ☐	☐ ☐ ☐ ☐
Warning Point personnel has authority to activate Warning System (written)	☒	☐	☐
Spotter Roster and Training Record	☒	☐	☐
Last Visit by Emergency Manager to NWS Office	07/2008	☐ Biennial	☐
Last Visit by NWS Officials to Community	03/2009	☐ Annual	
Last NWS Spotter Training for Spotters and Dispatchers	01/2009	☐ Biennial	
Last NWS Spotter Training Hosted/Co-Hosted *(For populations >40,000)*	02/2009	☐ Annual	
Exercises — Topic(s): Ice Storm (10/08)	Date:	☐	Date: Date:

List any additional descriptions, narratives, or documentation on a separate sheet

Verification Team Notes:

Renewal Comments:

Date: Initials:

Signature of Applying Official

Application Submitted by: (print name):	
Office:	Title:
Signature:	Date:
NWS Personnel Receiving Application (print name):	
Date Received:	

Note: Please do not write in shaded areas.

Chapter 7
The Shanghai Multi-Hazard Early Warning System: Addressing the Challenge of Disaster Risk Reduction in an Urban Megalopolis

Xu Tang, Lei Feng, Yongjie Zou, Haizhen Mu

Abstract On 27 June 2008, Mr Hu Jintao, President of China, declared that China needs to develop a sustainable Multi-Hazard Early Warning Systems (MHEWS) with a focus on multi-agency coordination and cooperation at all levels of government, to develop a regional joint response capability, and to build an optimum multi-hazard monitoring and emergency response warning process by actively involving multi-agency participation in the decision-making process. This process should also encourage active public participation in response to climate change. Responding to the President's direction and with the support of the China Meteorological Administration (CMA), the Shanghai Meteorological Bureau (SMB) enhanced its efforts to establish a comprehensive weather disaster prevention management system under the leadership of the Shanghai Municipal Government (SMG).

The CMA and SMG jointly support the Shanghai Multi-Hazard Early Warning System (MHEWS) as a WMO demonstration project with 'Multi-agency Response' as the core. The project integrates diversified advanced technologies into a multi-hazard warning process, advancing improved multi-agency coordination and cooperation through a multi-link communication platform with responsible emergency response and rescue agencies. The MHEWS is organized around its "4+1" technical platforms and three-level standard system on multi-agency coordination and cooperation. The technical platforms are: Multi-Hazard Detection & Monitoring, Forecast & Warning Information Generation, Multi-agency Coordination and Cooperation Support, Dissemination and User Application Platforms, and the Multi-Hazard Information Database. The three-level standard systems comprise: a Multi-agency Coordination and Cooperation Standard System, Safety Community Standard System and Regional Joint Defence Standard System. The MHEWS provides technical support to the Shanghai Emergency Response Platform and has been introduced into the Emergency Response Headquarters of the Shanghai Municipal Government (SMG). It provides forecast and warning services to the SMG's emergency response command centre, which is responsible for public emergency response actions and the delivery of emergency related information. The network will be fully operational to provide emergency response services to Shanghai Expo2010. Following Expo2010, WMO will conduct an assessment of the MHEWS.

Remarkable progress has been made. The dissemination platform has entered the testing stage; the forecast and warning information generation and multi-agency coordination and cooperation platforms are entering the development stages, with some modules already in operation. Warning subsystems for city traffic safety, heat wave and human health, power and energy security, and bacterial food poisoning are operational. There has been significant progress with grass roots level delivery of warning messages and with integration of information into the city grid management. Breakthroughs have been made in multi-agency coordination and cooperation. "The emergency response plan of Shanghai Municipality for rain, snow and freezing weather disasters", and "The emergency response plan of Shanghai Municipality for heavy fog disasters" have been distributed by the general office of SMG.

7.1 Overview

Shanghai is one of the world's densest cities in terms of population and economy. Urban public safety and successful operations during Expo 2010 require more efficient hazard early warning information. Moreover, Shanghai has rich experience in disaster prevention and mitigation (DPM) and has advanced urban infrastructures as well. Consequently, Shanghai was selected to conduct a Demonstration Project of the Multi-Hazard Early Warning System, for a mega-city.

Xu Tang, Director-General, Shanghai Regional Meteorological Center of China Meteorological Administration
Lei Feng, Deputy Director-General of Shanghai Meteorological Bureau
Yongjie Zou, Deputy-Director of Shanghai Emergency Response Management Office
Haizhen Mu, Deputy-Director of Shanghai Climate Center

M. Golnaraghi ed., *Institutional Partnerships in Multi-Hazard Early Warning Systems*,
DOI 10.1007/978-3-642-25373-7_7,

7.1.1 Background on the Establishment of EWS in Shanghai

In 2003, the Mayor of Shanghai signed the Shanghai Disastrous Weather Early Warning Signal Issuance Regulations. In April 2006, some necessary amendments were added to the above-mentioned regulations. On 26 October, 2006, the Standing Committee of the Twelfth Shanghai People's Congress approved "Shanghai implementation measures for Meteorological Law of the People's Republic of China", which defined the obligations and responsibilities of the Shanghai Meteorological Bureau (SMB) in disaster emergency response and prevention. In accordance with the above mandate, the city of Shanghai has integrated its emergency response and preparedness activities into the construction of the Multi-Hazard Early Warning System (MHEWS) with a focus on strengthening multi-agency coordination and cooperation as well as disaster prevention and mitigation.

Shanghai has developed a risk management framework in accordance with the administrative infrastructure of a two-level government, a three-level management structure and four-levels of networks and based on the National Overall Emergency Response Readiness (EmRR) Document and 25 special EmRR Documents. This consists of one overall city-level EmRR Document, 19 district/county-level EmRR Documents, seven city-level special EmRR Documents, 40 department/ agency EmRR Documents, and a number of key risk management units and event EmRR Documents. These sub-plans fulfil the dual requirements of being both all-reaching and grid-based and ensure that risk management reaches residential communities, factories, villages and schools. The Shanghai emergency response/ preparedness sub-plan on weather disaster is one of the department/agency EmRR Documents. The Shanghai Meteorological Bureau's (SMB's) responsibility is also embodied in other relevant sub-plans.

Shanghai is a mega-city with more than 19 million people and is the largest economic centre in China. It is also characterized by a small area, huge population, intensive new construction and a high concentration of economic activities. National public emergencies, such as those resulting from disasters caused by natural hazards, security accidents, and public health emergencies, require closer multi-agency coordination and cooperation including the meteorological department's response to disasters. The Shanghai Municipal Government has established a Multi-Hazard Information Integration, Multi-Agency Preparedness and Coordinated Multi-Phase Response Mechanism to fulfil the needs of emergency response. This has been accomplished through the following measures:

- Information Integration – an information communication platform has been built which connects the various districts/counties, departments and other response units, providing the Public Emergency Management Agency with strong technological support.
- Facility Integration – an integrated emergency management platform with unified leadership has been established to integrate emergency rescue facilities, law enforcement, fire protection, traffic management, public sanitation, water affairs, municipal administration, electric power company, seaports and airports.

In recent years, the SMB has enhanced its capacity for disseminating meteorological information, ensuring, to a substantial degree, that meteorological disaster warning information can be disseminated "to the community, to enterprises, to schools, and to rural areas". As part of the Shanghai MHEWS, the government of Shanghai has completed the development of a grid platform approach for management of meteorological services. At the same time, the meteorological department has improved its linkage with the relevant departments involved in MHEWS activities, focusing on partnership practice and better service delivery in disaster prevention and mitigation (DPM). It has, moreover, expanded its participation in relevant research activities and shared in the construction of a MHEWS operations centre. The centre is used in cooperation with all partner agencies having warning responsibility, such as the Flood Control Headquarters, Environmental Protection Departments, the Public Health Bureau and the Shanghai Municipal Centre for Disease Control and Prevention, and has established a partnership with the Shanghai Municipal Food and Drug Supervision Administration to issue early warnings of bacterial food poisoning.

7.2 Governance and Institutional Arrangements

7.2.1 Policy, Institutional and Legal Frameworks to Support Emergency Planning and Response

The emergency preparedness and response planning regulated by the government covers many disciplines such as public emergency response, meteorology, flood prevention and more. For example, there are two sections in the Law of Public Emergency Response – "Prevention and Emergency Response Preparedness" and "Monitoring and Warning" – in which preparedness, emergency shelter, registration and examination of risk sources, emergency response management training, professional emergency search and rescue operations, and emergency response materials are addressed. When a public emergency occurs, the local government will, initially, shoulder the direct responsibility though it will need to call on higher authority if the seriousness of the emergency reaches a level which overwhelms the local administration. Until this latter situation occurs, the local government will determine and implement all required response actions.

The Meteorology Law of China defines the roles, responsibilities and authorities of weather services as well as their operational functions such as construction and management of meteorological facilities, observation systems, weather forecasts and weather warning processes, weather disaster prevention, climate resource exploitation and their implementation and maintenance (Box 7.1). Another element of legislation addresses flood control in China, item 31 of which requires the establishment of China's flood control system, addressing hydrology, weather, communications, warning and flood disaster monitoring systems. According to item 43 of the Law, meteorological services should provide weather forecasts to the flood control headquarters during flood season. Item 17 of the Geology Disaster Prevention Law, moreover, explicitly directs that geology disaster forecasts should be jointly disseminated by China's land resources authority and Meteorological Administration.

The focus of the Shanghai Municipal Government is on weather disaster prevention and response. The Shanghai Implementation Plan, based on the Meteorology Law of China, was developed in 2006 and has been approved. The plan contains specific implementation measures aimed at adopting a more direct and practical approach to early warning and emergency response. The Shanghai Meteorological Bureau is a member of the Emergency Response Committee and Multi-agency Emergency Response Team. In addition, the dissemination platform of the MHEWS in the emergency response headquarters is used as the warning dissemination platform for public emergencies (Fig. 7.1).

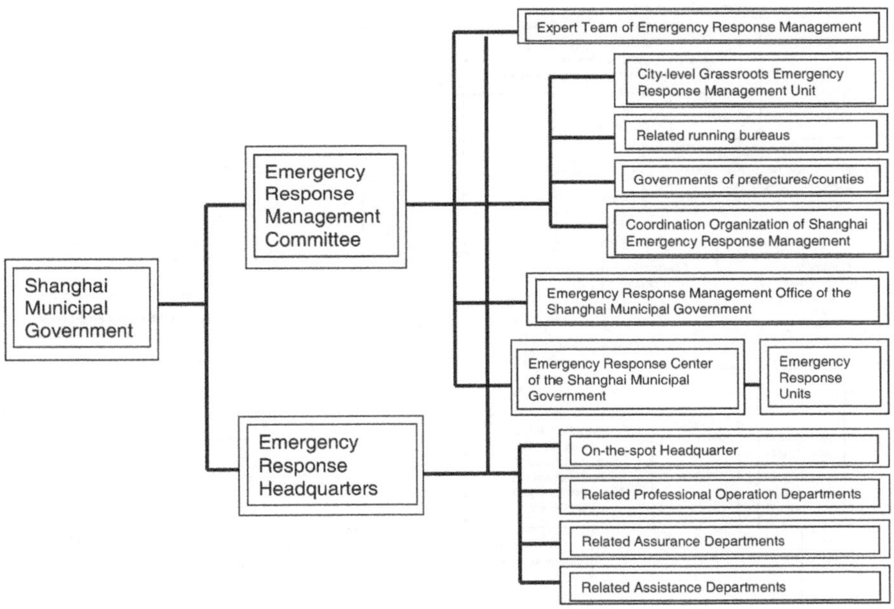

Fig. 7.1 Shanghai Institutional Framework for Emergency Preparedness and Response Planning

7.2.2 *National to Local Emergency Planning and Related Linkages to EWS*

The Shanghai municipal government integrates the monitoring information resources of various departments and forecasts the potential or predictable emergency event according to a scale of severity. The levels in the scale are – **Very Serious; Serious; Relatively Serious; or Moderate**. These levels correspond, respectively, to **Red, Orange, Yellow and Blue** warning signals. The Emergency Response Center is responsible for addressing public emergency events of general or greater scale as well as for the early handling of major or extraordinarily large emergencies. If an emergency is still beyond control after the initial handling, the municipal emergency response commission will determine the appropriate response level and scope, launch the corresponding preparedness measures and, if necessary, set up headquarters to deal with the emergency. In addition, the units and communities where the emergencies occur have full responsibility for initial response and should direct the people in undertaking self rescue and mutual rescue. Furthermore, the relevant departments should implement initial response actions as soon as possible.

Shanghai initiated active research and established an integrated disaster prevention and mitigation system at the beginning of 2001, in order to improve the city's capability to deal with emergency events. Since then it has established a complete emergency management mechanism, developed a preliminarily system framework, and implemented an emergency response system (Fig. 7.2). As a result, Shanghai has accumulated some experience in the construction of a cross-agency information facility, chain response actions and disaster mitigation. The agencies responsible for issuing warnings are members of the Shanghai Emergency Management Response Committee who actively participate in the MHEWS planning process. This Committee consists of more than 50 members from various government agencies and departments concerned with the issues of flooding, severe weather, fire, traffic accidents, chemical accidents, nuclear power accidents, public health, earthquakes, and marine emergencies. SMB is a member of the Committee and has as its primary role "Early Warning Generation and Dissemination". Early warning includes original disaster warnings and secondary-level disaster warnings that require cooperation with other departments, such as flood warnings caused by typhoons.

As an example of this approach, in December 2008, Shanghai city printed and distributed preparedness information on dispositions for heavy fog, low temperature, rain, snow and ice in response to a potential winter weather disaster. This stipulated that responsible departments must take appropriate measures when heavy fog, rain, snow and ice were forecast and also requested that all related departments implement all required actions in response to the threat of extreme weather.

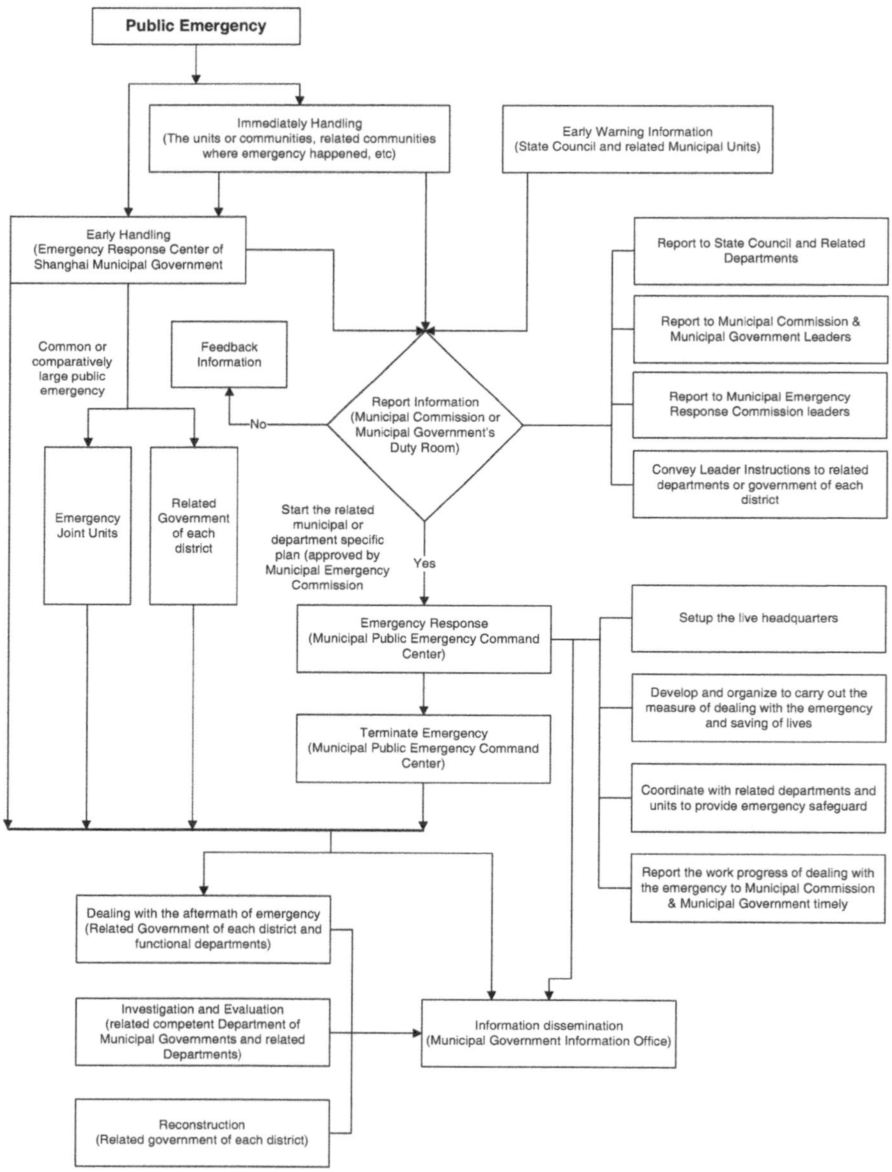

Fig. 7.2 The Emergency Management Workflow Chart for Public Emergency Response in the City of Shanghai

Box 7.1 A solid legal and institutional framework supports Shanghai's EWS.

The Meteorology Law of China defines the roles, responsibilities and authorities of weather services as well as their operational functions.

- *According to item 43 of the Law, meteorological services should provide weather forecasts to the flood control headquarters during flood season.*
- *In 2003, the Mayor of Shanghai signed the Shanghai Disastrous Weather Early Warning Signal Issuance Regulations. Some amendments were added in April 2006.*
- *On 26 October, 2006, the Twelfth Shanghai People's Congress approved "Shanghai implementation measures for Meteorological Law of the People's Republic of China" which defined the disaster-related responsibilities of the Shanghai Meteorological Bureau (SMB).*
- *The focus of the Shanghai Municipal Government is on weather disaster prevention and response. The Shanghai Implementation Plan, based on the Meteorology Law of China, contains specific implementation measures related to early warning and emergency response.*
- *Another element of legislation addresses flood control in China, item 31 of which requires the establishment of China's flood control system.*

Item 17 of the Geology Disaster Prevention Law directs that geology disaster forecasts should be jointly disseminated by China's land resources authority and meteorological administration.

7.2.3 *Organizational Structure for Implementing the Plans*

If there is no serious emergency, the departments should undertake their own actions. When a serious event occurs, the Shanghai Municipal Government will establish a unified command headquarters. Both the emergency management committee and headquarters are supported by district governments, related departments, supporting institutes and grass-root units. In addition, they will seek advice from expert teams if necessary. These expert teams consist of specialists in various fields. As a member of the Standing Headquarters, the SMB participates in emergency planning and also provides support to the expert teams as needed. In addition, SMB is a member of the related departments (Fig. 7.3).

Non-governmental organizations participate in emergency response activities when defined response levels are reached. For example, when Level 1 is exceeded and Level 2 is activated, agencies such as the Red Cross, Philanthropic Foundations and other related organizations will organize their members to carry out activities in disaster areas, including rescue, donation and salvation activities.

The MHEWS operation centre, as one of the technical centres in the Shanghai emergency management system, provides technical support to the Shanghai Emergency Management Platform and supplies guidance to decision-makers for emergency events in Shanghai (Fig. 7.4). In addition, the MHEWS warning dissemination platform also acts as the municipal-level platform for disseminating emergency response information.

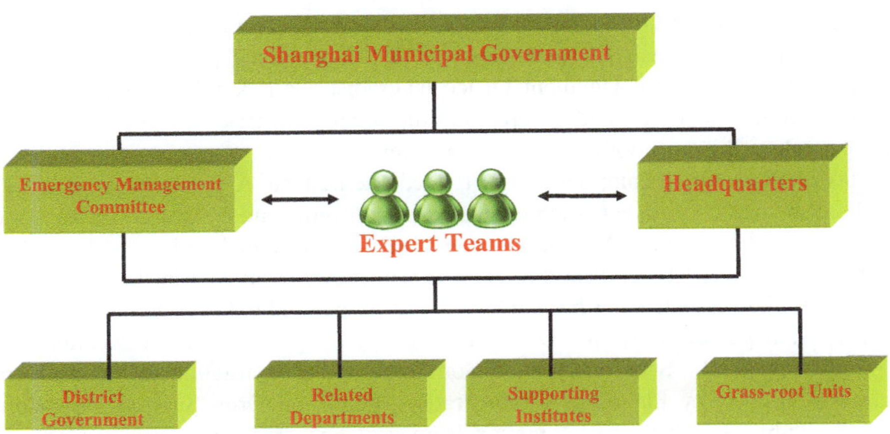

Fig. 7.3 Organizational Structure of Shanghai Emergency Responses System

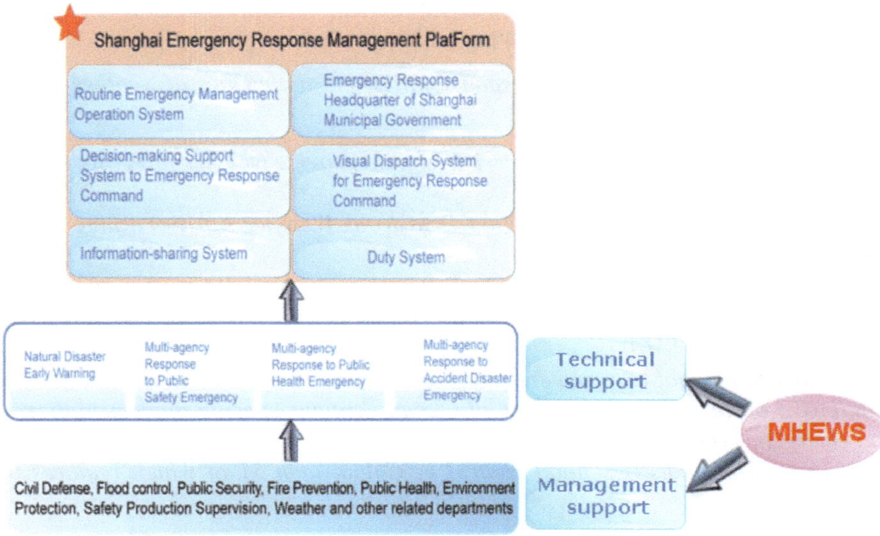

Fig. 7.4 MHEWS role in the Shanghai emergency management system

7.2.4 *Institutional Capacities and Concept of Operations (Standard Operating Procedures)*

As mentioned in the second paragraph of Sect. 2.1, Shanghai has developed a risk management framework consisting of one overall city-level Emergency Response Readiness (EmRR) Document, 19 district/county-level EmRR Documents, 7 city-level special EmRR Documents, 32 department/agency EmRR Documents, and a number of key risk management unit and event EmRR Documents. These sub-plans define the responsibilities of related departments, fulfil the requirements of being all-reaching and grid-based, and ensure that risk management reaches residential communities, factories, villages, and schools. The "Shanghai Emergency Response/Preparedness Sub-Plan on Weather Disaster" is one of the department/agency EmRR Documents. SMB's responsibility is also embodied in other relevant sub-plans.

Emergency response preparedness regulates post-disaster actions such as – what involved departments need to do; timely summarization and assessment of emergency response disposition actions; and amendment of preparedness plans in light of assessment results. The Shanghai Emergency Response Committee (SERC) periodically assesses and summarizes the status of the Shanghai emergency response actions during the past year, identifies improvement measures and initiates their implementation. Meetings on emergency response are, furthermore, convened every three months with the aim being to identify insufficiencies and define the focus of work for the next stage. In addition, an annual review meeting is held with attendance by SERC member department leaders being required.

The weather department's responsibilities include weather warnings, deciding when to activate emergency response preparedness actions, and identifying key impact areas requiring response measures. The Shanghai Meteorological Bureau (SMB) is also a member of the Shanghai Emergency Response Committee, the primary group that reviews, plans and directs the implementation of emergency response activities. The SMB has developed an early consultation mechanism with the Municipal Emergency Management Office to provide pre-warning information prior to warning issuance. This keeps decision-makers fully informed and assists in optimizing the decision-making process. By means of this consultation mechanism, the Municipal Emergency Management Office takes disaster risk information, rescue facilities and social situations into comprehensive consideration and ensures that the organizing tasks for disaster prevention, mitigation, response and recovery are conducted in a smooth and orderly manner.

Severe weather events often are often compounded by related phenomena. A typhoon, for example, brings both heavy rains and wind as well storm surge. Moreover, such events can occur quite suddenly and the impact can be multiplied in urban areas by such things as traffic etc. SMB has to communicate and collaborate closely with the MHEWS stakeholders/users so that the Multi-hazard Early Warning System will play a vital role in the city's disaster prevention and mitigation mechanism. Up to the present, SMB has contacted and cooperated with dozens of departmental users including the Municipal Emergency Management Response Office, Emergency Response Center, Education Commission, Development and Reform Commission, Public Security Bureau, Commission of Construction and Administration, Water Affairs Bureau, Harbour Administration, City Planning Administration, Food and Drug Supervision Administration, Environmental Protection Bureau, Public Heath Bureau, Aviation Bureau of Eastern China, Disease Control Center, Agriculture Commission, Economic Commission and so on. These departments are the key users of the MHEWS.

The SMB has established a fast, efficient, and unified MHEWS operation centre focusing on partnership practice, with the Chief Service Officer (CSO) as the core position. The goal here is to implement a "face-to-face" meteorological approach and enhance the level and efficiency of meteorological service. The main functions of the centre are as follows – implement multi-agency response actions; design service products, coordinate response actions and service delivery monitoring; dissemination to public, special users and government agencies; collect feedback information (Fig. 7.5). In this centre, the Chief Service Officer is the vital link to the user community, being responsible for dealing with comprehensive daily data flows, collecting the detailed requirements of the agencies, and providing specialized information to the right people. The CSO translates professional meteorological language into user-oriented language for different users through standard operating procedures (SOPs).

At present, Shanghai Meteorological Bureau (SMB) has formalized cooperative relationships by signing cooperation agreements with many departments including the Shanghai Municipal Food and Drug Supervision Administration (for joint dissemination of bacterial food poisoning warning information), Shanghai Health Bureau and the Shanghai Municipal Centre for Disease Control and Prevention (for joint establishment of a Heat Wave and Human Health Monitoring Warning System), Urban Management, and the Shanghai Municipal Agriculture Commission and so on. In essence the implementation of an effective Multi-hazard Early Warning System (MHEWS) aims to transform weather forecasting into disaster forecasting.

In overview, the SOPs that have been used in coordination and cooperation between SMB and the key users can be divided into the following types:

Joint Response Warnings and Standard Response Actions

Arrangements of this type are in place with the Shanghai Flood Control Headquarters, Shanghai Flood Prevention Department, Shanghai Water Affairs Bureau, Construction and Transportation Committee, Communications Authority and other related departments (Fig. 7.6).

Joint Production and Joint Dissemination

Arrangements of this type are in place with the Environmental Protection Bureau, Food and Drug Supervision Administration, Flood Prevention Department, Public Health Authority and others.

The Shanghai Municipal Food and Drug Supervision Administration and the SMB have jointly issued notices aimed at enhancing the ability to prevent and handle threats to food safety during periods of high temperature. Following the 'prevention first' principle, the two departments have established a joint forecast and warning mechanism and to take actions based on standardized response levels, following an agreement on work specifications for dealing with food safety during high temperature events. A warning model for bacterial food poisoning has been in regular operation since Jun 1st, 2009 and the 'Workflow for bacterial food poisoning warning' has been formalized to facilitate multi-agency coordination and cooperation.

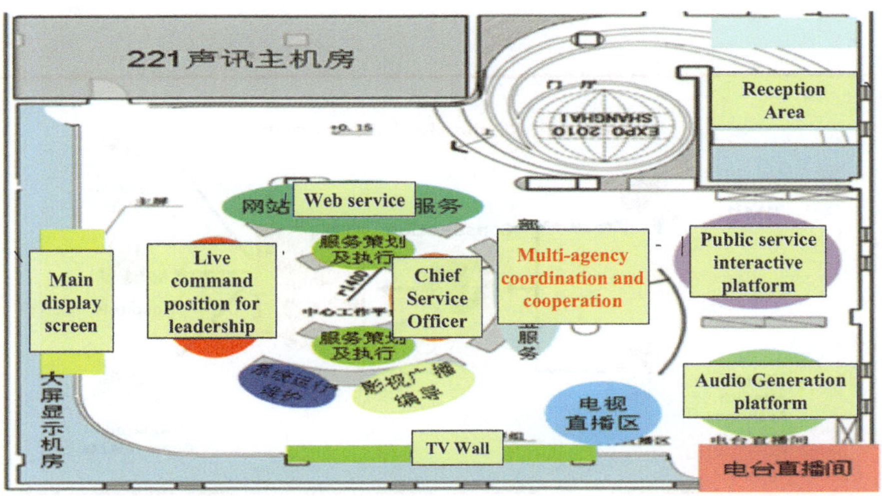

Fig. 7.5 Layout of MHEWS operation centre

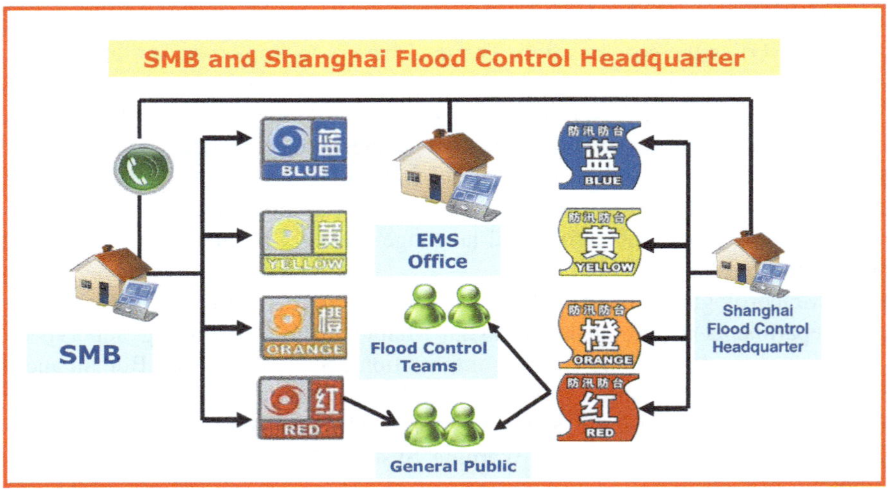

Fig. 7.6 Joint response warnings and standard respond actions

The SMB and the Shanghai Public Health Authority and Shanghai Disease Control Centre have jointly established the Shanghai Heat Wave and Human Health Monitoring and Warning System. Warnings about heat strokes will be jointly generated and disseminated through this system. As illustrated in Fig. 7.7, the SMB undertakes the responsibility of providing and disseminating meteorological forecasts on heat strokes during periods of extreme heat.

Joint Dissemination

The SMB partners with Broadcasting and Television departments, Education Committees, Agricultural Communities, the Urban Grid Management Centre and others to ensure the widest possible dissemination of warnings and other disaster related information.

The SMB and the Shanghai Media Group have established a collaborative agreement on the following:

- Ensuring that weather warning signals can reach all TV channels and broadcast frequencies.
- Establishing the mechanism for broadcasting weather warnings and real-time forecasts as well as, promptly, posting weather icons or broadcasting a rolling caption describing the threat.
- Enhancing the frequency of the weather broadcasts and establishing an 'inter-cut mechanism' for providing more timely updates of weather information.

The Meteorological Bureau of Songjiang district and the Grid Management Centre have developed a joint implementation plan that defines work flow, content and responsibilities to integrate meteorological early warning information into the city grid management system. The integration of meteorological early warning information into city grid management broadens the transmission and dissemination channels for meteorological early warning signals, ensuring broader information coverage.

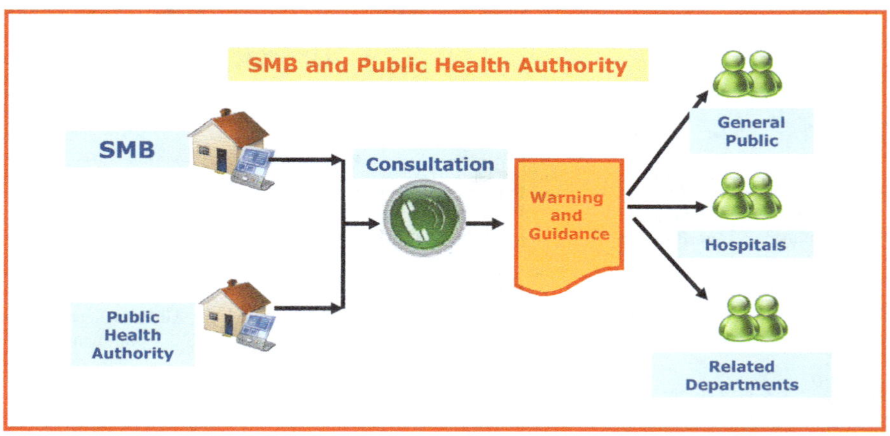

Fig. 7.7 Joint Production and Joint Dissemination.

Special Weather Office:

In special circumstances, the SMB works with an organization such as the Shanghai Municipal Electric Power Company to establish a special weather office to address the partner's specific needs for information and products.

The SMB and the Electric Company have undertaken a joint production initiative, pooling data and resources to develop a predictive electricity load and consumption forecasting system. In addition, an Electricity weather service department will be established.

Other Types of Coordination and Cooperation:

Other arrangements for coordination and cooperation also exist between the meteorological service and particular stakeholders. Examples include arrangements with the marine, aviation and agricultural sectors.

A SMB-SMSA Collaborative Framework has been created and agreements have been reached on the establishment of a sharing mechanism for meteorological and marine data and resources. This contributes to enhancing the capability for maritime search-rescue and disaster evaluation, developing the operation of coastal weather broadcasting, and improving observation networks to strengthen capacity for monitoring, forecasting, and warning on marine-related meteorological hazards.

In recent years, SMB and the Shanghai Air Traffic Management Bureau have engaged in comprehensive collaboration in several areas including detection data sharing, joint development of aviation-specific forecasting in eastern China, communication and training.

The development of an agro-meteorological disaster early warning system for Songjiang district is being led by the Meteorological Bureau of Songjiang district and the Agricultural Commission of Songjiang, in coordination with the Shanghai Climate Centre and Shanghai Meteorological Information Centre.

Certain deficiencies were identified during cooperation with key users. In some instances, SMB staff was not sufficiently acquainted with the users' requirements or were unfamiliar with relevant details such as with the partner agency's size, scope, operational workflow and the definition of the warning information. A MHEWS requires a standard information-sharing mechanism, a departmental consultation mechanism, an information dissemination mechanism, and a joint response mechanism to ensure smooth and orderly multi-agency cooperation and coordination. Consequently, the Shanghai Municipal Emergency Response Management Office and SMB are preparing a draft of "The warning and cooperation mechanism for Shanghai emergency events", together with the Shanghai Municipal Safe Production Supervision Administration, Shanghai Municipal Public Health Bureau, Shanghai Municipal Civil Affairs Bureau and Emergency Response Center of Shanghai Municipal Government. This draft document is expected to be published before the end of 2009 and addresses warning information, joint generation and dissemination, multi-agency joint response, social responsibility, safeguard measures, supervision and management mechanism for disaster emergencies arising from natural hazards as well as public health emergencies and accident disaster emergencies. It is anticipated that multi-agency coordination and cooperation will become formalized and routine with the help of this draft mechanism. SMB distributed a survey questionnaire to investigate current warning dissemination and related information for relevant departments in Shanghai and used the results of the survey as the basis of the draft mentioned above.

Obviously, user and public satisfaction are among the most important ways to judge the quality of the meteorological service. In order to assess the quality of meteorological production and service provided by MHEWS, users are divided into 3 groups for survey. The detailed survey methods are as follows:

- For governmental decision-makers: – by telephone, video conference, leadership and expert's conference, live consultation, collection of feedback information, etc.
- For departmental partners: – by expert consultation, expert forums, regular summary meetings, interviews, live surveys, website feedback information, etc.
- For the public: – by questionnaire survey, website survey, hotline, public supervisor, etc.

7.2.5 Financial and Budgetary Aspects

The development, maintenance, sustainability and improvement of early warning systems are resource intensive. The daily operation of the MHEWS is funded through both the CMA and Shanghai Municipal Government (SMG). Both of the above organizations applied for MHEWS-related key project development funding. In addition, the cooperating stakeholders will provide necessary financial resources to ensure the smooth operation of joint projects.

7.3 Utilization of Risk Information in Emergency Planning and Warnings

7.3.1 Organizational Responsibilities and Arrangements for the Development of Risk Information

The Shanghai Civil Administration is responsible for collection and statistical analysis of disaster risk information. The Shanghai Statistics Bureau is responsible for statistics information for population, economic and social development. The Shanghai Land Planning Administration is responsible for land use plans. The Shanghai Municipal Public Health Bureau is responsible for the collection of information on people's health. The Shanghai Municipal Power Company is responsible for information collection on electricity consumption and load factors. The Shanghai Municipal Food and Drug Supervision Administration is responsible for information collection on food poisoning. The Shanghai Municipal Water Affairs Bureau is responsible for monitoring and warning of water related disasters and the SMB for monitoring and warning of weather disasters.

A comprehensive examination of "hidden risks" in Shanghai is in progress. The Safety Administration has the responsibility for the distribution of data and information related to dangerous chemical sources. The Real Estate Department maintains data on buildings and houses that are vulnerable to disasters. SMB is also carrying out a weather disaster distribution project. Through multi-agency coordination and cooperation, these departments effectively share monitoring data and technical advances. These efforts have increased the number of disaster data categories to 65, addressing meteorological, hydrologic and marine, traffic, food, epidemic diseases and agriculture related hazards. Since 2006, the SMB has shared in more than 17,000 releases of disaster-related information and data through the Shanghai Emergency Response Centre of SMG. The SMB's transformation from weather monitoring to comprehensive monitoring has benefited in the establishment of a multi-hazard and early monitoring information database available to all agencies participating in the Shanghai MHEWS. The development of the MHEWS database is the basis for all disaster risk assessment operations and research.

The Meteorology Law of China stipulates that governments above county level should enhance the development of weather disaster monitoring and warning systems, coordinate related departments to document weather disaster prevention plans, and take effective measures to improve the capabilities of weather disaster prevention agencies. Disaster assessment standards, such as those found in the "Emergency Preparedness Plan for Shanghai Meteorological Disasters", can be used to asses damage and response levels.

7.3.2 Hazard Assessment, Quantification and Mapping

Based on historical data and expert analysis, the following disasters are the most likely to impact Shanghai: – typhoons, heavy rain, storm surges, red tides, tornados, fog, high temperatures, lightning, road traffic accidents, fires and chemical accidents (Table 7.1). Weather disasters not only lead to casualties but also influence city operations and the impact of weather disasters is magnified due to the rapid development of the economy. Such disasters can seriously impact city operations, social activities, agriculture, energy supply, transportation, price stability and countless others. In recent years, with the acceleration of global warming, changes in climate have brought more extremes in temperature and heavy precipitation and this trend will most likely continue. Rapid urbanization has compounded the effects of such weather extremes.

7.3.3 Assessment of Vulnerabilities and Exposure

The Meteorological Service undertakes the assessment of meteorological disaster risk. Through collection of weather disaster related data (typhoon, rainstorm and floods, strong wind, hail, thunder and lightning, heavy fog, cold wave and high temperature) and by statistical probability analysis, it calculates the probable occurrence of different levels in various areas. Meteorological disaster risk-mapping was undertaken in accordance with comprehensive disaster-causing factors, potential disaster environment, frequency of disasters and capacity for disaster resistance. SMB has conducted research on climate change and prediction of possible future situations in both Shanghai and Eastern China, based on global and regional climate modelling that integrates a comprehensive database of climate and climatic change in Shanghai. SMB has analyzed the decadal-scale evolution of extreme weather and climate in Eastern China (extreme rain, high temperature, drought and flood, cold events in winter, and typhoons affecting this area), along with the relationship between global warming and regional climate change, so as to predict the probable occurrence of important weather and climate events

Table 7.1 Meteorological and hydrological hazards in Shanghai

Hazards	Introduction	Dangerous Condition Occurring Frequency	Yearly Disasters Occurring Probability
Typhoon	Almost every year Shanghai has suffered Pacific tropical cyclone attacks. During 1949–2002 the number of tropical cyclones affected Shanghai was 186, bringing strong winds, heavy rains, storm surges and other disasters.	2~3/a	0.57
Heavy rain, flood	The average annual rainfall is 1123 mm, with 70 percent concentrated in the period from April to September. Low-lying regions may lead overflowing rivers and fields under water. Uneven distribution of urban drainage capacity requires further strengthening.	3.2d/a	0.61
Strong winds	Storm surge disasters caused by typhoons often occur along the Yangtze River and coastal regions and produce serious damage on seawalls, dams, and flood control river walls.	12.5d/a	0.65
Tornado	Annual average of Tornados is 2–3. They occur mainly in the outskirts (county) and are sudden and highly destructive.	2.1d/a	0.42
Lightning	Lightning is very common in Shanghai, with the average annual number of thunderstorm days reaching to 53.9. Lightning incidents happen every year bringing increasing economic losses.	28.3d/a	0.50
Heavy Fog	Mainly occurs in spring and winter and has great impact on the air quality and traffic.	39d/a	0.31
Heat Wave	About 9 high temperature days (higher than 35C) in normal years and about 20–30 days in abnormal years, may affect the city water and electricity supply, agricultural production and livelihood of the people.	9 d/a	
Red Tide	Frequent large-scale (an area of more than 1000 square km) occurrences near the Yangtze River estuary area cause severe damage to the marine biological resources and to human health and safety, due to red tide's toxicity.		

over the next ten years. SMB has already completed risk-mapping for heavy fog, high temperatures and typhoons. The Municipal Headquarters for Flood Prevention has mapped and created schematic diagrams of the Shanghai terrain. The Shanghai Municipal Water Affairs Bureau has completed a map of the distribution of drainage and dike facilities in Shanghai while the Shanghai Municipal Statistics Bureau provides the economic and social information for each district in the annual "Statistical Yearbook for Shanghai". The primary tool for review and feedback is to conduct a comparison of risk mapping with historical disaster data. If there is large difference between maps and the historical data then further assessment of disaster events should be conducted to improve the accuracy of the maps and statistical data.

7.3.4 *Storage and Accessibility of Disaster and National Hazard Risk Information*

The disaster information provided by the various MHEWS is housed in a unified management system operated by the Civil Affairs Department. The Civil Affairs department, routinely, collects information about disasters and checks the quality of data. The NMHS has no responsibility for storage of national risk information. The public and the international community can obtain hazard and vulnerability data from the Civil Affairs Department and relevant departments. The information is normally disseminated through media, publications and government websites.

7.3.5 *Development and Utilization of Hazard/Risk Information to Support Emergency Planning and Warnings*

The Emergency Response Management Plan utilizes disaster risk information in carrying out appropriate preventive actions. Special attention is given to high-risk areas. In view of the decision-making requirements of disaster prevention and mitigation, the government established a communication mechanism for responding to major meteorological disasters through early warnings in cooperation with the Emergency Response Management Office of the Shanghai Municipal Government. The mechanism ensures that prospective joint information for the Emergency Response Management Office takes into account the situation at a societal level and enhances the capacity for early detection and early decision-making in support of the government's emergency response and recovery actions. The assessment of disaster risk is the basis for the development of the risk management plan. When general disasters occur, SMB will provide risk assessment suggestions for all departments after becoming acquainted with the situation of each department. If severe disaster occurs, the risk assessment can then be integrated into the risk management plan. The risk information is used by the Shanghai Municipal Government in its decision-making process.

The Meteorological Service has integrated risk assessments into the early warning information it provides. Authorized by the government, the Meteorological Service disseminates not only weather disaster warning information but also information on the possible impact, intensity, scope of influence and possible losses, in addition to issuing disaster prevention guidelines. As an example, in view of the distribution of severe convective weather in Shanghai, SMB has formulated prevention measures for high impact weather and established a central warning facility as well as several district-level severe convective weather warning centres. These units are responsible for forecasting severe convective weather and providing information for the joint emergency response plan.

7.4 Hazard Monitoring, Forecasting, and Mandates for Warning Development

7.4.1 Organizational Responsibilities for Monitoring, Forecasting and Development of Hazard Warnings

In Table 7.2 the agency responsible for hazard warning is indicated by hazard and type, where:

- Type I hazards – SMB has sole responsibility for development of the warnings for the hazards.
- Type II hazards – SMB has a joint mandate for the development of the warnings for the hazards
- Type III hazards – SMB provides information to other agencies that have the mandate for the development of the warnings for the hazards.

Table 7.2 List of hazards, type and the responsible agency for warnings in Shanghai

Hazards	Type	Agency
Typhoon	Type I	SMB
Rainstorm	Type I	SMB
Snowstorm	Type I	SMB
High Temperature	Type I	SMB
Low Temperature	Type I	SMB
Cold wave	Type I	SMB
Gale	Type I	SMB
Heat wave	Type I	SMB
Lightning	Type I	SMB
Hail	Type I	SMB
Frost	Type I	SMB
Heavy fog	Type I	SMB
Haze	Type I	SMB
Road icing	Type I	SMB
Ozone	Type I	SMB
Storm Surge	Type II	SMB, SHB and Marine
Safety of Electricity and Power Supply	Type II	SMB and Economic Commission
Human Health	Type II	SMB and Public Health Bureau
Air Pollution	Type II	SMB and Environmental Protection Bureau
Bacterial Food Poisoning	Type II	SMB and Shanghai Municipal Food and Drug Supervision Administration
Flood	Type II	SMB and Shanghai Flood Prevention Department
Earthquake	Type III	Seismological Bureau
Tsunami	Type III	Oceanic Administration

7.4.2 *Organizational Collaboration and Coordination for Monitoring, Forecasting and Development of Hazard Warnings*

Shanghai continues to learn from its broad experience with the Grid Management Strategy, building on this experience to launch a safety approach for residential communities (Fig. 7.8). The community safety programme aims to provide "End-to-End-to-End" multi-hazard early warning services for decision-makers, the public and special users.

In Shanghai, there are, in total, more than 150 land-based automatic weather stations (AWSs), 20 AWSs located on islands, nearly 300 rain gauges, 2 wind profilers, 3 lightning localizers, 3 mobile vehicles, 7 remote video monitors and 18 GPS/MET. A Doppler radar has been operating for 10 years and a backup radar will be deployed this year. In addition there are other instruments including 2 wind profilers, a lightning positioning system, a video monitoring system, automobiles equipped for meteorological data collection and other resources. This comprehensive observation system emphasizes monitoring for strong mesoscale convective weather hazards and its construction provides the observation network needed for urban environmental meteorology. The system enhances real-time observation and information acquisition for the Expo site and other sensitive districts, achieving high precision observations with high spatial and temporal resolution. It improves the time margin for early warning of sudden meteorological hazards, satisfying the requirements for seamless weather forecasts and rapid response to hazards. Detailed information about organizational collaboration and coordination has been provided in Sect. 2.4.2.

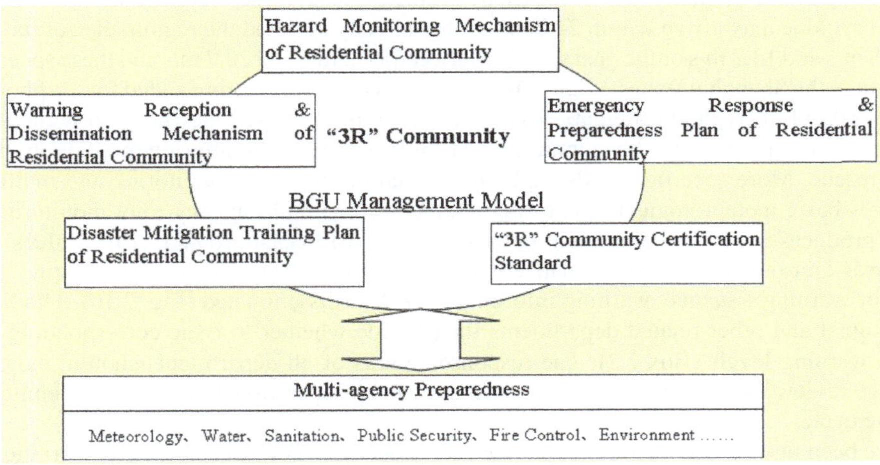

Fig. 7.8 The "3R" community strategy

7.5 Development of Timely, Authoritative, Recognizable and Understandable Warnings

7.5.1 *Warning Message Development Cycle*

Examples of hazard warning signals can be seen in Fig. 7.9. The warning levels and corresponding response action levels implemented by the Shanghai Flood Control Headquarters are shown in the associated Table 7.3.

Fig. 7.9 Meteorological and hydrological warning signals issued by Shanghai Meteorological Bureau (SMB)

Table 7.3 Shanghai Flood Control warning levels and response actions

Warning level	Response Action Levels (Implemented by Shanghai Flood Control Headquarters)
Blue	Organize a consultation meeting to strengthen flood monitoring and flood control supervision. Report the situation to Municipal Flood Control Headquarters and notify the Emergency Response Center of the Shanghai Municipal Government.
Yellow	Organize a consultation meeting to strengthen flood monitoring and flood control supervision, propose specific preparation requirements, report the situation to the Shanghai Municipal Government and notify the Emergency Response Center.
Orange	Deputy commander-in-chief for Municipal Flood Control Headquarters will chair the consultation meeting to strengthen flood monitoring and flood control supervision, to direct counter-measures, and report the situation to the leaders of the Shanghai Municipal Government. If necessary, an emergency response meeting will be held to plan flood and typhoon control more efficiently.
Red	Commander-in-chief of the Municipal Flood Control Headquarters will chair the consultation meeting, participating in planning flood and typhoon control guidelines. If necessary, the leaders of the Shanghai Municipal Government will deliver a speech on television or radio to encourage the military and civilians to devote themselves to disaster resistance and rescue. Report the situation to the State Council and related state-level departments according to National General Emergency Response Readiness and other related regulations.

Consider the "Blue" warning signal for a typhoon as an example. When the meteorological conditions reach a specific standard, the Blue warning signal for a typhoon will be issued by the meteorological service. The standards for the Blue level are as follows – the tropical cyclone may arrive within 24 hours or has already affected the region, the coastal or land surface wind speed is between 10.8 m/s and 13.8 m/s or the gust speed is between 17.2 m/s and 20.7 m/s and these speeds are likely to persist for some time. As soon as the Shanghai Water Affairs Office receives the Blue warning signal for a typhoon from the SMB, the Blue warning signal for flood and typhoon prevention will be issued immediately by the Shanghai Water Affairs Office.

The warning and response process can be summarized in several words – monitoring, prediction, preparedness, mitigation, assistance and rescue. More specifically, through operational observation, monitoring and multi-agency information sharing, SMB acquires basic meteorological monitoring data as well as multi-agency joint monitoring data. By means of analysis, SMB then produces the afore-mentioned multi-agency joint forecast product. If the values of these products do not reach the standards laid out for warning issuance, the system will simply continue monitoring. However, if they do meet the standards for warning issuance, warning information will be disseminated (Fig. 7.10). The SMB issues the meteorological warning signal and other related departments then decide whether to issue corresponding warning signals and determine their own warning levels (Box 7.2). The response actions of all departments should, moreover, be carried out immediately. Then we review the response actions and go back to the operational observation monitoring to complete the warning and response cycle.

Many studies have been undertaken regarding how the public accesses and interprets early warning messages. These results of these studies are used to help avoid confusion and panic during potential disaster events. Message format and dissemination processes are reviewed following each of the events and all relevant corrective actions are taken. The MHEWS has developed mechanisms for regular exercise and training to ensure that the public understands disaster response warning messages. These studies have been developed through the SMG based on risk analysis and agreed upon response actions. It is important to review and constantly improve the readiness response. The formulation of the warning level threshold value is based on historical statistical data. SMB is committed to the establishment of a variety of means (Fig. 7.11) to achieve early warning information distribution for a wider range of hazards.

The primary mechanism for coordination of warning level development among members is the Shanghai Emergency Response Management Committee. By means of signing departmental cooperation agreements, regularly scheduled and unscheduled inter-departmental joint communication and cooperation are being carried out to determine the early warning level and cooperative measures.

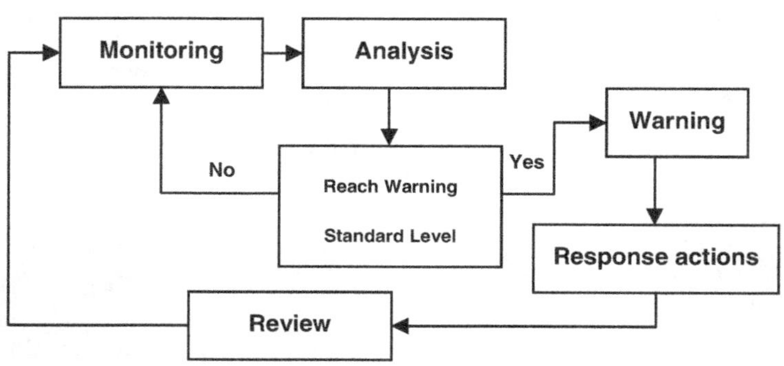

Fig. 7.10 The warning and response cycle

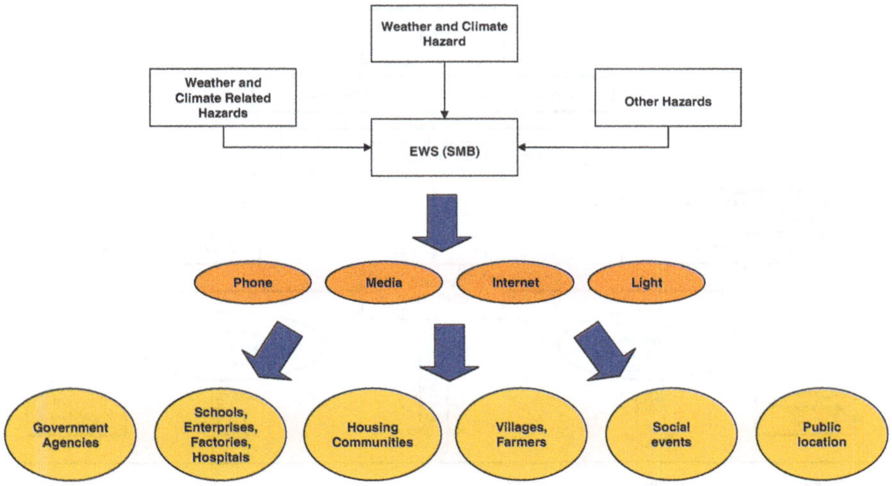

Fig. 7.11 Readiness levels communication flow

Box 7.2 Warning message of the Shanghai MHEWS

Meteorological hazard signals

Flood warning signals hazard

Blue: Municipal emergency headquarters arranges the whole resources for emergency to deal with the event under the lead and command of municipal government, and takes charge of information reporting.
Yellow: Municipal emergency headquarters takes responsibility to manage the emergency response, and asks for support of the municipal government if it necessary.
Orange: The related agencies take charge of disposal for hazard event. They need to take action to prevent and control the extension and ask for supports from municipal emergency headquarters if needed.
Red: The Basic Grid Units (BGU) takes charge of the emergency disposal of light hazard; the public takes suitable defensive action based on the actual need, and ask for help from related agencies according to the BGU mechanism.

7.5.2 Warning Message Improvement Cycle

The Shanghai Emergency Response Management Committee can identify and correct problems encountered at all levels through careful analysis of exercises and actual events. The weak links in the early warning process can be identified by means of regular and irregular review and assessment and measures will be formulated for improvement through inter-agency communication.

7.6 Warning Dissemination Mechanisms

Through the development of a community safety plan and implementation of urban grid management, three-class coverage of multi-hazard information can be implemented (Fig. 7.12).

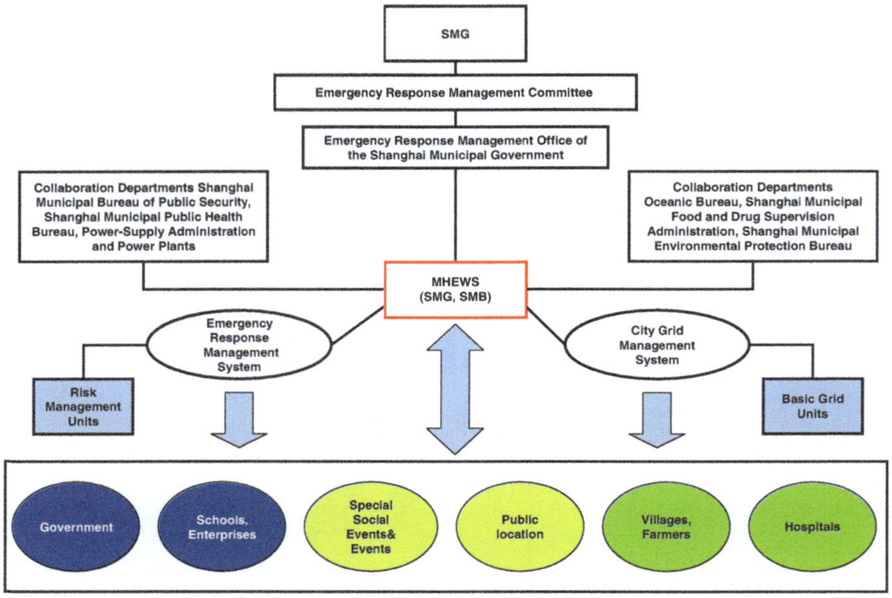

Fig. 7.12 Forecast and Warning Work Flow

- First, the initial establishment of municipal-level platform – the Emergency Response Management Office of the Shanghai Municipal Government relies on and is supported by the MHEWS Operations Centre in setting up a platform that covers city wide third-level government and 76 departments. In total, about 3,000 administrators and staff are involved. The Emergency Response Management Office of the Shanghai Municipal Government, and any other departments involved, will thus disseminate multi-hazard emergency response information and instructions.
- Second, the departmental-class platform has played a role – Our message platform has disseminated hydrological, flood prevention, heat stroke, food poisoning, community management and other relevant types of hazards early warning information. The city-wide information dissemination system has been established for communities, campuses and agriculture and covers 8,000 residential communities, 1,780 elementary and secondary schools as well as 300 agricultural departments.
- The third class in "3 class coverage" refers to the public-class platform that has been expanded effectively. A network-wide dissemination mechanism for early warning information on severe disasters has also been established for mobile phone users.

The Municipal Emergency Management Committee is responsible for organizing, coordinating and improving the warning information dissemination system. From its Operation Centre, the Shanghai Meteorological Bureau (SMB) operates and maintains the early warning information dissemination platform for Shanghai. As noted earlier, the Meteorological service is responsible for disseminating meteorological disaster warnings while the hydrological service is responsible for tidal-flood warnings. The Ministry of Health is responsible for pandemic and infectious disease warnings and the Public Security Department has a similar role relating to security warnings. The dissemination platform constructed by the meteorological service can be used to support the relevant departments in disseminating the required emergency response information and work instructions. The grid management system, thus, ensures widespread dissemination of hazard warnings to all residents in Shanghai. There is a mechanism for establishing confirmation of receipt within the Shanghai Emergency Warning Platform system.

The following dissemination mechanisms are being used in Shanghai:

- Cell-Phone Dissemination – The subsystems have been built and supported by the layered user database. The public receive warning information through Short Message Service (SMS), Multimedia Messaging Service (MMS), and Wireless Application Protocol (WAP).
- Radio Broadcasting Dissemination Subsystem – FM Subsidiary Communication Authorization (SCA) is used to broadcast warning information. The marine radio broadcast can cover the coastal waters surrounding Shanghai within a 1000 km radius.
- Basic Grid Unit System – Shanghai utilizes a Basic Grid Unit (BGU) management method for event monitoring and management in residential communities. The area of an average BGU is approximately 10,000 square meters. All BGUs are monitored and managed by supervisors who are responsible for collecting information and forwarding it to the city and district response centre through the BGU network. Based on the information obtained, the response centre will send

operational demands to related agencies and departments to deal with events within the individual grid or grids. SMB and other government agencies have provided many management and service products based on the BGU management framework. Examples of such SMB products are GIS-based dynamic rainfall-runoff simulations and chemical accident simulations.

- Public Electronic Screen Dissemination Subsystem – Early warnings are issued that include guidance for prevention and mitigation via display screens in the streets and parks. At present, there are 22,000 public electronic screens, 2,000 Digital TVs and 1,000 electronic screens in the streets.
- Mobile TV Dissemination Subsystem – You can catch the early warnings via Mobile TV in public transportation vehicles such as taxis, buses and subways.
- Warning Call Dissemination Subsystem – The system is based on the existing weather hotlines. The numbers are 12121, 969221, and 969222. There is also an active outgoing call service, named "Warning Call", which will be added to the subsystem (150 calls every minute).
- Dedicated Line Dissemination Subsystem – Texts, images, and voice calls can be transmitted by means of dedicated lines to the most important MHEWS partners and stakeholders – a local version of the cold war era "red telephone". The terminal display software can display the simulation images of the disaster's impact, development, and evolution for government agencies.
- Community Weather Light Warning System – SMB has established the first 'Community Weather Light' in the nation. The warning system in Baoshan district will be extended and expanded to the entire city and will include each district, in addition to well known regional landmarks.
- Outdoor Electronic Dissemination System – Electronic displays were set up in key areas, disseminating weather warnings, forecasts, real-time weather information and action guidelines. For example, after coordinating with the Information Commission of Songjiang district, a dissemination system of outdoor electronic display screens was developed for that district, focusing on the establishment of new screens (one has been completed) and the integration of older ones (41 have been completed). Information was disseminated using this system, with resource sharing and expanded utilization being realized with relevant departments.

7.7 Emergency Preparedness and Response Activities (National To Local)

7.7.1 Disaster Preparedness, Response Planning and Emergency Response Activation

The Community Safety Strategy not only includes all-reaching warning information but also information on emergency response preparedness, safety audit (integrated into residential communities and basic unit authentication system), disaster prevention training and rehearsals. Together these components form part of the disaster prevention scheme for the Shanghai MHEWS. The Community Safety Strategy also addresses schools, rural areas, enterprises, and grass-roots management units, with the participation of the Department of Civil Administration, Safety Supervision, Flood Prevention, Civilization Commission and other departments.

The Shanghai Municipal Government's and the Shanghai Meteorological Bureau schedule large-scale training and exercise sessions several times a year in an effort to, periodically, validate their EWS operations plan and develop and strengthen collaboration and relationships with EWS stakeholders.

An urban information interaction mechanism has been established on the basis of the Shanghai Urban Grid Management System. This mechanism enables grid managers to become warning information receivers and feedback providers. Each grid manager is responsible for ensuring that every ten thousand meter grid within the Shanghai MHEWS is supported. Shanghai MHEWS has taken into account the individual needs of vulnerable communities. To meet these responsibilities, emergency management relies on teams of grid managers and volunteers. Risk maps are used by SMG in establishing and prioritizing emergency preparedness and response actions and to carry out intensive preparedness and prevention in regions of high disaster risk and vulnerability.

The Shanghai Municipal Emergency Management Committee invites experts, relevant departments and representatives from vulnerable areas to participate in the development of emergency response plans and issues the plans in the form of government regulations. The Committee regularly organizes large-scale comprehensive training and exercises once or twice each year, and every department organizes training and assessment as needed according to the actual situation. The meteorological services actively participate in emergency response planning. The Municipal Emergency Response Management Committee and meteorological services jointly organize and conduct training sessions on an annual basis. A personnel evacuation system related to typhoons has been set up and has been activated several times, utilizing established evacuation routes. Other smaller scale evacuation exercises are also periodically carried out.

7.7.2 *Community Response Capacities*

Mechanisms and standards are being developed to assess communities' ability to respond effectively to early warnings through the following programs:

- Community disaster prevention certification;
- Community disaster prevention regulations development;
- Community disaster risk assessment;
- Community disaster and emergency response exercise;
- Community disaster information.

The following example illustrates how the Shanghai Emergency Response Committee reviews multi-hazard emergency response incidents to identify problem areas, providing an opportunity to update the Shanghai Municipal Government's (SMG's) response plans to respond to such events in the future. Low-temperature, freezing rain and snow weather disasters in early 2008 resulted in a wide ranging impact on society. Subsequently, SMG developed a specialized emergency response plan based on the idea of a multi-hazard early warning issued by the General Office of the Shanghai Municipal Government. At the direction of the government, organizations at the district-level, neighbourhood, grid-level, and block-level, such as community residents' committees, act for the government in addressing response and preparedness actions with the local population. This requires organizations to prepare and train these communities in multi-hazard incident response. This is accomplished with assistance of the relevant MHEWS partners who, routinely. provide such training when needs exist. The training programs were jointly developed by MHEWS partners and such training is conducted as required. The meteorological services provide routine and multi-hazard related information services to local communities, including jointly sponsored safety information and preparedness related training.

7.7.3 *Public Awareness and Education*

The local communities are educated by means of the Internet, the media, distribution of multi-hazard brochures and posters and training and exercises conducted by representatives of meteorological services and other partners as part of their public outreach programs. The Emergency Management Committee has established a public awareness program with the assistance of its media channels in order to improve public awareness of multi-hazards. There are dedicated weather broadcast channels that incorporate multi-hazard related programs. At this time, there are no specific public awareness and education campaigns tailored to the specific hazards in Shanghai but days such as Meteorological Day and Fire Prevention Day are observed as a means of maintaining public awareness. With that goal in mind, the Committee periodically reviews/surveys the level of public understanding and awareness of the multi-hazard process. Through this process, the Committee will identify and make necessary changes to maintain the highest possible public awareness level. Metrological services are a partner, actively participating in the Emergency Management Committee's public awareness program.

Joint activity between the MHEWS partners and the Shanghai Municipal Educational Committee will determine how public awareness education programs are built into school curricula. In 2007, Shanghai carried out image education in nearly 60 primary schools with total participation of about 50,000 students. The aim of the education was safety and awareness for meteorological disasters. Meteorological services printed extra-curricular reading material about weather for children which attracted children's interest with their cute cartoon characters, brilliant illustrations and lively language. These materials focused on safety techniques for children such as "In case of floods caused by heavy rains, they should escape into strong buildings, wave bright red scarves to draw the attention of rescuers".

7.8 Improvement of Overall Operational Framework of the EWS

The Municipal Emergency Response Management Committee holds quarterly working group meetings of MHEWS partners to review feedback on system operations and to improve the work flow mechanism. In addition, the SMB has established a professional position to collect feedback from disaster risk management agencies, other stakeholders, and the public for the purpose of improving MHEWS products.

7.9 Examples of Previous Events Where the Operational EWS Has Led to Improvements in

The following illustrate situations where improvements were implemented in Shanghai's MHEWS as a result of experience gained during these severe or disastrous events.

The August 2001 Heavy Rainfall Drew Attention to Needs for an Effective EWS

On the 5th of August 2001, a rainstorm impacted Shanghai and led to serious economic losses. The public was unprepared as there had been no warning and the media had insufficient protocols for weather disaster information dissemination. This event highlighted needs for greater awareness, improved weather warning information and coordination by agencies involved in disaster response activities.

The Cold Spell in 2008 Reinforced the Importance of the Shanghai Multi-Hazard Early Warning System Project

The cold weather that hit 10 provinces in central and southern China in January and February 2008 was the worst in decades. About 100 million people were affected by the severe weather. The storm spiralled into a crisis in some areas, with potential loss of critical supplies such as power, water, and communications as well as the threat of coal and rice running out in a matter of days. It is estimated that the direct economic loss was 15 to 16 billion Yuan during this weather related disaster. SMB issued road ice, cold front and heavy fog warnings during that period. The Emergency Response Management Office demanded that 300 government agencies should take the warnings issued by SMB as the starting point for multi-agency actions for risk response and prevention. During this period, it became apparent that that there were still many difficulties and weaknesses in prevention and preparedness when severe weather disasters occurred in China. For example, pre-evaluation was inadequate regarding the influence weather might have on transportation and electric power supply. Furthermore, when confronted by a severe disaster, the multi-agency collaboration and information sharing mechanism was seen to be in urgent need of strengthening. This was especially true following the dissemination of warnings when it became clear that multi-agency coordination and cooperation as well as effective disaster risk management needed to be improved. With the concept of "Multi-agency Teamwork" as the core consideration, the Shanghai Multi-hazard Early Warning System Project became the object of future planning, as it became obvious that such an approach was essential in local disaster risk management.

The August 2008 Rainstorm Provided Useful Lessons

A rainstorm on August 25, 2008 broke a 100-year record and caused severe social problems and economic losses to the mega-city. More than 100 mm of precipitation fell on urban areas in about a 1–2 hour time period and almost 150 streets and 11,000 houses were flooded. Some low lying road intersections were shut down and more than one hundred flights were delayed at Pudong and Hongqiao airports. During the course of the disaster, the SMB issued several early warnings responding to the meteorological emergency. However, the existing updating mechanism for warning information proved inadequate with respect to its ability to adapt to the rapid changes that occurred in the weather system. It was realized that early warning requires not only the release of warning information but also earlier consultation and early notification. In addition, feedback information from multi-agency cooperation and coordination is extremely important for updating warning levels in the whole process of disaster prevention and mitigation. It is very important to change the process from simple weather forecasting into disaster forecasting.

Shanghai's MHEWS Was Effective During a Severe Cold Wave in January 2009

In January of 2009, the city of Shanghai experienced a cold front that caused unusual freezing weather. On January 23rd, the low temperature reached minus 5.9 degrees Celsius, the lowest temperature recorded in the urban core during the previous 18 years. The Shanghai meteorological service disseminated early warnings of the cold front and frost. "The Shanghai Municipality emergency response planning for rain, snow and freezing weather disasters", published by the General Office of the Shanghai Municipal Government at the end of 2008, played an important role. The idea of multi-agency early warning embodied in the multi-agency coordination and cooperation guidelines facilitated an effective response to this low-temperature and freezing weather disaster. It proved remarkably effective and has been highly praised by the Shanghai Municipal Government. Examples of actions taken are as follows:

- Municipal government – the vice-mayor deployed the response action for each department on 21st January (two days ahead of time).
- Shanghai Municipal Economic Commission – enhanced distribution and coordination to guarantee material supply and ensure the supply of electricity, gas, water, coal, heat and oil.
- Municipal Agriculture Commission – took steps to protect crops; to harvest vegetables early and enter them into market; adopt multilayer-covering and electric heating and any other measures. And do a good job in cold and freezing prevention for all types of seedlings, as well as for livestock and aquaculture seedlings.
- Public traffic, police and municipal department – changed vehicle fuel (instead of number 10 fuel, use of number 0); strengthened road safety precautions, especially for highways.
- Housing authority – undertook check-up of safety and freeze protection for old houses.
- News department – released forecasting of disaster information and public self-protection knowledge.
- Emergency Response Management Office of the Shanghai Municipal Government – convened relevant departments to outline "The emergency response planning for rain, snow and freezing weather disasters", jointly with the Meteorological Bureau, and to urge all departments to master the actions required by planning through live simulations and develop perfected and specific plans.
- Each district or county – emergency response team of militia increased on-duty strength. Neighbourhood Committees and property managers inspect water pipes and natural gas infrastructure in their communities.

7.10 Overall Lessons Learned and Future Steps for Improving NMHS Contribution in EWS With a Particular Focus on Institutional Coordination and Cooperation

In terms of the city's comprehensive disaster prevention and mitigation system, the Shanghai Multi-Hazard Early Warning System gained significant success in disaster prevention and mitigation (DPM) through the CLAP Approach (Critical locations, Line, Area and People) and the SMART mechanism (Standardized Multi Agency Response and in-Time Coordination Mechanism).

In the CLAP approach:

- "*Critical Locations*" means embedding the MHEWS into a number of safety-sensitive hot spots related to the safety of the city, such as the Shanghai South Railway Station, International Airports, Yangshan Deep-sea Harbour and the Shanghai Petrochemical Department. For example, a special meteorological warning service facility would be set up at the International Deep-sea Harbour of Yangshan and the establishment of a marine observatory will provide early warning service for shipping, loading and unloading, logistics and, in the event of an emergency, timely evacuation.
- The "L" in CLAP refers to "Line" and is conceived as a reliable chain of information linking together all DPM assets and decision makers from the governmental and agency level down to the grass roots level. An effective "Line" can be achieved by strengthening the capacity of standardized inter-departmental and interdisciplinary coordination and cooperation. For example, the Flood Control Department and the Meteorological Service have established standard emergency response procedures (SOPs) and a joint mechanism for typhoons, rainstorms and other urban disasters through the whole process of DPM. Another example is how the Food and Drug Supervision Administration and the Meteorological Service have established a joint mechanism for information-sharing, development of technology and dissemination of early warning information for food poisoning. By reinforcing the capacity of standardized inter-departmental and interdisciplinary coordination and cooperation, a more effective defensive chain has been formed for responding to the city disasters.
- The so-called "Area" means combining the urgent emergency response mechanism with the daily urban grid management mechanism. In other words, the MHEWS is integrated with local level managers. Relying on the city grid management mechanism covering 19 districts around the whole city, social coverage of early warning information and cooperation of grass-root units have been realized. When the Meteorological Service disseminates early warning information, city grid administrators are responsible for checking the operation of important facilities in the grid area (10,000 square meters) for which they are responsible and providing timely reporting of related disaster information to the meteorological services, as feedback. Such critical information includes whether drainage is sufficient, if flooding has occurred, whether low-lying squares have accumulated stagnant water, whether substation facilities vulnerable to lightning strike are damaged, and so on.

- The "People" component requires effective application of MHEWS for decision-makers, social groups and individuals. MHEWS provides decision-makers with critical support for decision-making through comprehensive and effective disaster warnings and response measures as one component of the emergency response platform. Where social groups are concerned, the disaster prevention capacity of basic social units will be enhanced through the establishment of a security system within the local community. Moreover, awareness of and knowledge about disasters by different groups will be improved in collaboration with the departments of Education, Association for Science and Technology, the Red Cross, youth and women's organizations.

By successfully integrating the above four components – Critical locations, Line, Area and People – the Shanghai Multi-Hazard Early Warning has successfully achieved a more comprehensive, more effective, more targeted, disaster prevention capability.

7.10.1 The SMART Mechanism

By using the SOPs, the SMART mechanism can be used to translate meteorological (scientific) language into user language, which ensures that the stakeholders (users) receive the necessary information at the right time, enabling them to take timely action to avoid unnecessary catastrophic consequences. It should be stressed that the mechanism must be actionable based on operational weather service delivery capability. Furthermore, SMART practices should be extended to the community level.

List of Acronyms

AWS	Automatic Weather Stations
BGU	Basic Grid Unit
CLAP	Critical locations, Line, Area and People
CMA	China Meteorological Administration
CSO	Chief Service Officer
DPM	Disaster Prevention and Mitigation
EmRR	Emergency Response Readiness
EWS	Early Warning System
FM	Frequency Modulation
GPS/MET	Global Positioning System/Meteorology
MHEWS	Multi-hazard Early Warning System
MMS	Mulitimedia Messaging Service
SCA	Subsidiary Communications Authorization
SERC	Shanghai Emergency Response Committee
SMART	Standardized Multi Agency Response and in-Time Coordination Mechanism
SMB	Shanghai Meteorological Bureau
SMG	Shanghai Municipal Government
SMS	Short Message Service
SMSA	Shanghai Maritime Safety Administration
SOP	Standard Operating Procedure
WAP	Wireless Application Protocol
WMO	World Meteorological Organization

Chapter 8
Multi-Hazard Early Warning System in Japan

Naoyuki Hasegawa, Satoshi Harada, Shotaro Tanaka, Satoshi Ogawa, Atsushi Goto, Yutaka Sasagawa, Norihisa Washitake

Abstract Japan is located on the "Pacific Ring of Fire" where seismic and volcanic activity is high. As a result, the country has a relatively large number of earthquakes and volcanoes in relation to the very small area of its territory which occupies only 0.25% of the world's land mass. Due to its geographic, topographic and meteorological conditions, Japan is also prone to natural disasters such as typhoons, torrential rains and heavy snow.

The impact of natural disasters can be considered to be determined by two factors. One is a natural factor – the magnitude of natural phenomena (e.g. the scale and intensity of earthquakes and typhoons). The other is a social factor – vulnerability to hazards than can cause disasters. While it is almost impossible for people to control the natural factor, the social factor can be controlled or modified by human intervention. Consequently, Japan has made a special effort to reduce society's vulnerability to disasters since the end of the World War II. In order to reduce Japanese society's vulnerability to disasters and enhance its ability for disaster countermeasures, Japan has established and strengthened a disaster prevention scheme, undertaken protective measures, enhanced weather forecast technology, and installed communications infrastructure to disseminate disaster prevention information, based on lessons learnt from disasters that have occurred almost every year.

The basic scheme for disaster prevention in Japan was established following the Ise-wan Typhoon in 1959, which resulted in more than five thousands victims. Comprehensive disaster countermeasures were established and implemented based on the Disaster Countermeasures Basic Act, enacted in 1961 just after the Ise-wan typhoon disaster. Subsequent to the Ise-wan typhoon disaster, Japan did not experience any disasters which resulted in more than a thousand victims until 1995 when the Great Hanshin-Awaji Earthquake occurred. This earthquake not only caused more than six thousand casualties but also resulted in complete and unprecedented paralysis of urban cities. After the earthquake, Japan modified its various Acts and plans, established new systems and reinforced initial response capability for disaster prevention. In particular, observation systems that can accurately detect disaster risks in real-time have been progressively improved in order to establish early warning systems that facilitate the early evacuation of residents and the response activities of disaster management organizations, thereby reducing disaster damage. Organizations involved in disaster reduction, particularly the Japan Meteorological Agency (JMA), use 24-hour systems to carefully monitor various natural phenomena and weather conditions. In addition to disseminating observed information related to natural phenomena, the JMA also issues a wide range of forecasts, warnings and advisories regarding severe weather events, earthquake ground motions, tsunamis and volcanic eruptions.

Naoyuki Hasegawa, Head, Office of International Affairs of the Japan Meteorological Agency
Satoshi Harada, Senior Scientific Officer, Office of International Affairs of the Japan Meteorological Agency
Shotaro Tanaka, Senior Scientific Officer, Office of International Affairs of the Japan Meteorological Agency
Satoshi Ogawa, Scientific Officer, Office of International Affairs of the Japan Meteorological Agency
Atsushi Goto, Officer, Office of International Affairs of the Japan Meteorological Agency
Yutaka Sasagawa, Officer, Office of International Affairs of the Japan Meteorological Agency
Norihisa Washitake, Officer, Office of International Affairs of the Japan Meteorological Agency

M. Golnaraghi ed., *Institutional Partnerships in Multi-Hazard Early Warning Systems*,
DOI 10.1007/978-3-642-25373-7_8, © Springer-Verlag Berlin Heidelberg 2012

8.1 Overview of Early Warning Systems (EWS) and the Role of the National Meteorological and Hydrological Service (NMHS)

8.1.1 Background

Japan is an island country located in the western Pacific Ocean, with a total land area of about 378,000 square kilometres. More than 70 percent of the country is mountainous. Being situated along the circum-Pacific volcanic belt, Japan has several volcanic regions and is frequently affected by earthquakes and tsunamis. In spite of its rather small area, the climate differs regionally from a subarctic climate to a subtropical climate with a major feature being the clear-cut temperature changes between the four seasons. Most regions have a damp, rainy, season from May to July, accompanied by seasonal winds from the Pacific Ocean. The side of the country that faces the Sea of Japan has a climate with much snow in winter, associated with seasonal winds from Siberia. From July to September, Japan frequently suffers from typhoons.

Japan is an earthquake prone nation due to its geological structure in proximity to the plate boundaries of the Pacific Plate, the Philippine Sea Plate, the Eurasian Plate, and the North American Plate. Every year there is a great loss of life and property due to natural disasters. Almost every year, Japan is affected by volcanic disasters triggered by eruptions and volcanic earthquakes, in addition to the disasters caused by typhoons. Until the 1950s, numerous large-scale typhoons and earthquakes caused extensive damage and thousands of casualties. However, disaster damage has shown a declining trend with advances in society's capabilities to address disasters and the mitigation of vulnerabilities by developing disaster management systems, promoting nationwide land conservation, improving weather forecasting technologies, and upgrading disaster information communications systems. In spite of such efforts, however, more than 6,400 people became casualties of the Great Hanshin-Awaji Earthquake in 1995. Furthermore, in 2004, ten typhoons, the largest number recorded in a single year, crossed over Japan, causing damage throughout the nation. There is, in addition, a high probability of occurrence of large-scale earthquakes during the coming decades and, in consequence, natural disasters remain a menacing threat to the safety and security of the country.

8.2 Governance and Institutional Arrangements (National to Local Levels)

8.2.1 Policy, Intuitional and Legal Frameworks to Support Emergency Planning and Response

Just after World War II, Japan enacted various laws for disaster management, especially for the purpose of disaster emergency response. For instance, the Disaster Relief Act was enacted in 1947 and the Flood Control Act in 1949, triggered by the 1946 Nankai Earthquake which caused more than 1,400 deaths. In addition, the Building Standard Law was enacted in 1950, triggered by the 1948 Fukui Earthquake which caused more than 3,700 deaths. Prior to 1961, however, there was no comprehensive and strategic disaster management system. The turning point in strengthening Japan's disaster management system followed the immense damage caused by the Ise-wan Typhoon in 1959. This led to the enactment of the Disaster Countermeasures Basic Act in 1961, which formulates a comprehensive and strategic disaster management system. This disaster management system has been further strengthened following the lessons learned from subsequent large-scale disasters such as the Great Hanshin-Awaji Earthquake in 1995. The evolution of disaster management laws and systems is detailed in Table 8.1.

The main contents of the Disaster Countermeasures Basic Act are as follows:

- Definition of responsibilities for disaster management;
- Disaster management organizations;
- Disaster management planning system;
- Disaster prevention and preparedness;
- Disaster emergency response;
- Disaster recovery and rehabilitation;
- Financial measures;
- State of Disaster Emergency.

Disaster countermeasures are undertaken based on the Disaster Countermeasures Basic Act and various disaster management related laws described in Appendix 1.

Table 8.1 History of major natural hazards and disaster countermeasures in Japan

	Events	Disaster Management Acts	Disaster Management Plans and Systems
1940	46 · Nankai Earthquake 48 · Fukui Earthquake	47 · Disaster Relief Act 49 · Flood Control Act	
1950	59 · Typhoon Ise-wan	50 · Building Standard Law	
1960	61 · Heavy Snowfalls 64 · Niigata Earthquake	60 · Soil Conservation and Flood Control Urgent Measures Act 61 · Disaster Countermeasures Basic Act 62 · Act on Special Financial Support to Deal with Extremely Severe Disasters · Act on Special Measures for Heavy Snowfall Areas 66 · Act on Earthquake Insurance	61 Disignation of Disaster Reduction Day 62 Establishment of Central Disaster Management Council 63 Basic Disaster Management Plan
1970	73 · Mt. Sakurajima Eruption 76 · Seismological Society of Japan's report about the possibility of Tokai Earthquake 78 · Miyagi-ken-oki Earthquake	73 · Act on Special Measures for Active Volcanoes 78 · Act on Special Measures for Large-scale Earthquakes	79 Tokai Earthquake Contermeasures Basic Plan
1980		80 · Act on Special Financial Measures for Urgent Earthquake Countermeasure Improvement Projects in Areas for Intensified Measures 81 · Amendment of Building Standard Law	
1990	95 · Great Hanshin-Awaji Earthquake 99 · Torrential Rains in Hiroshima · JCO Nuclear Accident	95 · Act on Special Measures for Earthquake Disaster Countermeasures · Act on Promotion of the Earthquake-proof Retrofit of Buildings · Amendment of Disaster Countermeasures Basic Act · Amendment of Act on Special Measures for Large-scale Earthquakes 96 · Act on Special Measures for Preservation of Rights and Profits of the Victims of Specified Disasters 97 · Act on Promotion of Disaster Resilience Improvement in Densely Inhabited Areas 98 · Act on Support for Livelihood Recovery of Disaster Victims 99 · Act on Special Measures for Nuclear Disasters	95 Amendment of Basic Disaster Management Plan Disignation of Disaster Reduction and Volunteer Day
2000	00 · Torrential Rains in the Tokai Region 04 · Niigata-ken-Chuetsu Earthquake 05 · Typhoons and Torrential Rains ※10 typhoons, the largest number in a single year on record, landed on Japan (an average of 2.6 typhoons per year)	00 · Act on Promotion of Sediment Disaster Countermeasures for Sediment Disaster Prone Areas 02 · Act on Special Measures for Promotion of Tonankai and Nankai Earthquake Disaster Management 03 · Specified Urban River Inundation Countermeasures Act 04 · Act on Special Measures for Promotion of Disaster Management for Trench-type Earthquakes in the Vicinity of the Japan and Chishima Trenches 05 · Amendment of Act on Promotion of the Earthquake-proof Retrofit of Buildings · Amendment of Flood Control Act · Amendment of Act on Promotion of Sediment Disaster Countermeasures for Sediment Disaster Prone Areas	01 Establishment of the Cabinet Office 03 Policy Framework for Tokai Earthquake Policy Framework for Tonankai and Nankai Earthquakes 04 Tonankai and Nankai Earthquake Countermeasures Basic Plan 05 Tokai Earthquake Disaster Reduction Strategy Tonankai and Nankai Earthquake Disaster Reduction Strategy Policy Framework for Tokyo Inland Earthquakes 06 Policy Framework for Trench-type Earthquakes in the Vicinity of the Japan and Chishima Trenches Countermeasures Basic Plan for Trench-type Earthquakes in the Vicinity of the Japan and Chishima Trenches Tokyo Inland Earthquake Disaster Reduction Strategy Basic Framework for promoting a Nationwide Movement for Disaster Reduction

伊勢湾台風，1959
写真提供：岐阜県
Ise-wan Typhoon, 1959
Photo: Gifu Prefecture

長崎豪雨災害，1982
写真提供：長崎市
Torrential Rains in Nagasaki, 1982
Photo: Nagasaki City

The Japan Meteorological Agency (JMA) is one of the 23 Designated Government Organizations described in the Disaster Countermeasures Basic Act. These Designated Government Organizations have the following obligations for disaster countermeasures:

- Cooperating in the formulation of the Local Disaster Management Plans of prefectures and local municipalities;
- Formulating/Reviewing/Revising their own operational disaster prevention plans;
- Delegating officials to the headquarters for major/emergency disaster control;
- Dispatching officials for emergency measures or rehabilitation efforts;
- Maintaining organizations for disaster prevention;
- Implementing disaster prevention;
- Implementing emergency measures for disasters;
- Gathering/Transmitting information related to disasters;
- Reporting conditions of disasters;
- Implementing the work of rehabilitation after disasters.

JMA's work and responsibilities are prescribed in the Meteorological Service Law, which is one of the laws on Disaster Prevention and Preparedness.

8.2.2 National to Local Emergency Planning and Related Linkages to EWS

Japan has three administrative levels of governance – national, prefectural and municipal. Each level of government has its own disaster management organization, policy framework and budget. When disasters occur, municipalities respond first. In the event that disasters are large in scale and beyond municipal capacity, the national and prefectural governments provide all possible support. Figure 8.1 outlines the disaster management system in Japan.

The Disaster Management Planning System in Japan is formulated as follows:

- *Basic Disaster Management Plan:* This plan is a basis for disaster reduction activities and is prepared by the Central Disaster Management Council, based on the Disaster Countermeasures Basic Act (Fig. 8.2).
- *Disaster Management Operation Plan:* This is a plan prepared by each designated government organization and designated public corporation, based on the Basic Disaster Management Plan.
- *Local Disaster Management Plan:* This is a plan prepared by each prefectural and municipal disaster management council, subject to local circumstances, and based on the Basic Disaster Management Plan.

The Basic Disaster Management Plan addresses comprehensive and long-term disaster reduction issues such as disaster management related systems, disaster reduction projects, early and appropriate disaster recovery and rehabilitation, as well as scientific and technical research.

Based on the Disaster Management Basic Act, the Central Disaster Management Council formulates and reviews the Basic Disaster Management Plan. This Plan was revised completely in 1995, based on the experiences of the Great Hanshin-Awaji Earthquake. It now consists of various plans for each type of disaster in which tangible countermeasures to be taken by each stakeholder, such as the national and local governments, public corporations and other entities, are described for easy reference, in accordance with the disaster phases of prevention and preparedness, emergency response, and recovery and rehabilitation.

The Japan Meteorological Agency (JMA) is involved in the above scheme as one of the twenty-three Designated Government Organizations described in the Disaster Countermeasures Basic Act. The Heads of these Designated Government Organizations are the members of the Secretariat Organization under the Central Disaster Management Council. Each Designated Government Organization also formulates its own Disaster Management Operation Plan. The Plan has been reviewed every year and revised several times as required.

The Disaster Management Operation Plan of the JMA prescribes the following measures to be taken by the JMA in support of disaster preparedness and emergency response:

Fig. 8.1 Outline of the Disaster Management System

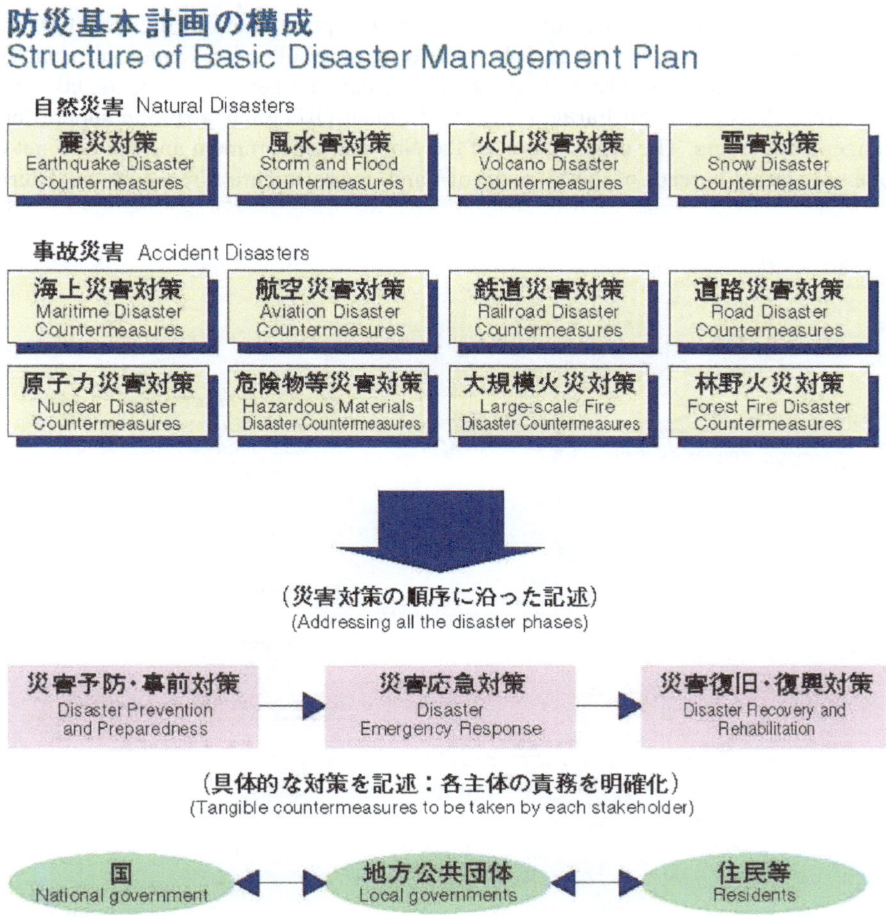

Fig. 8.2 Structure of Basic Disaster Management Plan

Disaster Preparedness

- Implementing meteorological services for disaster preparedness;
- Implementing a cooperation scheme with national disaster authorities, prefectures and mass media;
- Conducting drills and training for disaster mitigation;
- Conducting surveys and research on disaster management and providing knowledge to the public.

Emergency Response

- Establishing an emergency response scheme and activities;
- Setting priorities for services during emergency response;
- Cooperating with national disaster authorities, prefectures and mass media;
- Reporting results of surveys after disasters.

8.2.3 Organizational Structure for Implementing the Plans

Japan's disaster management system addresses all of the disaster phases of prevention, mitigation and preparedness, emergency response and recovery and rehabilitation. With the roles and responsibilities of the national and local governments being clearly defined, the relevant public and private sector stakeholders cooperate in implementing various disaster countermeasures.

As part of a series of reforms of the central government system, the post of Minister of State for Disaster Management was created in 2001 to integrate and coordinate disaster reduction policies and activities of ministries and agencies. The Director-General for Disaster Management in the Cabinet Office, which is responsible for securing cooperation and collaboration on wide-ranging issues among related government organizations, is mandated to undertake the planning of basic disaster management policies and response to large-scale disasters, as well as to conduct overall coordination (Fig. 8.3).

The Cabinet Secretariat system was also strengthened through the appointment of the Deputy Chief Cabinet Secretary for Crisis Management and the establishment of the Cabinet Information Collection Center, to strengthen risk management functions in order to address emergencies such as large-scale disasters and serious accidents, taking into account the lessons learned from the Great Hanshin-Awaji Earthquake. The Cabinet Office has a role in supporting the Cabinet Secretariat on disaster management matters. The organization of the National Government and the Cabinet Office are illustrated in Fig. 8.4. JMA is one of external bureaus of the Ministry of Land, Infrastructure, Transport and Tourism (MLIT).

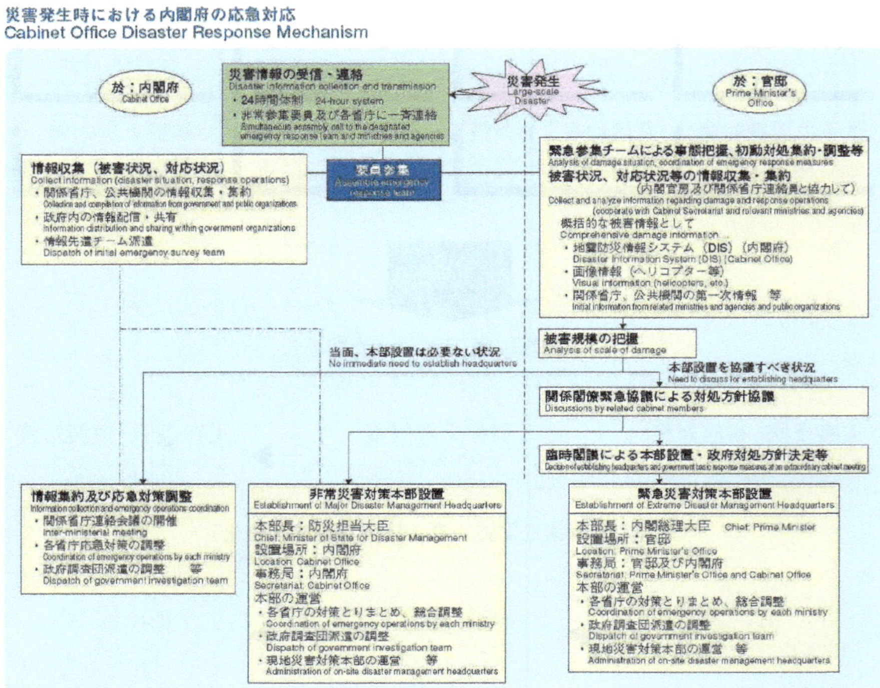

Fig. 8.3 Disaster Response Mechanism

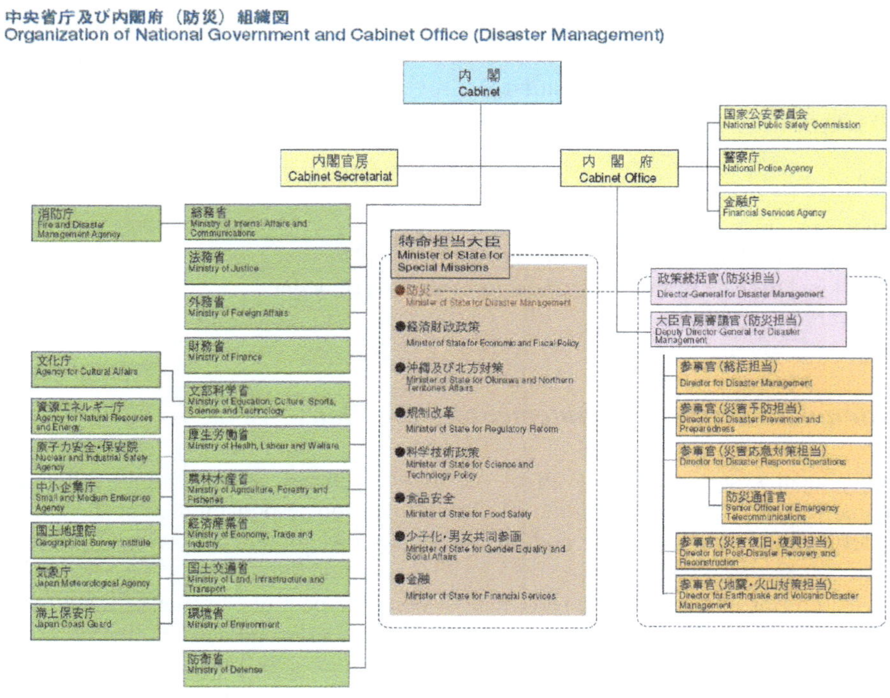

Fig. 8.4 Organization of the National Government and Cabinet Office

The national and local governments need to collect and disseminate disaster and damage information quickly and to have secure communications so that they can carry out effective disaster emergency activities such as search and rescue and medical operations. Based on such information, local governments set up a disaster response headquarters (headed by mayors of the cities and/or governors of the prefectures concerned) while related organizations establish their own operations mechanisms.

The national government collects disaster information at the Cabinet Information Collection Center 24 hours a day. At the time of a large-scale disaster, the designated emergency response team, comprised of the director-generals of the respective ministries and agencies, assembles immediately at the Crisis Management Center in the Prime Minister's Office to analyze and assess the disaster situation and report to the Prime Minister. If necessary, inter-ministerial meetings are held, at ministerial or high-ranking senior official level, to decide on basic response policies. Depending on the degree of damage, the government may establish a Major Disaster Management Headquarters, headed by the Minister of State for Disaster Management, or an Extreme Disaster Management Headquarters, headed by the Prime Minister. In addition, a government investigation team, headed by the Minister of State for Disaster Management, may be dispatched or an on-site disaster management headquarters may be established.

In the case of large-scale disasters that exceed the response capabilities of the affected local government, various wide-area support mechanisms are mobilized by the National Police Agency (Inter-prefectural Emergency Rescue Unit), Fire and Disaster Management Agency (Emergency Fire Rescue Team), and Japan Coast Guard. In addition, the Self-Defense Forces can be dispatched to undertake emergency response activities on request from the governor of the affected prefectural government. A wide-area medical transportation system is being developed for dispatching disaster medical assistance teams (DMAT) and ambulance parties to transport seriously injured people to disaster management base hospitals outside of the disaster-stricken area.

As part of the Disaster Response Mechanism, the JMA provides timely and accurate 'Disaster Information' to central and local disaster management authorities, in order to supply such information to the public through these authorities. This 'Disaster Information' consists of following warnings, advisories and information:

- Warnings and advisories on weather, high tides, high waves and flooding;
- Earthquake Early Warnings;
- Tsunami Warnings and Advisories;
- Volcanic Warnings;
- Information regarding typhoons, heavy rain, tornadoes, earthquakes, tsunamis and volcanic activities.

8.2.4 Institutional Capacities and Concept of Operations (Coordination and Operational Collaboration)

The Central Disaster Management Council was established under the Disaster Countermeasures Basic Act, with its brief being to ensure the comprehensiveness of disaster risk management and to discuss matters of importance with regard to disaster management. The council consists of the Prime Minister, who is the chairperson, the Minister of State for Disaster Management, all ministers, heads of major public institutions and experts (Fig. 8.5).

The duties of the Central Disaster Management Council are to:

- Formulate and promote implementation of the Basic Disaster Management Plan and Earthquake Countermeasures Plans;
- Formulate and promote implementation of the urgent measures plan for major disasters;
- Deliberate on important disaster reduction issues, in response to requests from the Prime Minister or Minister of State for Disaster Management (basic disaster management policies, overall coordination of disaster countermeasures and declaration of state of disaster emergency);
- Offer opinions regarding important disaster reduction issues to the Prime Minister and Minister of State for Disaster Management.

Prefectural and municipal Disaster Management Councils are established in prefectures and local municipalities, with membership comprised of representatives of local government organizations, including police and fire management departments, and designated local public corporations. Implementation of disaster risk management measures is based on the Local Disaster Management Plans drafted by the Councils. These Disaster Management Councils and Disaster Management Plans, at each level from central government (Table 8.2) to local municipalities, are prescribed in the Disaster Countermeasures Basic Act. This Act requires the Disaster Management Council at each level to review its Disaster Management Plan every year and amend it in order to ensure that the capacities of all early warning system stakeholders are utilized in the most effective and efficient manner. Each Disaster Management Council should communicate developments and/or amendments of their Disaster Management Plan to the Prime Minister, governors of prefectures and local municipalities.

Table 8.2 The History of the Basic Disaster Management Plan

防災基本計画の策定・修正経緯
History of Basic Disaster Management Plan

年 Year	内　容 Contents
昭和38年 1963	作成 Initial plan drawn up
昭和46年 1971	地震対策、石油コンビナート対策等に係る修正 Revision of Earthquake Disaster Countermeasures and Petroleum Industrial Complexes Disaster Countermeasures
平成 7 年 1995	自然災害対策編の全面的な修正 Overall Revision of Natural Disaster Countermeasures
平成 9 年 1997	事故災害対策編の追加 Addition of Accident Disaster Countermeasures
平成12年 2000	原子力災害対策編の全面的な修正 Overall Revision of Nuclear Disaster Countermeasures 省庁再編に伴う修正 Revision along with a series of reforms of the central government system
平成14年 2002	風水害対策編、原子力災害対策編の修正 Revision of Storm and Flood Countermeasures and Nuclear Disaster Countermeasures
平成16年 2004	震災対策編の修正 Revision of Earthquake Disaster Countermeasures
平成17年 2005	自然災害対策に係る各編の修正 Revision of Natural Disaster Countermeasures

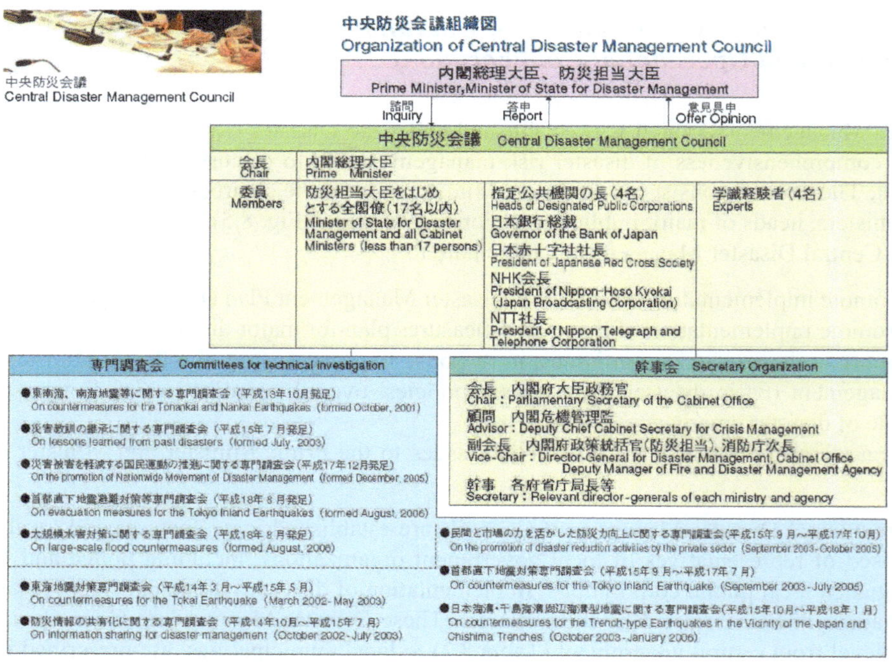

Fig. 8.5 Organization of the Central Disaster Management Council

8.2.5 *Utilization of a Multi-Hazard Approach*

Collaboration with other agencies is promoted to optimize resources for improvement of early warning systems, with the fields of earthquake and tsunami monitoring providing good examples. The Act on Special Measures for Earthquake Disaster Countermeasures was enacted in July 1995, to promote a comprehensive national policy on earthquake disaster prevention following the 1995 Great Hanshin-Awaji Earthquake. Based on this Act, the Headquarters for Earthquake Research Promotion was established in the Ministry of Education, Culture, Sports, Science and Technology (MEXT). In the Earthquake Research Promotion scheme, JMA has been assigned responsibility for collecting all of the seismic data from seismometers which have been installed by JMA and other organizations, analyzing these data and determining hypocentres of earthquakes occurring in and around Japan. The number of hypocentres identified has been dramatically increased by utilizing the data from the seismometers of other organizations. These hypocentre data are compiled and provided, as a national earthquake catalogue, for use by earthquake researchers, disaster management authorities and the public.

8.2.6 *Key Factors Considered for Sustainability of Different Components of Early Warning Systems With a Multi-Hazard Approach*

Disaster risk reduction is funded in the budgets of national and local governments. The budget is requested by each organization independently. From 1995 to 2004, the average annual national budget for disaster management was approximately 4.5 trillion yen, accounting for approximately 5% of the total budget for general accounts. The percentage breakdown of the disaster management budget is as follows: Scientific Technology Research – 1.3%; Disaster Prevention and Preparedness – 23.6%; National Land Conservation – 48.7%; and Disaster Recovery and Rehabilitation – 26.4%. Following each severe disaster, a specified budgetary amount has been approved for disaster restoration and disaster prevention and the JMA has improved its systems by utilizing these budgetary allocations. For example, a supplemental budget was allocated for the installation of more than 150 remote seismic observation stations following the earthquake and tsunami in the north-west of Hokkaido area in 1993. As another example, a budget supplement was allocated for the installation of automatic seismic intensity meters at more than 600 sites following the Great Hanshin-Awaji Earthquake in 1995.

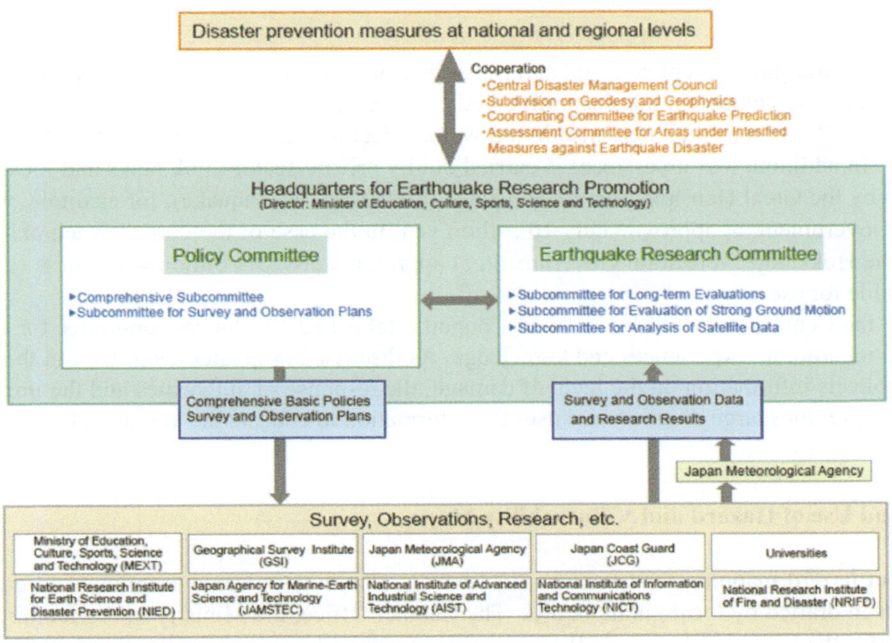

Fig. 8.6 Structure of the Headquarters for Earthquake Research Promotion

As the National Meteorological Service of Japan, the Japan Meteorological Agency (JMA) has a mandate to enhance its institutional capacities to support the early warning system. In this context, the JMA delivers a series of internal training courses to various levels of staff to build capacity in disaster management. These courses are conducted in JMA headquarters, district meteorological observatories and the Meteorological College. In addition, the Government of Japan is striving to make work processes and systems in each government agency more efficient and rational through initiatives such as:

- Review of the system and operational aspects of work;
- Computerization and standardization of paperwork;
- System unification and concentration, in order to become more streamlined and efficient.

This promotion of optimization of work and systems is being applied to the installation of meteorological instruments and central analysis/communication systems whose value exceeds a certain budgetary figure.

8.3 Utilization of Risk Information in Emergency Planning and Warnings

8.3.1 Organizational Responsibilities and Arrangements for the Development of Risk Information

Key government agencies involved in hazard and vulnerability assessments are as follows:

- The Fire and Disaster Management Agency, Ministry of Internal Affairs and Communications (MIC) has drawn up a procedure enabling local public bodies to make an objective assessment of their own disaster risk reduction and crisis-management systems. Assessment is carried out using results based on replies to roughly 800 questions, each offering 2 to 4 alternatives. The Agency is responsible for utilizing available facilities, human resources and fire services to protect people's lives and property from fire and to take precautions against disasters such as storms, floods, fires and earthquakes and mitigate the damage resulting from these.
- The Cabinet Office provides an efficient methodology to self-evaluate and enhance community-based capability to collect information, foresee probable disasters and ensure rapid evacuation. This self-evaluation would be relayed to the Director for Disaster Preparedness and the Cabinet Office for Disaster Management. The Cabinet Office is responsible for ensuring cooperation and collaboration among related government organizations on wide-ranging issues.
- The Central Disaster Management Council and local governments have anticipated the damage that would be caused in the event of a major earthquake and have assessed vulnerability by, among other measures, evaluating the earthquake resistance of buildings and infrastructure and by assessing the awareness of community members with regard to disaster risk reduction.

Although Japan's central government does not carry out systematic socio-economic and environmental impact and loss analysis for all disasters, the competent administrative bodies assess disaster damage in relation to rivers, roads, forestry and fisheries facilities, educational facilities, health and welfare facilities and other public facilities, in order to expedite their rapid recovery. In addition, loss assessment is carried out by private-sector think tanks and local government bodies. The damage caused by the Great Hanshin-Awaji Earthquake (1995 Kobe Earthquake), for example, was estimated, by the Hyogo Prefectural Government, at approximately 10 trillion yen. In the case of the 2003 Tokachioki Earthquake, damage was estimated, by the Hokkaido Prefectural Government, at approximately 25.4 billion yen. These results have been published and are available for use.

In October 2003, the Central Disaster Management Council established a technical committee for Disaster Lessons Assessment in order to incorporate experiences and knowledge. As discussed in greater detail later in this report, the committee systematically collects information on the level of damage, the response of authorities and the impacts on people's living conditions during previous large disasters and uses the information to educate the young and old.

The Preparation and Use of Hazard and Vulnerability Maps

Japan has carried out hazard mapping for tsunamis, tidal waves, flooding, volcanic eruptions and earthquakes and many hazard maps have been drafted by local public bodies. The Cabinet Office, the Ministry of Agriculture, Forestry and Fisheries of Japan (MAFF), the Fisheries Agency, MLIT and other agencies have, moreover, drawn up manuals on the subject. Many of these maps have been made available to the general public on the Internet and elsewhere. Provision for promoting hazard mapping and keeping the public informed regarding disaster countermeasures is included in the country's basic disaster-risk reduction plan. In addition to the country's disaster-risk reduction plan, some legislation also mandates municipalities with the preparation of hazard maps. Municipalities are, for example, mandated to distribute flood mapping, by the Flood Control Act, and sediment disaster hazard mapping, by the Act on Promotion of Sediment Disaster Countermeasures for Sediment Disaster Prone Areas. Those hazard maps prepared by municipalities are posted to the web portal site. Hazard maps have not, however, been prepared for all of Japan.

The hazard and risk maps are used in emergency planning and response operations. Volcano hazard maps, for example, show areas predicted to be influenced by volcanic eruptions, evacuation facilities, etc and, as of 2006, maps had been prepared for 37 volcanoes. At the time of the Mt. Usu eruption in 2000, residents were familiar with the hazard map. As a result, the evacuation of about 16,000 residents was conducted in an orderly manner in accordance with the volcano alert issued by JMA two days prior to the eruption and no one was killed or injured. Moreover, the Mt. Fuji Volcano Disaster Management Conference, consisting of relevant ministries and local governments, created a Mt. Fuji hazard map and reviewed wide-area disaster countermeasures in view of the possible enormous influence of an eruption. In addition, the "Basic Framework for Wide-area Countermeasures against Mt. Fuji Volcano Disaster" was formulated by the Central Disaster Management Council in 2006.

Examples of the use of hazard maps with early warnings for earthquakes and tsunamis, volcano early warning information, early warnings for severe weather disasters and issuance of early warning information for sediment-related disasters are shown at: http://www.bousai.go.jp/kyoryoku/pdf/soukikeikai.pdf.

Systematic Collection, Sharing and Assessment of Hazard and Vulnerability Data

Based on the experiences of the Great Hanshin-Awaji Earthquake, the Cabinet Office has been developing an integrated disaster management information system that assists in grasping the disaster situation at an early stage and promotes information sharing among relevant organizations, thereby enabling rapid and appropriate decision-making for emergency response operations. Its components are the:

- *Earthquake Disaster Information System* (DIS): The DIS is automatically activated to estimate, within 30 minutes, the approximate distribution of seismic intensity and scale of damage, human suffering and building damage, upon the receipt of information from JMA indicating earthquake intensity level of 4 or greater;
- *Real Damage Analysis System by Artificial Satellite* (RAS): The RAS uses satellite images to assess actual disaster damage when it is otherwise difficult to determine the disaster situation due to the disruption of transportation and communications networks;
- *Disaster Information Sharing Platform* (PF): The PF is a common information sharing system, with a standardized information format, where various disaster information provided by ministries and agencies, local governments, relevant organizations and residents, can be posted and freely accessed by all.

Determination of the Risks Faced by Each Region or Community

When a disaster occurs or is imminent, residents may start evacuating on their own and the mayor of the municipality may also issue an evacuation order or instruction. It is effective for municipalities to prepare a manual outlining the criteria for disaster situations that require the issuance of evacuation orders or instructions, thereby facilitating a rapid decision by the mayor. In 2005, the Cabinet Office, in cooperation with relevant ministries, published "Guidelines for Producing a Decision and Dissemination Manual for Evacuation Orders and Instructions" and is promoting its implementation.

NMHS's Role in the Provision of Hazard Information

The NHMS plays the following roles:

- *Supporting the preparation of Manuals on the Decision and Dissemination of Evacuation Orders and Instructions*: When municipalities prepare Manuals on the Decision and Dissemination of Evacuation Orders and Instructions, the corresponding Local Meteorological Observatory of JMA supports the municipality's decision-making on the criteria for disaster situations that require the issuance of evacuation orders by providing observational data and forecasts. In addition, the Local Meteorological Observatory of JMA advises municipalities' decision-making on evacuation orders by commenting on the relationship between the meteorological characteristics and historical data for disasters.
- *Providing information through the Information Network for Disaster Prevention*: As described earlier, JMA and its Local Meteorological Observatories disseminate information rapidly and effectively to municipalities through the In-

formation Network for Disaster Prevention. They have also established the framework for information exchange with municipalities.

- *Supporting decision-making and dissemination for Evacuation Orders and Instructions*: The Local Meteorological Observatory of JMA supports mayors' decisions on and dissemination of Evacuation Orders and Instructions through the provision of meteorological alerts and advisories. JMA introduced a new index closely related to disasters in June 2008 to provide more precise support for mayors' decision-making.

8.3.2 *Hazard Assessment, Quantification and Mapping (National to Local)*

Hazards in Japan

As indicated earlier, there is great loss of lives and property in Japan every year as a result of natural disasters (Table 8.3). Until the 1950s, numerous large-scale typhoons and earthquakes caused extensive damage and thousands of casualties (Fig. 8.7). Disaster damage has, however, shown a declining trend with the advances in society's capabilities to address disasters and the mitigation of vulnerability arising from progress in the development of disaster management systems, promotion of national land conservation, improvements in weather forecasting technologies and upgrading of disaster information communications systems.

Table 8.3 Major disasters in Japan

日本の主な自然災害　Major Disasters in Japan

年 月 Date	災 害 Disaster	死者・行方不明者数 Number of Deaths and Missing
明治21. 7.15 1888	磐梯山噴火 Mt. Bandai Eruption	461
明治24.10.28 1891	濃尾地震 (M7.9) Nobi Earthquake	7,273
明治29. 6.15 1896	明治三陸地震津波 (M7.1) Meiji Sanriku Earthquake Tsunami	21,959
大正 3. 1.12 1914	桜島噴火 Mt. Sakurajima Eruption	58
大正12. 9. 1 1923	関東大地震 (M7.9) Great Kanto Earthquake	約105,000
大正15. 5.24 1926	十勝岳噴火 Mt. Tokachi Eruption	144
昭和 2. 3. 7 1927	北丹後地震 (M7.5) Kita-Tango Earthquake	2,925
昭和 8. 3. 3 1933	昭和三陸地震津波 (M8.3) Showa Sanriku Earthquake Tsunami	3,064
昭和 9. 9.20~22 1934	室戸台風 Typhoon Muroto	3,036
昭和13. 6.28~ 7. 5 1938	豪雨 (神戸大水害) Torrential Rains	925
昭和18. 9.10 1943	鳥取地震 (M7.2) Tottori Earthquake	1,083
昭和19.12. 7 1944	東南海地震 (M7.9) Tonankai Earthquake	1,223
昭和20. 1.13 1945	三河地震 (M6.8) Mikawa Earthquake	2,306
昭和20. 9. 17~18 1945	枕崎台風 (広島、西日本) Typhoon Makurazaki	3,756
昭和21.12.21 1946	南海地震 (M8.0) Nankai Earthquake	1,443
昭和22. 8.14 1947	浅間山噴火 Mt. Asama Eruption	11
昭和22. 9.14~15 1947	カスリーン台風 (東海以北) Typhoon Catherine	1,930
昭和23. 6.28 1948	福井地震 (M7.1) Fukui Earthquake	3,769
昭和23. 9.15~17 1948	アイオン台風 (東北等) Typhoon Ion	838
昭和25. 9. 2~ 4 1950	ジェーン台風 (四国以北) Typhoon Jane	539
昭和26.10.13~15 1951	ルース台風 (中国、九州等) Typhoon Ruth	943
昭和27. 3. 4 1952	十勝沖地震 (M8.2) Tokachi-oki Earthquake	33
昭和28. 6.25~29 1953	大雨 (前線：九州、四国、中国) Torrential Rains	1,013
昭和28. 7.16~24 1953	南紀豪雨 (和歌山、東北以西) Torrential Rains	1,124
昭和29. 5. 8~12 1954	風害 (低気圧：北日本、近畿) Storm damage	670
昭和29. 9.25~27 1954	洞爺丸台風 Typhoon Toyamaru	1,761
昭和32. 7.25~28 1957	諫早豪雨 Torrential Rains	722
昭和33. 6.24 1958	阿蘇山噴火 Mt. Aso Eruption	12
昭和33. 9.26~28 1958	狩野川台風 Typhoon Kanogawa	1,269
昭和34. 9.26~27 1959	伊勢湾台風 Typhoon Ise-wan	5,098
昭和35. 5.23 1960	チリ地震津波 Chile Earthquake Tsunami	142
昭和38. 1. 1963	※雪害 (北陸) Heavy Snowfall	231
昭和39. 6.16 1964	※新潟地震 (M7.5) Niigata Earthquake	26
昭和40. 9.10~18 1965	※台風第23、24、25号 (徳島、兵庫、福井等) Typhoons 23, 24, 25	181
昭和41. 9.23~25 1966	※台風第24、26号 (静岡、山梨等) Typhoons 24,26	317
昭和42. 7. ~ 8. 1967	※7、8月豪雨 (中部以西、東北) Torrential Rains	256
昭和43. 5.16 1968	※十勝沖地震 (M7.9) Tokachi-oki Earthquake	52
昭和47. 7. 3~15 1972	※台風第6、7、9号及び7月豪雨 Typhoons 6,7,9 and Torrential Rains	447
昭和49. 5. 9 1974	伊豆半島沖地震 (M6.9) Izu-hanto-oki Earthquake	30
昭和51. 9. 8~14 1976	※台風第17号及び9月豪雨 (香川、岡山等) Typhoon 17 and Torrential Rains	171
昭和52. 1. 1977	雪害 (東北、近畿北部、北陸) Heavy Snowfall	101
昭和52. 8. 7~53.10. 1977	※有珠山噴火 Mt. Usu Eruption	3
昭和53. 1.14 1978	※伊豆大島近海地震 (M7.0) Izu-Oshima-kinkai Earthquake	25
昭和53. 6.12 1978	※宮城県沖地震 (M7.4) Miyagi-ken-oki Earthquake	28
昭和54.10.17~20 1979	※台風第20号 (東海、関東、東北) Typhoon 20	115
昭和55.12.~56. 3. 1980	雪害 (東北、北陸) Heavy Snowfall	152
昭和57. 7. ~ 8. 1982	※7、8月豪雨及び台風第10号 Torrential Rains and Typhoon 10	439
昭和58. 5.26 1983	※日本海中部地震 (M7.7) Nihon-kai-chubu Earthquake	104
昭和58. 7.20~ 29 1983	※梅雨前線豪雨 (山陰以東) Torrential Rains	117
昭和58.10.3 1983	※三宅島噴火 Miyake Is. Eruption	—
昭和58.12.~59.3. 1983	雪害 (東北、北陸) Heavy Snowfall	131
昭和59. 9.14 1984	※長野県西部地震 (M6.8) Nagano-ken-seibu Earthquake	29
昭和61.11.15~12.18 1986	伊豆大島噴火 Izu-Oshima Is. Eruption	—
平成 2.11.17~ 1990	※雲仙岳噴火 Mt. Unzen Eruption	44
平成 5. 7.12 1993	※北海道南西沖地震 (M7.8) Hokkaido-nansei-oki Earthquake	230
平成 5. 7.31~ 8. 7 1993	※平成5年8月豪雨 (全国) Torrential Rains	79
平成7. 1.17 1995	※阪神・淡路大震災 (M7.3) Great Hanshin-Awaji Earthquake	6,437
平成12. 3.31~13. 6.28 2000	※有珠山噴火 Mt. Usu Eruption	—
平成12. 6.25~17. 3.31 2000	※三宅島噴火及び新島、神津島近海地震 Miyake Is. Eruption and Niijima and Kozushima Is. Earthquake	1
平成16. 10.18~21 2004	※台風第23号 (近畿、四国地方) Typhoon 23	98
平成16. 10.23 2004	※平成16年 (2004年) 新潟県中越地震 (M6.8) Niigata-ken-Chuetsu Earthquake	67
平成17. 12~18. 3 2005	平成18年豪雪 (北陸、日本海側) Heavy Snowfall	152

注) 風水害は死者・行方不明者500人以上、地震・津波・火山噴火は死者・行方不明者10人以上のもののほか、災害対策基本法による非常災害対策本部が設置されたもの (※) 等。

資料：気象年鑑、理科年表による。死者・行方不明者については消防庁の調べによる。平成16年新潟県中越地震については平成18年6月8日現在の数値。

Note : Regarding the damage caused by storms and floods, data is included for disasters in which 500 or more people were killed or reported missing. Regarding the damage caused by earthquakes, tsunamis or volcanic eruptions, data is included for disasters in which 10 or more people were killed or reported missing. The data also includes disasters if a Major Disaster Management Headquarters was established based on the Disaster Countermeasures Basic Act.

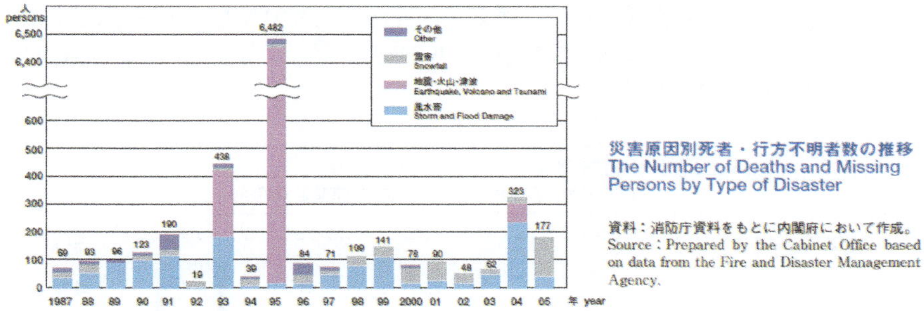

Fig. 8.7 The Number of Deaths and Missing Persons by Type of Disaster

Nevertheless, in spite of such efforts, more than 6,400 people became casualties of the 1995 Great Hanshin-Awaji Earthquake and the ten typhoons in 2004 caused damage throughout the nation. There is, moreover, a high probability of the occurrence of large-scale earthquakes in the coming decades. In consequence, natural disasters remain a menacing threat to the safety and security of the country.

Japan is prone to a variety of disasters including flooding, landslides, tidal waves and storm hazards. These arise from meteorological conditions such as typhoons and active weather-front systems, geographical conditions such as precipitous terrain and steep rivers, and social conditions where many of the cities are built on flood plains. There is an increasing tendency for downpours to occur throughout the country, as a long-term trend (Fig. 8.8). One-half of the population is concentrated in possible inundation areas which account for about 10% of the national landmass. Although there has been a large reduction in the area inundated by floods as a result of soil conservation and flood control projects undertaken over many years, the value of general assets damaged in flooded areas has increased rapidly in recent years (Fig. 8.9).

Disasters Led to Improvements in the Early Warning System

As mentioned earlier, the disaster management system in Japan has been developed and strengthened following the bitter experiences of large-scale natural disasters and accidents. The 1959 Ise-wan Typhoon and the Great Hanshin-Awaji Earthquake were the two specific disaster events which led to significant improvements in the Early Warning System in Japan, as discussed in greater detail in later sections of this report.

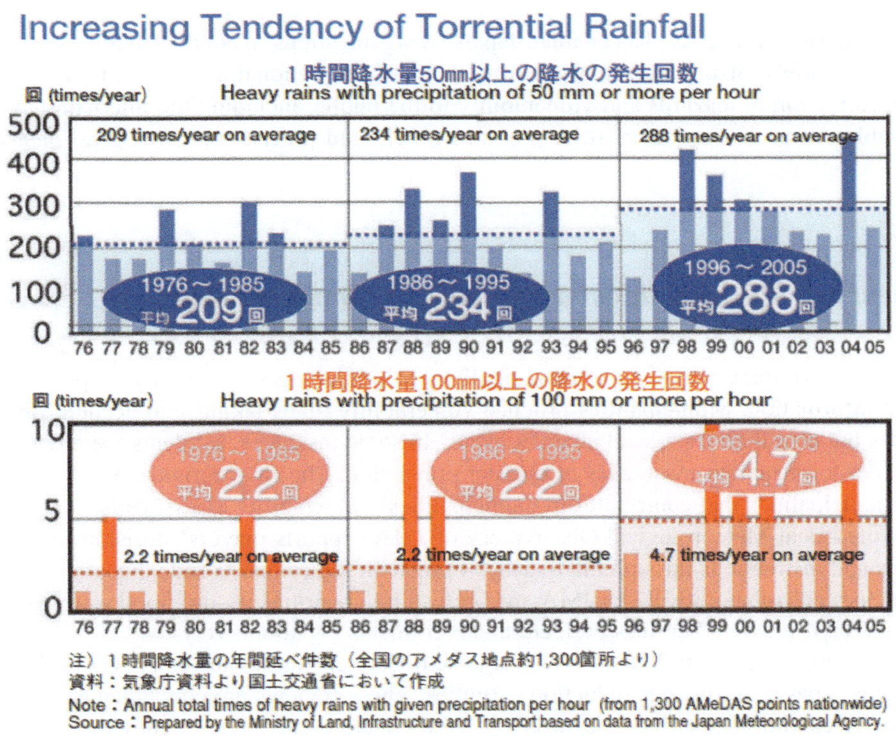

Fig. 8.8 The Increasing Tendency of Torrential Rainfall

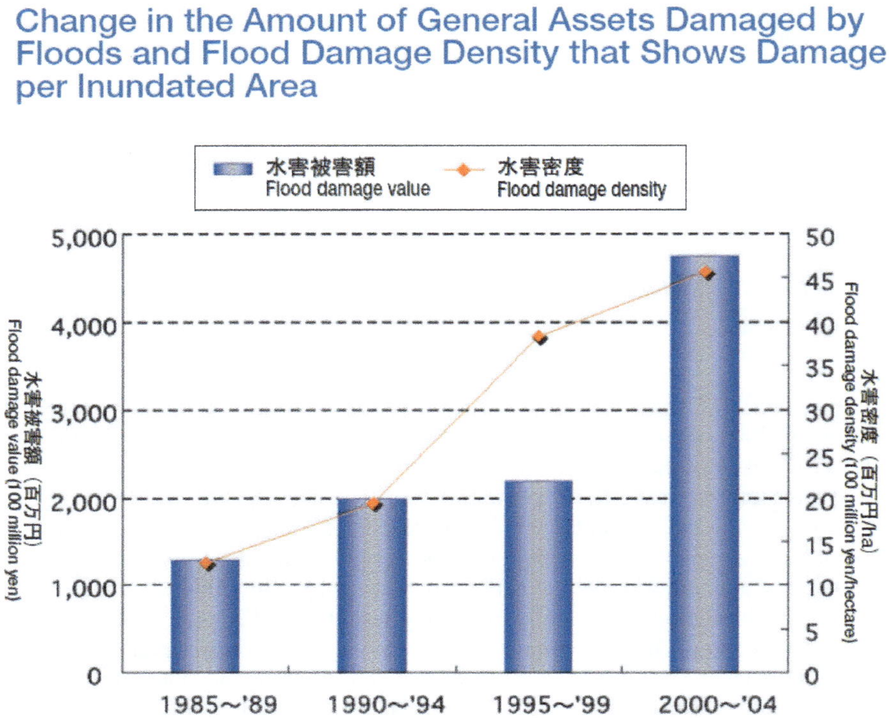

注）水害密度：水害面積（水害による「宅地その他」の浸水面積）当たりの一般資産被害額
資料：国土交通省河川局「水害統計」より内閣府作成
Note：Flood disaster density: General assets damage value per inundated area.
Source：Prepared by the Cabinet Office using the "Flood Disaster Statistics" of the Ministry of Land, Infrastructure and Transport.

Fig. 8.9 Changes in the Amount of General Assets Damaged by Floods and Flood Damage Density (i.e. damage per unit area)

The Dissemination and Sharing of Risk Information

The Disaster Countermeasures Basic Act has enabled Japanese organizations to conduct synthesized and systematic disaster countermeasures. A variety of organizations, such as those of the national government and research institutes, disseminate data and information on hazards and vulnerability to the public, including the international community, through web sites and by publications. In particular, JMA provides timely and precise meteorological data as well as historical data and statistics.

8.3.3 *Assessment of Vulnerabilities and Exposure (National to Local)*

The relevant Ministries encourage and support municipalities in producing vulnerability maps through provision of related guidelines and information. Municipalities produce vulnerability maps taking into account not only the ministries' advice and guidelines but also their own local characteristics. In some instances, residents are advised to assist in producing these maps in order to raise their awareness of disaster prevention. The JMA provides timely and precise information on meteorology to municipalities that contributes to municipalities' vulnerability assessments.

As noted earlier, the Local Meteorological Observatory of JMA supports mayors' decision-making and dissemination for evacuation orders and instructions through the issuance of meteorological alerts and advisories. Sediment disaster alerts for municipalities are issued collaboratively by JMA and prefectural governments and flood warnings for designated rivers are issued by JMA and Hydrological Services/prefectural governments collaboratively.

As also mentioned earlier, the Central Disaster Management Council formulates and reviews the Basic Disaster Management Plan, which is a basis for disaster reduction activities. These reviews are conducted each year in the light of research findings, disaster conditions that have occurred, and the effectiveness of emergency measures that have bee taken. The Plan is revised if deemed necessary.

8.3.4 Storage and Accessibility of Disaster and National Hazard Risk Information

A variety of organizations, such as those of the national government and research institutes, disseminate data and information on hazards and vulnerability to the public and the international community through web sites and by publications. (Websites introduced in this section are all Japanese only.) As an example, the Ministry of Land, Infrastructure, Transport and Tourism (MLIT) centralizes various types of information for disaster prevention that are held by organizations within MLIT, including the River Bureau, Road Bureau and JMA. MLIT provides such information to disaster management organizations and to the public in an easy-to-understand format through an "Information Service Center for Disaster Prevention (ISCDP)" web site. The disaster information available on the ISCDP includes hazard information on rivers and sediment provided by the River Bureau and real-time rainfall information based on observations by the River Bureau, Road Bureau and JMA.

http://www.mlit.go.jp/saigai/bosaijoho/index.html

The following are some other websites which provide citizens with disaster and natural hazard risk information:

Hazard Map Portal by MLIT – This site provides hazard maps for floods, high tides, tsunamis, sediment disasters and volcanic disasters, as well as maps for earthquake disaster prevention made by cities in Japan:

http://www1.gsi.go.jp/geowww/disapotal/index.html

Natural hazards database "SAIGAI" – The Natural Disaster Science Data Center in each district collects materials on natural disasters and provides them in a unique format mainly for scientists in Japan:

http://maple.dpri.kyoto-u.ac.jp/saigai/info.html

Typhoon hazards database system – The National Research Institute for Earth Science and Disaster Prevention provides information on typhoon disasters since 1951 on this website:

http://ccwd05.bosai.go.jp/DTD/search_jsp/login.jsp

Disaster legend information database – The Fire and Disaster Management Agency collects lessons and legends for past disasters and provides them on this website:

http://www.saigaidensho.soumu.go.jp/

The Japan Meteorological Agency stores its historical observational data, including precipitation data, and provides this observational data to disaster management organizations and the public by various means, including through its website. JMA devotes a special page to disastrous natural phenomena such as typhoons, earthquakes, tsunamis and volcanic eruptions on its website. This page includes annual reports on natural hazards and actions taken by JMA in support of disaster countermeasures by relevant organizations:

http://www.jma.go.jp/jma/kishou/know/saigai_link.html

8.3.5 Development and Utilization of Hazard/Risk Information to Support Emergency Planning and Warnings

JMA provides hazard/risk information and contributes to the production of hazard/risk information by other agencies in order to support emergency planning. Hazard risk information obtained through early warning systems is utilized in supporting the early evacuation of residents and the response activities of disaster management organizations. JMA operates 24-hourly systems to monitor various natural phenomena and weather conditions and it issues a wide range of forecasts and advisories regarding earthquake-generated tsunami and severe weather events such as heavy rain. JMA incorporates the results of monitoring and forecasting into warning messages including, for example, sediment disaster alerts for municipalities, collaboratively issued by JMA and prefectural governments, and flood warnings for designated rivers, collaboratively issued by JMA and Hydrological Services/prefectural governments. Additional details are provided in later sections.

8.4 Hazard Monitoring, Forecasting, and Mandates for Warning Development

8.4.1 Organizational Responsibilities for Monitoring, Forecasting and Development of Hazard Warnings

Figure 8.10 illustrates the position of JMA in the operational disaster prevention system in Japan. JMA has the mandate for monitoring, forecasting and development of warnings for various kinds of high risk natural hazards such as earthquakes

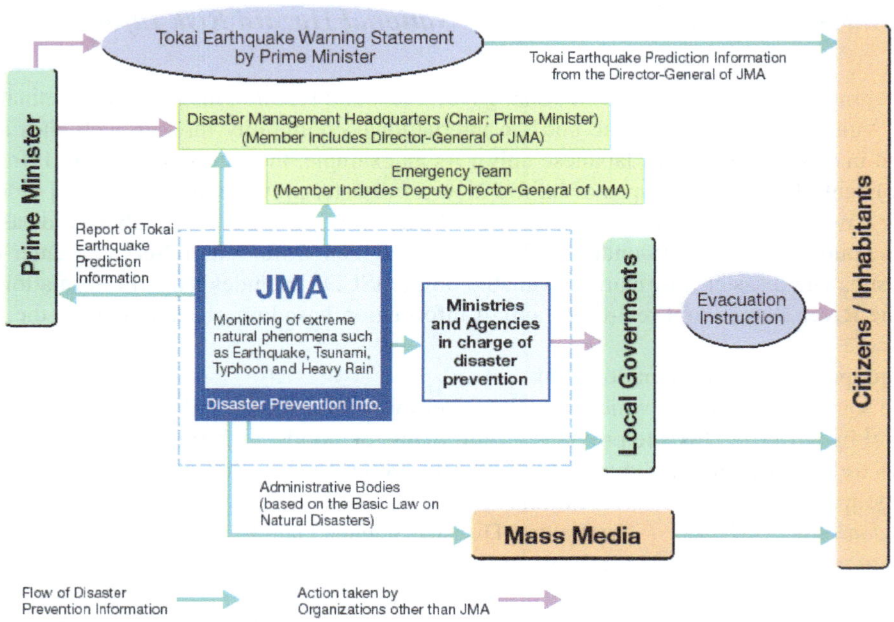

Fig. 8.10 Disaster prevention operation schemes (JMA Brochure 2006)

and severe weather conditions, including heavy rain, heavy snow and storm surge. JMA also has responsibility for issuing flood warnings, in collaboration with the River Bureau of MLIT or prefectural governments. Details of JMA's responsibilities for the development of hazard warnings are as follows:

Type I hazards – JMA has sole mandate for the development of warnings for the hazard.

- According to Article 13, paragraph (1) of the Meteorological Service Act (Act No. 165 of 1952), JMA "shall, pursuant to the provisions of Cabinet Order, give suitable forecasts and warnings for general use concerning meteorological phenomena, terrestrial phenomena (in the case of earthquakes, limited to earthquake ground motions; hereinafter, the same shall apply in this report), tsunamis, storm surges, high waves, and floods";
- According to Article 14, paragraph (1) of the Meteorological Service Act, JMA "shall, pursuant to the provisions of Cabinet Order, give suitable forecasts and warnings for use for aircraft and vessels concerning meteorological phenomena, terrestrial phenomena, tsunamis, storm surges, and high waves";
- According to Article 14-2, paragraph (1) of the Meteorological Service Act, JMA "shall, pursuant to the provisions of Cabinet Order, give suitable forecasts and warnings for use in flood prevention activities concerning meteorological phenomena, storm surges, and floods".
- In addition to other requirements arising from the preceding Articles, JMA issues seven types of meteorological warnings and sixteen types of meteorological advisories, as illustrated in Table 8.4. An overall listing of hazards for which warnings are issued in Japan is shown in Table 8.5.

Type II hazards – JMA has a joint (or shared) mandate for the development of warnings for the hazard.

- According to Article 14-2, paragraph (2) of the Meteorological Service Act, with respect to the rivers designated pursuant to the provision of Article 10, paragraph (2) of the Flood Prevention Act (Act No. 193 of 1949), JMA "shall, jointly with the Minister of Land, Infrastructure, Transport and Tourism who administers the affairs concerning flood prevention, give suitable forecasts and warnings of floods for use in flood prevention activities by indicating the water levels and flow volumes of the said rivers (or after overflow of rivers, water levels or flow volumes, or areas flooded by overflow and water depths therein)".
- According to Article 14-2, paragraph (3) of the Meteorological Service Act, with respect to the rivers designated pursuant to the provision of Article 11, paragraph (1) of the Flood Prevention Act, JMA "shall, jointly with prefectural governors, give suitable forecasts and warnings of floods for use in flood prevention activities by indicating the water levels or flow volumes thereof".

Type III hazards – JMA Provides information to other agencies that have the mandates for the development of warnings for the hazard.

In Japan, all warnings for hazards are issued by JMA either solely or jointly with other authorities.

Table 8.4 Meteorological Warnings and Advisories of JMA

Meteorological Warnings and Advisories of JMA

Warnings (7)	Advisories (16)	Disasters
Heavy Rain	Heavy Rain	Sediment Disaster; Inundation of houses, load, farming land etc.
Flood	Flood	Flood; Dike break; Washout of bank; Inundation etc. except rivers designated for joint flood forecast
Storm Surge	Storm Surge	Inundation; Damage of ships, shore facilities including ones for fishery etc.
Storm	Gale	Damage of houses, buildings, agricultural crops etc.; Maritime disaster
High Wave	High Wave	Maritime disaster; Inundation; Damage of ships, shore facilities including ones for fishery etc.
Heavy Snow	Heavy Snow	Traffic disturbance; Damage of houses etc.
Snow-storm	Snow-storm	Traffic disturbance; Damage of houses, buildings etc.; Maritime disaster
	Thunderstorm	Disasters caused by Thunderbolt, hail, gust (tornado, downburst)
	Dense Fog	Traffic disturbance, Maritime disaster
	Frost	Damage of agricultural crops in early/late winter
	Dry Air	Fire disaster; Forest fire
	Avalanche	Avalanche except steep mountain areas
	Low Temperature	Water pipe freeze-up and break; Damage of agricultural crops
	Snow-melting	Flood, Sediment disaster, Inundation
	Ice Accretion	Significant accretion of ice on electric power cable, communication wire, vessel's body
	Snow Accretion	Significant accretion of snow on electric power cable, communication wire

Table 8.5 Hazards for which warnings are issued in Japan

Hazard	Type	Authorities in collaboration with JMA to issue warnings
Heavy rain	I	
Flood	I	
River flooding	II	MLIT (national-managed rivers), prefectural governments (prefectural-managed rivers)
Landslide	II	prefectural government
Tropical cyclone		
Storm surge	I	
Strong winds	I	
High wave	I	
Heavy snow	I	
Snow storm	I	
Earthquakes	I	
Tsunami	I	
Volcanic events	I	

8.4.2 Organizational Collaboration and Coordination for Development of Hazard Warnings

JMA issues flood warnings in collaboration with the Hydrological Service (Fig. 8.11). JMA collects weather information and predicts the weather situation while the Hydrological Service has its own observation network of rain gauges and water level recorders for keeping an eye on the river situation. The information from the two organizations is combined and delivered to users in the form of river-specific flood warnings. Until 2001, the River Bureau was located within the Ministry of Construction while JMA was an external organ attached to the Ministry of Transport and, in consequence, establishing a good partnership was a challenge. However, the two Ministries have been combined to form the Ministry of Land, Infrastructure, Transport and Tourism (MLIT). This Government reform, combined with the efforts made by both agencies, has now resulted in the collaboration on flood warnings being one of the best practices in establishing partnerships between different governmental organizations. Furthermore, JMA has extended the range of rivers for which flood warnings are issued by including medium to small rivers that are managed by local governments. Consequently, collaborative flood warnings are now issued for about 350 rivers (252 rivers jointly with MLIT; 93 rivers jointly with prefectures). JMA devotes continuing efforts to making the products understandable to the end users and has also promoted the exchange of the related data with relevant organizations and users.

A partnership has also been established with prefectures for issue of sediment disaster alerts (Fig. 8.12). This partnership is very similar to that with hydrological services for flood warnings. Prefectures have their local networks of precipitation measurements and topographical and geological information including hazard maps. By combining these sources, JMA and prefectures issue collaborative sediment disaster alerts that specify cities, towns and villages affected by the alerts.

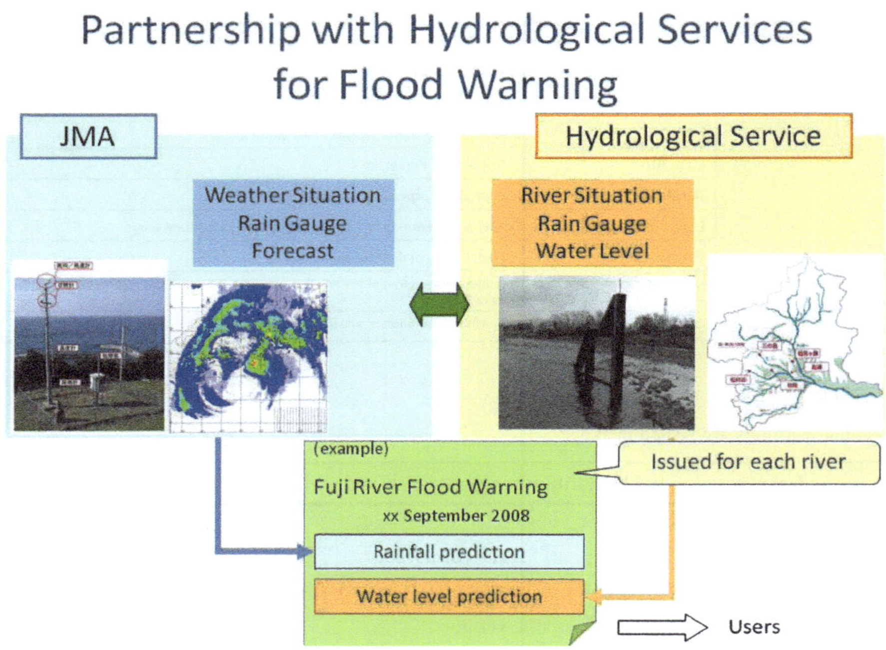

Fig. 8.11 Partnership with Hydrological Services for flood warning

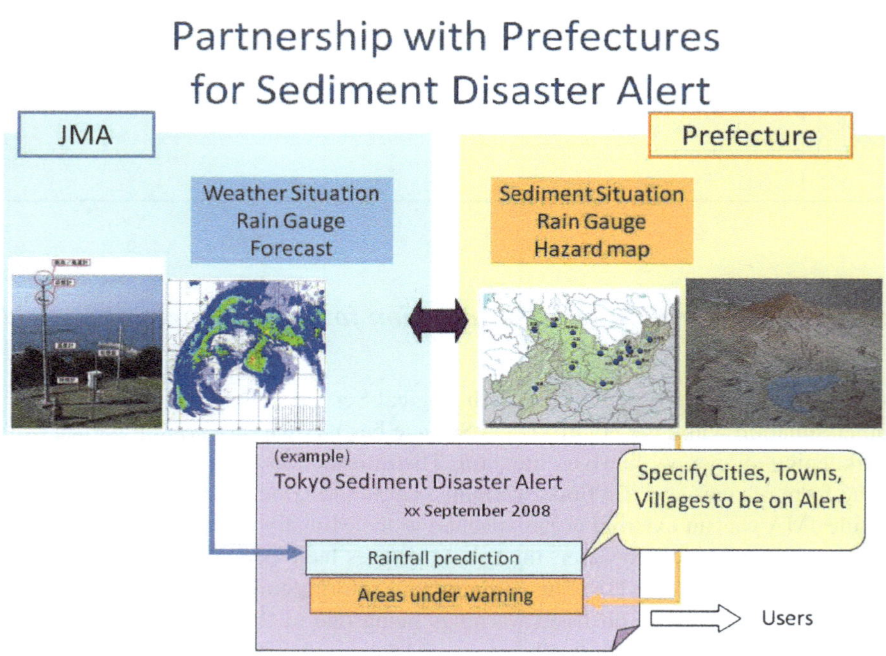

Fig. 8.12 Partnership with prefectures for sediment disaster alert

8.5 Development of Understandable, Authoritative, Recognizable and Timely Warnings

8.5.1 Warning Message Development Cycle

In Japan, warning and response cycles differ significantly between different types of disasters, as illustrated in the following examples.

Countermeasures Against Tokai Earthquake

The strain on the earth's crust along the Suruga Trough has been building up for approximately 150 years, ever since the 1854 Ansei-Tokai Earthquake. As a result, it is believed that there is a high possibility of a Tokai Earthquake occurring. The Tokai Earthquake is, at present, the only earthquake with a possibility of being predicted just before it occurs. Areas for intensified measures against the earthquake were designated under the Act on Special Measures for Large-scale Earthquakes. In these areas, the observation system has been reinforced and an earthquake response system has been developed in the event of a prediction report being announced. Data that are believed to be effective for the earthquake prediction are monitored in real-time by JMA. Upon the detection of any abnormality in the data, the observation, caution and prediction information will be announced, based on the level of the abnormality. The Prime Minister will then issue a Warning Declaration based on the prediction report and implement the required measures, including the establishment of the Earthquake Disaster Warning Headquarters (Fig. 8.13).

The Central Disaster Management Council drew up the Earthquake Countermeasures Basic Plan containing basic policies for actions to be taken in response to a Warning Declaration based on the Act. In addition, relevant organizations have their own corresponding plans. When local governments carry out urgent projects, based on their own plans, to improve facilities for mitigating possible damage resulting from the Tokai Earthquake, special measures will be taken such as increasing national government subsidies and fiscal support for the local governments. These measures are based on the Act on Special Financial Measures for Urgent Earthquake Countermeasure Improvement Projects in Areas for Intensified Measures.

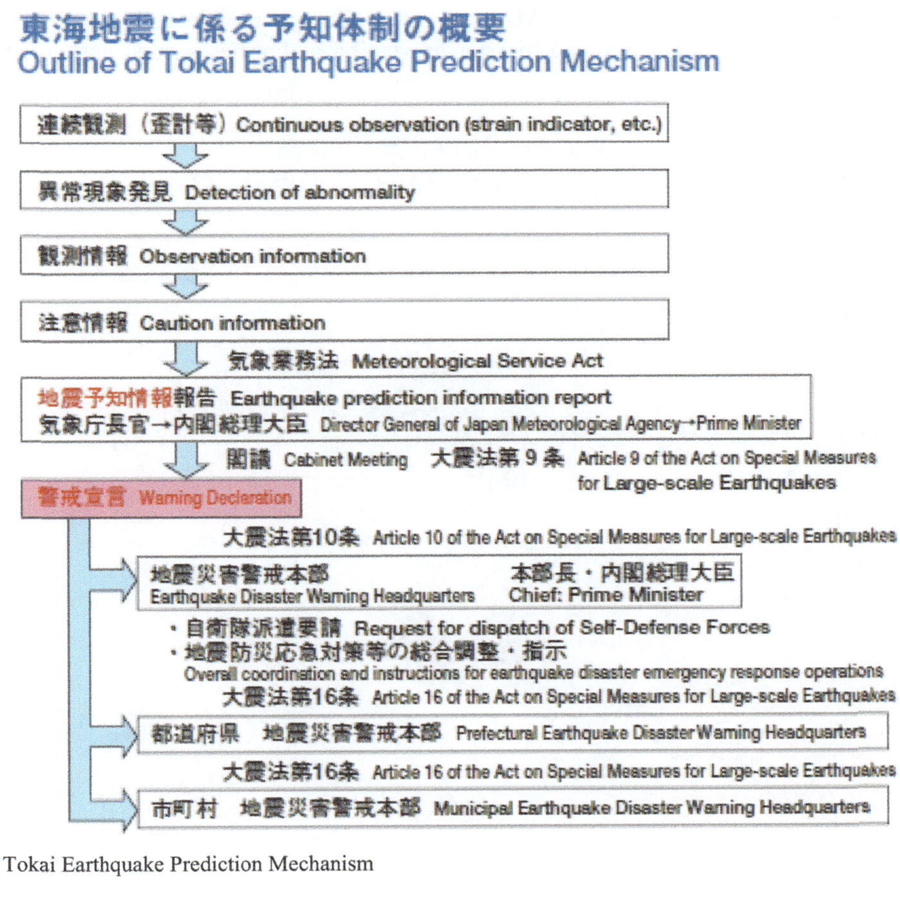

Fig. 8.13 Outline of Tokai Earthquake Prediction Mechanism

Tsunami Countermeasures

Being surrounded by water on all sides and with long and complex coastlines, Japan is highly vulnerable to earthquake-generated tsunamis. In the past, severe damage has been caused by various tsunamis including the Meiji-Sanriku Earthquake Tsunami (1896), Nihon-kai-Chubu Earthquake (1983), and Hokkaido Nansei-oki Earthquake (1993) Tsunamis. In addition to local tsunamis generated by earthquakes near the coast, Japan has also suffered major damage from the onslaught of distant tsunamis generated by open-sea earthquakes. In 1960, for example, a tsunami generated by the Chile Earthquake crossed the Pacific Ocean and reached the shores of Japan about 22 hours later, killing 142 people. When a tsunami is expected to cause coastal damage, JMA issues a tsunami warning or advisory within two to three minutes after the earthquake and then follows up with announcements concerning the estimated height and arrival time of the tsunami. This information is transmitted immediately to disaster management organizations and media outlets and forwarded onward to residents and maritime vessels.

Tsunami countermeasures have been undertaken such as expediting the announcement/transmission of tsunami forecasts and improving coastal embankments, tidal embankments and tide prevention gates. Moreover, the Cabinet Office, in cooperation with relevant ministries, has prepared guidelines for the creation of a tsunami hazard map (Fig. 8.14) and the designation/development of tsunami evacuation buildings by local governments and is working on disseminating these guidelines.

Countermeasures Against Volcanic Eruptions

On December 1, 2007, JMA began issuing Volcanic Warnings and Volcanic Forecasts (Fig. 8.15) for each active volcano in Japan in order to mitigate damage from volcanic activities. Volcanic warnings are issued for expected volcanic disasters and specify municipalities where people need to take actions. Volcanic Forecasts are issued for less active volcanoes or those that become less active.

Warnings and Responses to Weather Disasters

When a hazardous weather condition is expected, JMA delivers various plain language messages including Warnings, Advisories and Bulletins to the general public and disaster prevention authorities so that appropriate measures can be

Fig. 8.14 Example of a Tsunami Hazard Map (Suzaki City, Kochi Prefecture)

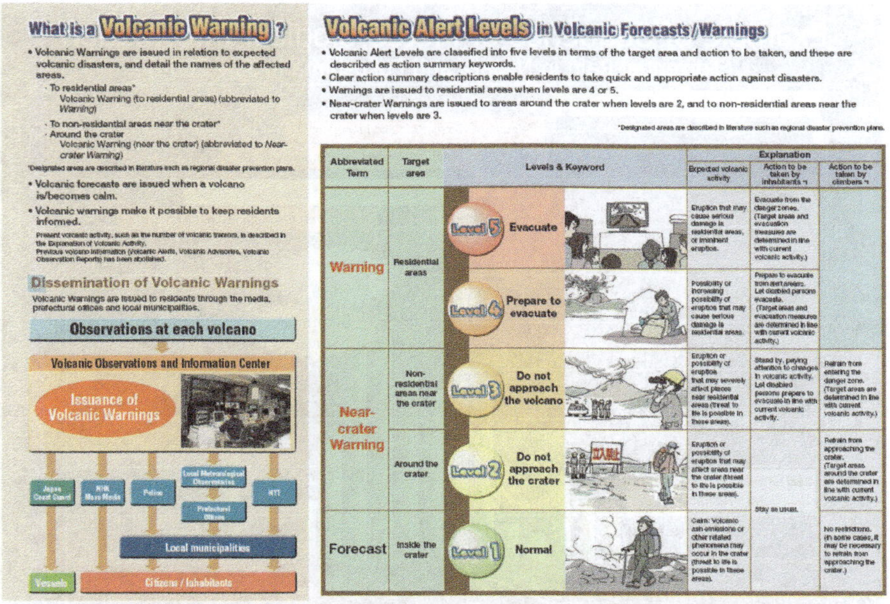

Volcanic Warnings/Forecasts for volcanoes where Volcanic Alert Levels are not applied.

Abbreviated Term	Target area	Explanation
Warning	Residential areas	Warnings in residential areas and non-residential areas near the crater
Near-crater Warning	Non-residential areas near the crater	Warnings in non-residential areas near the crater
	Around the crater	Warnings around the crater
Forecast	Inside the crater	Calm (normal conditions)

*Designated areas are described in literature such as regional disaster prevention plans.

Fig. 8.15 Explanation of Volcanic Warning/Forecast and Volcanic Alert Level

taken to mitigate possible hazards. Warnings are issued when weather conditions are expected to be catastrophic and meet warning criteria. Advisories are issued when weather conditions meet advisory criteria but are expected to remain below the warning criteria. Bulletins provide information to supplement the Warnings and Advisories. These messages are issued by Local Meteorological Observatories (LMOs) for each of the sub-divisions in their respective prefectures. As of March 2006, a total of 370 sub-divisions were designated across the country, taking into consideration their meteorological characteristics and local administrative boundaries.

Bulletins are also issued from District Meteorological Observatories (DMOs) or JMA Headquarters depending on the scale or severity of the disturbances of concern. They include information not only for severe weather but also for other possible disasters. Sediment disaster bulletins, as an example, are issued jointly with municipalities, in particular when sediment-related disasters due to heavy rain are expected, with high probability, in a few hours. All such information is delivered directly to disaster prevention authorities from JMA Headquarters and/or DMOs and to the public through the mass media. Figure 8.16 illustrates the sequencing of JMA's weather information issued immediately before and during the torrential rain in Niigata and Fukushima in July 2004.

Responsible Organization for the Development of Warnings

JMA is responsible for the development of Warnings in relation to weather (i.e. storm, snow-storm, heavy rain, heavy snow, storm surge, high waves and flood), volcanoes, tsunamis and earthquakes. In addition, JMA issues early warning information for extreme weather (e.g. extremely high or low seven-day average temperature). As mentioned earlier, JMA also has a joint or shared responsibility for issuing River Flood Warnings, in collaboration with Hydrological Services at national and municipal levels, and for Sediment Disaster Alerts, with sediment control authorities at a municipal level.

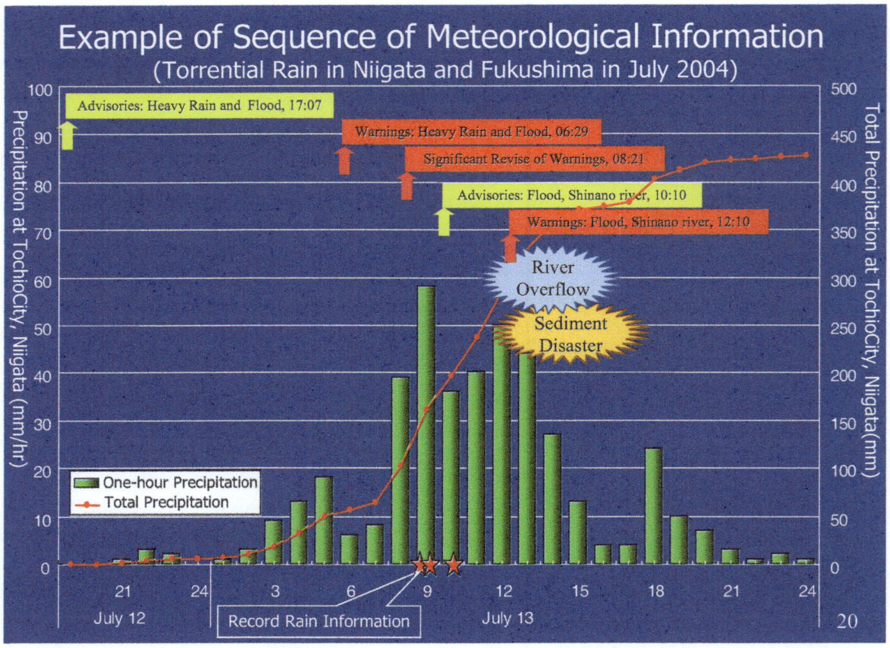

Fig. 8.16 Example of the Sequence of Meteorological Information (Torrential Rain in Niigata and Fukushima in July 2004)

Disaster Reduction Drills and Exercises

Disaster reduction drills and exercises represent good opportunities to review the effectiveness of the disaster management system in delivering rapid and appropriate emergency operations and to enhance public awareness through widespread participation. The Disaster Countermeasures Basic Act stipulates that disaster reduction drills should be undertaken. In order to promote various drills and exercises nationwide, the Central Disaster Management Council prepares a "Comprehensive Disaster Reduction Drills Plan" each year, which outlines the comprehensive disaster reduction drills to be carried out by the national government in cooperation with local governments and relevant organizations and stipulates the basic principles for executing the drills.

On September 1st, Disaster Reduction Day, wide-area, large-scale disaster reduction drills are conducted in every region across the country, in collaboration with disaster related organizations. Furthermore, drills based on the experiences during past disasters are conducted in every region throughout the year. In recent years, practical disaster reduction drill methods such as role-playing simulation systems have been introduced in which participants are not given any information in advance and are required to make decisions and respond to the situation based upon the information provided after the drill starts.

As pointed out earlier, the Central Disaster Management Council has a responsibility for overall coordination regarding disaster prevention programs among JMA and other disaster risk management agencies. In addition, JMA collaborates with relevant agencies at municipal levels to ensure that their needs are reflected in planning meteorological services.

Capacity Building and Training Programs

In many instances, JMA undertakes capacity building and training programs involving stakeholders and the public before and after it commences the issue of new disaster-related information. By doing so, JMA ensures that the recipients understand the new information correctly and take necessary actions. Moreover, as follow-up, JMA assesses the effectiveness of the new information in order to determine whether any improvements to the existing services are required. This is a form of PDCA-cycle ("Plan-Do-Check-Act") based service management, meaning that service providers can improve information step by step, taking into account feedback from the users.

Earthquake Early Warning (EEW) Service

A good illustration of the above approach is provided by a series of programs undertaken in relation to the Earthquake Early Warning (EEW) service which commenced in October 2007. As this is a quite innovative Alert, JMA, in collaboration with other relevant authorities, has conducted many PR activities as follows:

- *Task Team on the Launch of the EEW Operation* – In November 2005, prior to the official launch of the EEW operation, JMA established a task team consisting of scholars and representatives of other authorities to obtain recommendations on the following topics:

  i. EEW provision for advanced users;
  ii. EEW provision for the general public;
  iii. Guidelines on the use of EEW;
  iv. Methods to raise public awareness of EEW.

  Following a series of discussion sessions, the team finalized its recommendations in March 2007.

- *Meetings of Representatives of Ministries Responsible for Public Relations Activities on EEW* – Since March 2007, JMA together with other ministries and agencies has held meetings to exchange views on how EEW can be effectively utilized in related fields as well as to establish necessary coordination on related government programs such as:

  i. Strengthening public relations activities on EEW;
  ii. More effective use of EEW.

- *Task Team on the Assessment and Improvement of EEW* – To further improve EEW in terms of its technology and operation, JMA established this task team in February 2009, involving scholars in relevant fields. The team is tasked with the following:

  i. Assessment of operation and information content;
  ii. Activities to raise public awareness of EEW;
  iii. Operational improvements to EEW in terms of issuance criteria, information content and delivery methods;
  iv. Technical improvements to EEW issuance.

- *Questionnaire surveys on EEW* – As part of annual satisfaction surveys in 2009, JMA conducted a questionnaire survey on EEW, targeting disaster-related organizations and residents nationwide. JMA also conducted a survey on the use of EEW on the occasion of the earthquake in Iwate and Miyagi in 2008. This survey targeted municipal governments in the region, schools and kindergartens that had installed EEW receivers, and private business offices that were reported by the media to utilize EEW. The survey found many good examples of effective uses of Earthquake Early Warnings.

Tornado Watch

Tornado Watch, which was launched by JMA in March 2008, represents another good example of interaction with user communities. The accuracy of tornado prediction is not very high even with the latest techniques. As the first step towards making the information as effective as possible, JMA needed to design the information product based on the users' requirements, in order to convey the risk clearly. JMA established an advisory committee for this purpose, composed of professors of meteorology, sociology and risk management and media people, and asked the committee for advice on information content and how to convey the risk. Opinions were also solicited from potential users such as construction firms, schools, and railway companies.

Following the committee's recommendation, JMA decided to name the information product "Tornado Watch", based on advice that the word "tornado" should be included because it attracts people's attention. It also decided to indicate that "This information remains in effect for one hour", based on the users' advice that an indication of the time period of validity helps users to understand exactly when they should pay attention. A further outcome of the discussions was the committee's encouragement to JMA to commence issuing the Tornado Watch despite the limited accuracy that could be achieved at that point in time. Here, the committee and JMA agreed that the information can save people's lives if it is properly used.

JMA developed a guideline on how to use the Tornado Watch, for practical application. The main purpose of this guideline is to outline the basic concept of "required action" and the scenario under which JMA's information on hazardous winds can be properly used. The basic concept of required action is quite simple. In view of the fact that about 90% of Tornado Watches are not followed by a real tornado or gust of wind, users are prompted to simply watch the sky and not to evacuate immediately when a Watch is issued. Users should take refuge only when there is a visible sign of a thunderstorm. All that users have to do, therefore, is pay more attention than usual to weather conditions. This approach reduces risk to human life, without needlessly taking refuge too frequently.

Raising public awareness about the information is undoubtedly a key consideration in preventing real damage. For this purpose, JMA published and distributed a brochure, which is shown in Fig. 8.17. This booklet introduces the Tornado

Fig. 8.17 Brochure on Tornado Watch

Watch, its meaning and required action, as well as providing basic knowledge about tornadoes. In addition, JMA delivered a number of lectures to media people, including weather persons, and local government staff in charge of disaster prevention and risk management. The mass media also presented some TV and radio programs to introduce this information and explain how to use it effectively. It is hoped that these initiatives help people to understand what the information is, and exactly what they should do to protect themselves when they receive a Tornado Watch.

8.5.2 Warning Message Improvement Cycle

JMA undertakes assessments of meteorological services with the aim of improving weather services, establishing efficient and quality meteorological administration, converting to result-based management, and achieving accountability to the general public. The following are some examples of such assessments.

Service Assessment Implementation Plan

Each year, JMA develops a service assessment implementation plan and conducts assessment based on this plan.

Program Assessment (Policy Review)

In this assessment, specific themes are selected in accordance with social and economic conditions and general and in-depth analyses/assessments are undertaken. This assessment enables the government to review relevant policies responsively and continuously, taking into account the achievement of goals and changes in social and economic conditions. The themes of this assessment are selected by MLIT each year. Examples of assessments related to JMA include:

- Enrichment of meteorological information – toward reducing disaster damage (March 2005);
- Countermeasures against volcanic eruptions – drawing on the experience of Usu and Miyake (March 2004).

Customer Satisfaction Surveys

Customer satisfaction surveys are carried out every year with the aim of understanding user needs for weather services, exploring future directions for service improvements and measuring the outcome of service improvements. Recent disaster-related topics surveyed were:

- Earthquake and volcano information for disaster prevention (FY2008);
- Weather information for disaster prevention (FY2006, 2004, 2001).

Advisory Meetings on Weather Services Assessment

JMA strives to improve its weather, earthquake, volcano and ocean services with the aim of contributing to disaster prevention and reduction, transportation safety and the development of the economy. The purpose of these advisory meetings is to receive opinions and advice from external experts including scholars, representatives of the media and telecommunication companies.

8.6 Warning Dissemination Mechanisms

The development of a rapid and accurate telecommunications system is essential for the effective use of early warning information. JMA has, therefore, implemented an on-line system linking JMA and various users including disaster management organizations of the national and local governments and media organizations. Figure 8.18 shows the outline of this system. Disaster management organizations have also been developing radio communications networks exclusively for disasters. The Central Disaster Management Radio Communications System (CDMRCS) connects national organizations (Fig. 8.19). The CDMRCS is composed of a telephone and facsimile hotline between designated government bodies and public corporations, an image transmission circuit for transmitting pictures of disaster situations from helicopters in real-time, and a satellite communications system as a backup. The Fire Disaster Management Radio Communications System connects fire fighting organizations across the country. Prefectural and municipal disaster management radio communications systems connect local disaster management organizations and residents. Local governments use simultaneous wireless communications systems including outdoor loudspeakers and indoor radio receivers to disseminate disaster information to residents. Tsunami and severe weather warnings are also widely disseminated to citizens via TV and radio broadcasts.

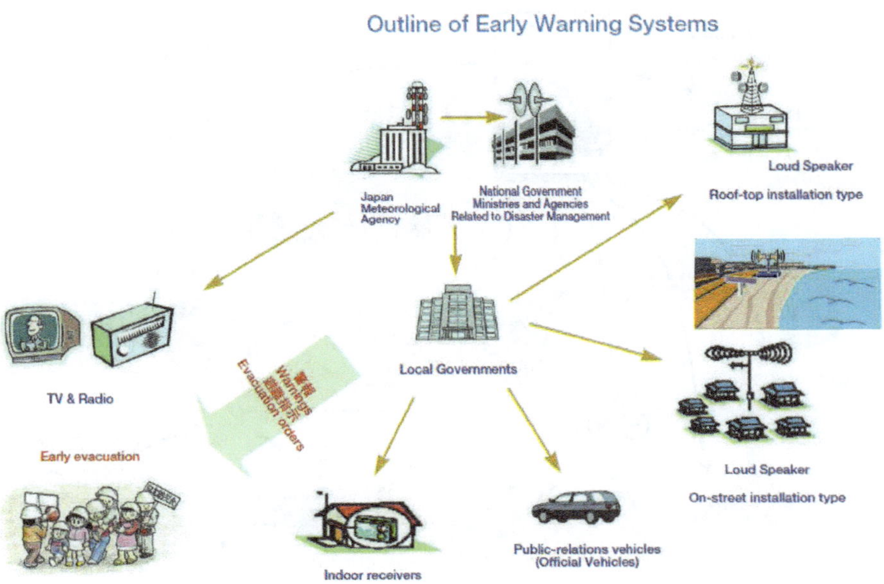

Fig. 8.18 Outline of Early Warning System

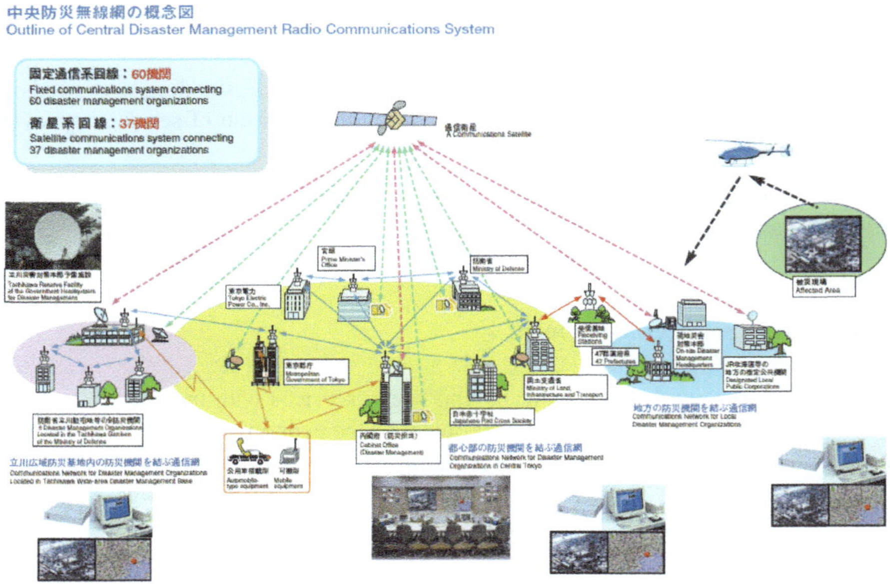

Fig. 8.19 Outline of the Central Disaster Management Radio Communications System (CDMRCS)

Within this dissemination framework, JMA provides disaster related information to national and local governmental bodies as well as the general public by the most appropriate means for each body (Fig. 8.20). Some national administrations and local governments are also responsible for issuing warnings and advisories, such as civil protection warnings issued by the Cabinet Office and photochemical smog warnings issued by each local government.

As indicated earlier, JMA delivers information to disaster-related departments of local governments using a dedicated communication system for disaster-related information. This enables JMA to deliver its products directly to the local governments and to support them, in the event of approaching typhoons, heavy downpours and earthquakes, by sending e-mail messages directly to each local government and by placing information on dedicated web pages prepared for individual users on the basis of their own preferences. JMA has further developed its warning products and is now able to issue warnings and advisories targeted to each city, town or village and tailored to suit the dedicated information system to each municipal government.

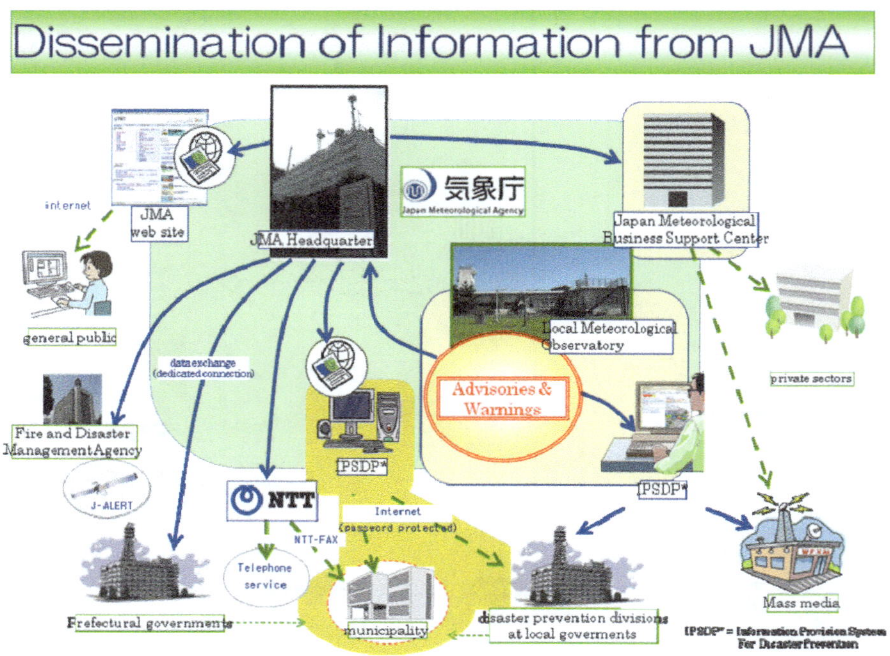

Fig. 8.20 Warning dissemination route from JMA

While its current targets are sub-prefecture zones, JMA plans to commence issuing weather warnings and advisories that target local governments, effective May 2010. One reason behind this change is administrative regulations specific to Japan. In Japan, heads of local governments such as cities, towns and villages have a legal responsibility to develop measures against the natural disasters, such as issuing evacuation directives and advisories. JMA, therefore, decided to target specific local governments when issuing warnings and advisories in order to strengthen the linkage between weather warnings and evacuation directives and advisories (Fig. 8.21).

In Tokyo, for example, warnings and advisories from JMA currently target five sub-prefecture zones. Beginning May 2010, the targeted recipients will increase to 53 individual local governments (Fig. 8.22). JMA anticipates that this change will make warnings more effective in supporting actions by local governments and can be accommodated by its dedicated communication system.

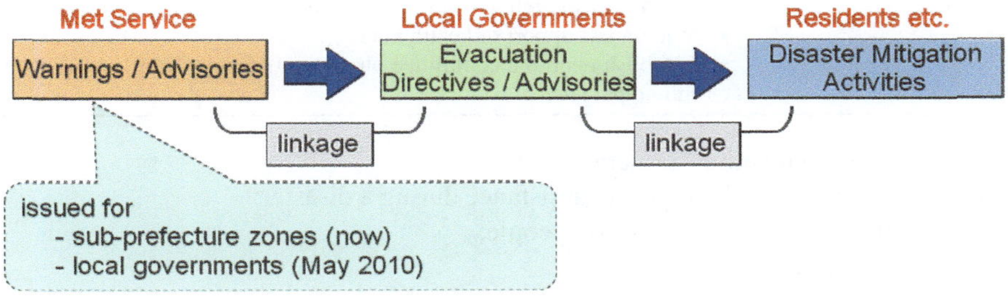

Fig. 8.21 Linkage between weather warnings and evacuation directives and advisories

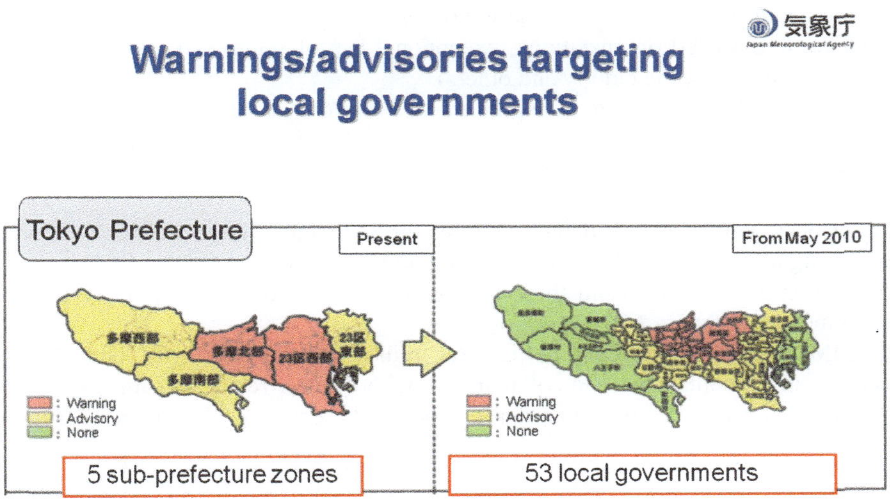

Fig. 8.22 Change in warnings/advisories targeting zones in Tokyo

8.7 Emergency Preparedness and Response Activities (National to Local)

8.7.1 Disaster Preparedness and Response Planning and Emergency Response Activation

Japan's basic Disaster Management Plan addresses comprehensive and long-term disaster reduction issues such as disaster management related systems, disaster reduction projects, early and appropriate disaster recovery and rehabilitation, and scientific and technical research. The Plan was revised entirely in 1995, based on the experiences of the Great Hanshin-Awaji Earthquake, and further modified as needed to meet the specific needs of vulnerable communities, authorities and emergency respondents. In view of Japan's aging society and the increasing number of the elderly being killed or injured by disasters, measures need to be reinforced to provide necessary assistance during disasters to those such as the elderly and physically impaired. The Cabinet Office in cooperation with relevant ministries published "Guidelines for Evacuation Support of People Requiring Assistance during a Disaster" in 2005 (revised in 2006), for implementation at the municipal level. As detailed in Table 8.6, these Guidelines address:

Table 8.6 Outline of Guidelines for Evacuation Support of People Requiring Assistance during a Disaster

Improving the information communications system	• Announcement of evacuation preparation information • Establishment of a support unit for people requiring assistance • Secure communications by making use of various means such as the Internet, emergency call message service, etc.
Sharing of Information concerning people requiring assistance during a disaster	• Collection and sharing of information on people requiring assistance in various ways • Promotion of exceptional use of social welfare-related personal information to prepare evacuation support systems
Creating a tangible evacuation support plan for people requiring assistance during a disaster	• Creation of an evacuation support plan for each individual requiring assistance • Recognition of the importance of making communities resilient to disasters
Assistance at evacuation centres	• Establishment of an information desk for people requiring assistance at evacuation centres • Establishment of welfare evacuation centres
Collaboration among related organizations	• Continuity of welfare services in disaster situation • Wide-area support of health nurses • Establishment of a committee on evacuation support of people requiring assistance at the municipal level

- Improving the information communications system;
- Sharing of information concerning people requiring assistance during a disaster;
- Creating a tangible evacuation support plan for those people;
- Assistance at evacuation centres; and
- Collaboration among related organizations.

Tangible countermeasures are included such as the issuance of evacuation preparation information that calls for the early evacuation of people requiring assistance and sharing of information on people requiring assistance between disaster management and social welfare-related organizations (exceptional use of social welfare-related personal information to prepare evacuation support systems for the elderly and others).

The Flood Control Act and other relevant Acts require the delivery of hazard maps to facilities used by people requiring assistance. The Ministry of Land, Infrastructure, Transport and Tourism (MLIT) launched the following web portal for hazard maps to assist these efforts to enhance preparedness: http://www1.gsi.go.jp/ geowww/disapotal/index.html (Japanese only)

In addition, the Act on Sediment Disaster Countermeasures for Sediment Disaster Prone Areas requires prefectural and city governments to investigate the geography, precipitation, land use and other relevant factors in vulnerable areas approximately every five years. Information from these investigations is needed to support appropriate preparedness and countermeasures for sediment disasters.

As indicated earlier, JMA also cooperates and participates in disaster reduction drills conducted by local governments and relevant organizations. In particular, JMA in collaboration with disaster related organizations conducts warning issuance drills in each region and across the country in order to expedite the issuance of Tsunami Warnings/Advisories.

8.7.2 Community Response Capacities

In May 2003, the Central Disaster Prevention Council of the Cabinet Office established an expert panel on the inheritance of lessons from past disasters. The purposes of the expert panel are to incorporate lessons from previous disasters and to improve emergency preparedness and public awareness through systematic collection of information on past catastrophic disasters such as the damage, governmental response and impacts on citizen's lives and social economy. Textbooks summarizing the information collected are published by the expert panel approximately every ten years and contribute to the inclusion of lessons from past disasters in compulsory education and lifelong learning. Disaster reduction drills and exercises are also a useful tool in developing effective community response capabilities and have been discussed previously.

When hazardous weather conditions are expected, the Japan Meteorological Agency (JMA), as outlined earlier, delivers a variety of plain language messages including Warnings, Advisories and Bulletins to the general public and disaster prevention authorities so that appropriate measures can be taken to mitigate possible hazards. JMA provides an automatic telephone answering service for the public and maintains direct communication links between meteorological offices and central/local governments and the media for the dissemination of warnings. The Agency is also active in introducing new information technologies into its meteorological information services such as, for example, the high-speed communications network known as the Information Network for Disaster Prevention (INDiP) that enables rapid and effective dissemination of information in both text and graphic forms.

8.7.3 Public Awareness and Education

Disaster education in schools is important in conveying vital disaster-related knowledge to people from childhood and this subject is, therefore, taught in various school curricula. Public education at the community level is also important, including town watching and hazard-mapping programs in which residents participate. The Cabinet Office promotes disaster-related education including sharing good examples of disaster education programs. The Ministry of Education, Culture, Sports, Science and Technology (MEXT) prepares and provides the educational tools for students which enable them to behave safely and appropriately under hazardous conditions. MEXT also holds "Safe School" promotion forums to enhance safety management in schools.

Government disseminates information relevant to disasters and disaster prevention in cooperation with mass media. Such information may include, for example, a review of routinely held disaster drills or the status of devastated areas and government efforts in those areas. Since 2001, the Cabinet Office has published the "Disaster Management News" in order to share current information relevant to disasters and their prevention between the disaster risk management agencies. Background information is, moreover, disseminated intensively prior to the issuance of new disaster prevention information products. In 2007, for example, many TV and radio programs and newspapers devoted increased coverage to the Earthquake Early Warning (EEW) program prior to the official commencement of the EEW operation for the public. The series of programs and campaigns for EEW, outlined earlier in Sect. 8.5.1, represents a model for public awareness and education campaigns tailored to specific hazards.

Some materials which describe useful tips for disaster countermeasures are provided in several languages. As a good example, Shizuoka Prefecture, which may be severely affected by a future Tokai Earthquake, has distributed brochures entitled, respectively, an "Earthquake Disaster Prevention Guidebook" and a "Life Passport" to provide foreigners in the prefecture with some advice on how to protect themselves from earthquake disasters. These brochures are published in English, Spanish, Portuguese, Chinese, Korean, Tagalong and Japanese.

The Japan Meteorological Agency holds seminars to raise public awareness of weather-related disaster prevention and disseminate knowledge of weather and climate. JMA also hosts explanatory meetings regarding the dissemination of forecasts, warnings and other relevant information for the disaster risk management agencies.

8.8 Improvement of Overall Operational Framework of EWS

Earlier sections of this report have described how JMA reviews its meteorological services on a regular basis, particularly its warning and advisory services, in pursuit of improvements. In addition, the Japanese government is investigating lessons from historical disasters and devastations that could lead to improvements in the overall operation of the Early Warning System such as, for example, enhanced awareness by the people of disaster prevention and mitigation. As noted earlier in Sect. 8.3.1, a Technical Investigation Group on Inheritance of Lessons from Disasters (TIGLD) was organized in 2003 under the Central Disaster Management Council (Fig. 8.23). This group aims to collect, systematically, historical disaster-related information such as damages, countermeasures by the central and local governments and impacts on society and the economy in order to convey traditional "know-how" to younger residents and make people aware of disaster prevention plans. The Council targets about 100 historical disasters and is investigating them over a 10 year period, examining 10 disasters each year. The targeted disasters include earthquakes, tsunamis, typhoons, volcanic activities and conflagrations that have occurred from the seventeenth century up to the present time.

A specific example of an issue which the Technical Investigation Group on Inheritance of Lessons from Disasters (TIGLD) has discussed is the Torrential Rains in Nagasaki in 1982. Nagasaki Prefecture is a local administrative area located in the west of Japan. It experienced continuous rains from 11 to 21 July 1982 because of the stalled Baiu front, a seasonal rain front in Japan in early summer. The local meteorological observatory issued heavy rain warnings for the prefecture as a whole five times during this period. At that time, warnings were issued when heavy rainfall was expected to occur somewhere in a prefecture but did not, necessarily, affect people in the remaining areas of that administrative unit. In the case of the 1982 Torrential Rains in Nagasaki, the first four warnings turned out to be targeted to the border areas of the prefecture and did not affect its capital city. Consequently, when the fifth and final warning was issued, most people in the prefecture did not take it seriously, including local government staff and mass media. However, the Baiu front moved northward and became very active, causing heavy precipitation in most areas of the prefecture. The heavy rainfall resulted in 299 casualties due to landslides and flooding, in spite of the warning having been issued. Drawing lessons from this disaster, Japan Meteorological Agency began to sub-divide the target areas for weather warnings and advisories from whole prefectures to sub-prefectural zones, in 1987 and 2003. As of May 2010, JMA is further sub-dividing the warning areas to target municipal governments such as cities and towns. It has also considered how to convey more effectively the severity of the warning and, to that end, begun to include an indication of the level of importance in the warning message.

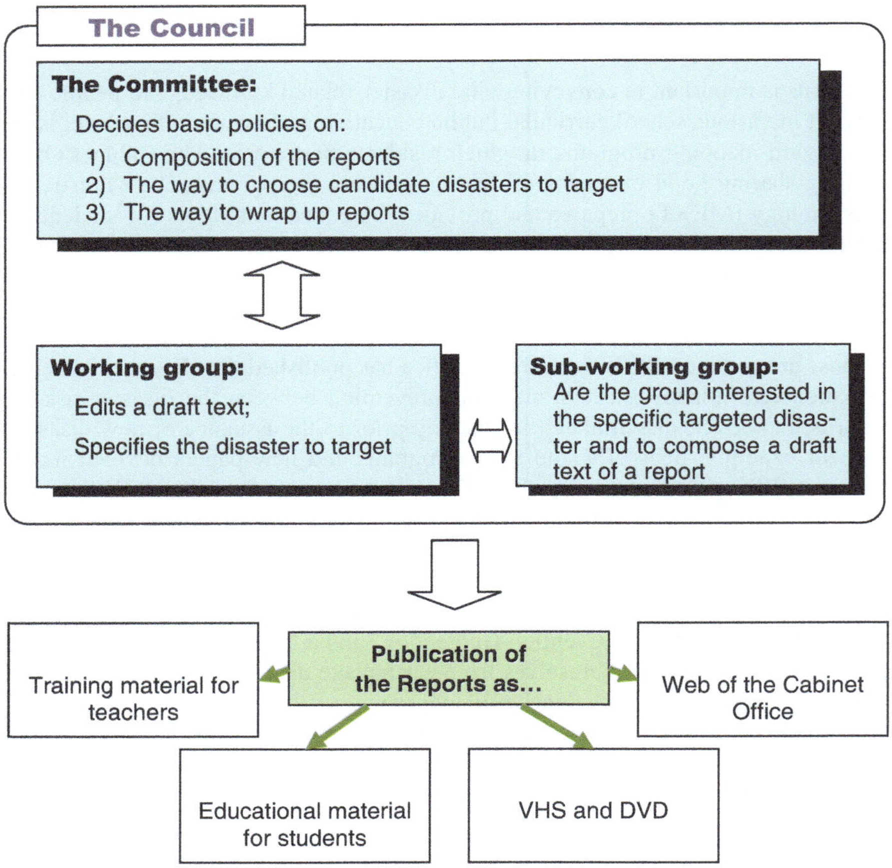

Fig. 8.23 Technical Investigation on Inheritance of Lessons from Disasters

In view of the fact that Japan is one of the countries most prone to natural disasters, the retention of experience and knowledge of disasters and the development of communication are important. Discussions by the Technical Investigation Group on Inheritance of Lessons from Disasters (TIGLD) cover all of the issues and measures mentioned above. The Group not only reviews the systems of JMA but also those of many other bodies including the mass media and medical organizations, as it is considered important to improve the nationwide disaster warning system as a whole by identifying problems and difficulties experienced following the damage caused by disasters and seeking improvements. Immediately after the 2004 Indian Ocean Earthquake, TV programs and newspapers in Japan often referred to the 1896 Meiji Sanriku Earthquake Tsunami that caused more than 220 thousand casualties in northern Japan. A historical record or a live image from a distant location on the earth, by itself, does not help us to depict a concrete image of a specific disaster. However, a combination of them would draw a striking image to people's attention and make them aware of disaster prevention and mitigation. Such countermeasures will assist not only Japanese citizens but also people all over the world who experience natural disasters.

As required by the Disaster Countermeasures Basic Act, the disaster-related services of the Japanese government are obliged to submit a White Paper to the Diet containing their plans for disaster prevention and a general account of measures undertaken for disaster prevention. This White Paper is composed of three parts:

- A description of disasters and countermeasures undertaken in the past year;
- A description of disaster prevention measures undertaken in the past year; and
- Plans for disaster prevention for the next year.

The JMA together with other ministries and agencies contributes to this White Paper every year in order to review measures and plans for disaster prevention and mitigation on a nationwide basis. In addition, JMA also publishes its own White Papers on its services and plans to provide useful information systematically. According to the White Paper for 2009, the risk of disasters has been higher than ever because of the recent change in the natural environment. As an example, local downpours in 2008 caused some casualties in Kobe and Okazaki cities and this trend is likely to continue. JMA points out in the White Paper that there were one and a half times as many heavy downpours during the last ten years as occurred in the 1980s and the risk seems to higher in the next 10 years. The White Paper also included results of a poll of the general public that revealed that about 74% of people think that the family connection is the most important one in the event of disasters. At the same time, 80% of people do not share with their family how to communicate with each other in the event of disaster.

8.9 Examples of Previous Events Where the Operational EWS Has Led to Improvements in Emergency Preparedness and Prevention

The disaster management system in Japan has been developed and strengthened following the bitter experiences of large-scale natural disasters and accidents. Among these disasters, the Ise-wan Typhoon in 1959 and the Great Hanshin-Awaji Earthquake in 1995, in particular, were two specific events that led to significant improvements in Japan's Early Warning System.

8.9.1 The Ise-wan Typhoon (1959)

The Ise-wan Typhoon, or Typhoon Vera (1959), struck the middle part of Japan (Tokai area) on September 26, 1959. A total of 4,697 people were killed, 401 people were lost and more than 38,000 people were injured. Such heavy losses were mainly caused by high tide in the Ise Bay (Ise-wan) and the name of this typhoon was taken from the name of the Bay. The Ise-wan Typhoon was the major turning point for disaster prevention in Japan. Following this typhoon, the government of Japan enacted the Disaster Countermeasures Basic Act in 1961. Two major policy changes were outlined in this Act. First, it changed the previous approach that emphasized post-disaster emergency response and initiated consistent disaster countermeasures including disaster prevention, emergency response and recovery from disasters. In the previous countermeasures approach there had, moreover, been no scheme to organize various types of individual actions for disaster mitigation. Secondly, therefore, based on the Disaster Countermeasures Basic Act, the government has promoted comprehensive and systematic countermeasures for disaster mitigation by organizing actions undertaken by each stakeholder through the establishment of the Central Disaster Management Council and the formulation of a Disaster Management Planning System. Consistent with these new principles, the Disaster Countermeasures Basic Act described the establishment of the Central Disaster Management Council with the Prime Minister as the chairperson and also mentioned the Disaster Management Planning System based on the Basic Disaster Management Plan. This Act enabled the conduct of synthesized and systematic disaster countermeasures.

Following the Ise-wan Typhoon, both observational networks and forecast techniques were significantly upgraded. In 1974, for example, JMA began installing automated weather stations (AWS) named AMeDAS (Automated Meteorological Data Acquisition System) and, today, JMA has more than 1,300 such systems across Japan. JMA has also installed 20 radar systems, while only 5 systems were in place before the Ise-wan typhoon. In 1978, JMA began to use Geostationary Meteorological Satellites. These satellites enable JMA to monitor the tracks of typhoons on a 24/7 basis, contributing to disaster prevention not only in Japan but also in East Asia and the western Pacific. Numerical weather prediction (NWP) in JMA commenced in 1959 and as the supercomputer system for NWP advances, it enables JMA to issue more accurate and longer range forecasts.

8.9.2 The Great Hanshin-Awaji Earthquake (1995)

On January 17, 1995, an earthquake of magnitude 7.3 occurred at the northern end of Awaji Island and struck the western part of Japan. Maximum intensities of 7 on the JMA intensity scale were recorded at Kobe, Ashiya, and part of Nishinomiya City and the tremor was felt by people across a wide area of Japan from Tohoku to Kyushu. This Great Hanshin-Awaji Earthquake directly affected large cities where social and economic functions were densely integrated. More than 6,000 people were killed and approximately 43,700 people were injured. Information networks were disrupted by the earthquake, local government functions were paralyzed and infrastructure such as roads, trains and ports was severely damaged. Where utilities are concerned, water supply to 1.3 million houses, electricity supply to 2.6 million houses and gas supply to 860 thousand houses were cut. Most city functions were completely paralyzed.

Following the Great Hanshin-Awaji Earthquake, the government of Japan had to reinforce its disaster prevention system. Consequently, the government enacted or revised relevant Acts, substantially modified the Basic Disaster Management Plan and established a series of information exchange systems. These latter included the Cabinet Information Collection Center, which enabled the government to collect disaster information on a 24/7 basis, and the Crisis Management Center in the Prime Minister's Office, where the designated emergency response team, comprised of the director-generals of the respective ministries and agencies, assembled.

Osaka District Meteorological Observatory, one of JMA's district branches, disseminated information, in a timely manner, on the recorded intensity at every station as well as the position and magnitude of the earthquake, while JMA Headquarters Office issued information on the earthquake to relevant ministries and the mass media in Tokyo. Several days after the earthquake, however, as a result of field investigations by JMA staff, it was revealed that part of Kobe City and Awaji Island were estimated to have been hit by tremors of intensity 7 – the maximum intensity in the JMA intensity table. Prior to the Great Hanshin-Awaji Earthquake, intensity values had been estimated by human observation. Furthermore, maximum intensity 7 was estimated only through subsequent investigations after earthquakes. Following this earthquake event, however, it was recognized that the immediate dissemination of intensity values was required to support initial responses for disaster prevention. Consequently, JMA installed about 600 intensity meters that can record intensity values immediately and transmit them automatically. These intensity meters have also been installed by local governments. As of 2010, more than 4,000 intensity meters have been installed across the Japan. Their data are transmitted to JMA and disseminated to the public within several minutes after the occurrence of an earthquake.

More than 80% of the deaths in the Great Hanshin-Awaji Earthquake occurred as a result of people being crushed or suffocated to death due to building collapse and falling furniture. In particular, wooden residences constructed in accordance with the former structural design criteria for earthquake resistance, prior to June 1981, were seriously damaged. As one of the projects to protect as many lives as possible from an earthquake, Shizuoka Prefecture, which may be severely affected by a future Tokai Earthquake, has been promoting the project "TOUKAI-0" (double meaning of the name of the region 'Tokai' and 'no collapse' in Japanese), in cooperation with cities and towns. In this project, the prefecture provides citizens living in wooden houses with a free "Structural Evaluation of Residences by Specialists" and also supports some part of the costs of reinforcement of wooden houses.

8.10 Overall Lessons Learned and Future Steps for Improving NMHS Contribution in EWS Particularly Focusing on Institutional Coordination and Cooperation

There are four key points to be made regarding the role of NHMS in contributing to an effective Multi-hazard Early Warning System:

- First, every NHMS should continue its efforts to improve its products. JMA has improved its products by introducing many kinds of technologies for observation and forecast production, such as AWS systems, radars, meteorological satellites and NWP systems. These advanced technologies enable JMA to provide more accurate observations and forecasts to disaster management organizations, local municipalities and citizens.
- Second, every NHMS needs to provide timely access to its products in order to ensure that NHMS products are used effectively. To that end, JMA strives to develop an information environment in which the public can access weather information as needed at any time, regardless of location. In this context, the latest Information and Communication Technologies (ICT) are the most powerful tools. For example, JMA exchanges information and views with private information providers for mobile phones and information technology (IT) companies. JMA also promotes new information provision channels using the latest ICT, such as car navigation systems. Furthermore, JMA develops a common and easy-to-process data format in issuing weather information.
- Third, it is also important to disseminate safety knowledge regarding disaster mitigation. For example, flash flooding observed in Kobe in 2008 was triggered by localized downpours and it killed five people. Where this type of hazard is concerned, there may be no time for people to receive hazard information from NMHSs and other authorities. Consequently, it is important to make citizens aware of disaster-vulnerable areas and the characteristics of expected hazards, as well as how to utilize NHMS's information, warnings and advisories to enable them to protect themselves from such hazards. JMA concentrates efforts on effective dissemination of safety knowledge. Many educational activities have been conducted by local meteorological observatories (LMOs). JMA sends meteorologists to give lectures on safety knowledge, including the use of JMA's weather information, to infrastructure managers and community organizations. JMA also distributes many kinds of leaflets illustrating how to use weather information and how to escape related dangers from natural disasters.

- Finally, every NHMS should promote collaboration with other organizations in order, to work better together in comprehensive disaster management programs. An example of such promotion in Japan is weather warnings and advisories that target local governments instead of sub-prefectural zones. As mentioned earlier, the intent is to strengthen the linkage between weather warnings and evacuation directives and advisories from local governments. Other examples of partnerships include the Flood Warnings produced through collaboration between JMA and the River Bureau and the Sediment Disaster Alerts produced through collaboration between JMA and prefectures.

As the National Meteorological Service in Japan, the Japan Meteorological Agency (JMA) will continuously devote its best efforts to improving its capabilities in observation and forecasting, providing timely access to weather information, disseminating safety knowledge, and strengthening partnerships with other authorities.

References

White Paper on Disaster Management.
http://www.bousai.go.jp/hakusho/hakusho.html

Disaster Management in Japan. (brochure:PDF).
http://www.bousai.go.jp/1info/pdf/saigaipanf.pdf

Disaster Countermeasures Basic Act.
http://law.e-gov.go.jp/htmldata/S36/S36HO223.html

Information on Disaster Risk Reduction of the Member Countries (Japan) in Asian Disaster Reduction Center.
http://www.adrc.asia/nationinformation.php?NationCode=392&Lang=en&NationNum=23

Japan's Natural Disaster Early Warning Systems and International Cooperative Efforts.
http://www.bousai.go.jp/kyoryoku/pdf/soukikeikai.pdf

JMA homepage.
http://www.jma.go.jp/jma/indexe.html

Japan Meteorological Agency (brochure:PDF).
http://www.jma.go.jp/jma/en/Activities/brochure200603.pdf

Disaster Management Operation Plan of JMA (2008).
http://www.jma.go.jp/jma/kishou/intro/gyomu/JMAdisaster_prevention_operating_plan.pdf

e-Gov http://www.e-gov.go.jp/index.html

Headquarters for Earthquake Research Promotion.
http://www.jishin.go.jp/main/index-e.html

Report of the Technical Investigation Group on Inheritance of Lessons from Disasters: "Torrential Rains in Nagasaki, 1982" (only in Japanese).
http://www.bousai.go.jp/jishin/chubou/kyoukun/rep/1982--nagasakiGOUU/index.html

National Report of Japan on Disaster Reduction for the World Conference on Disaster Reduction (Kobe-Hyogo, Japan, 18–22 January 2005).
http://www.unisdr.org/eng/country-inform/reports/Japan-report.pdf

Fire and Disaster Management Agency Webpage.
http://www.fdma.go.jp/en/pam02.html#introduction_4

The Guidelines for Producing a Decision and Dissemination Manual for Evacuation Orders and Instructions (available in Japanese only).
http://www.bousai.go.jp/3oukyutaisaku/pdf/04_shiryou2.pdf

The Guidelines for Evacuation Support of People Requiring Assistance During a Disaster (available in Japanese only).
http://www.bousai.go.jp/hinan_kentou/060328/index.html

Appendix 1 Disaster Management Related Laws in Japan

Basic Acts

1. Disaster Countermeasures Basic Act (1961)
2. Act on Prevention of Marine Pollution and Maritime Disaster (1970)
3. Act on Disaster Prevention in Petroleum Industrial Complexes and other Petroleum Facilities (1975)
4. Act on Special Measures for Large-scale Earthquakes (1978)
5. Act on Special Measures for Nuclear Disasters (1999)
6. Act on Special Measures for Promotion of Tonankai and Nankai Earthquake Disaster Management (2002)
7. Act on Special Measures for Promotion of Disaster Management for Trench-type Earthquakes in the Vicinity of the Japan and Chishima Trenches (2004)

Disaster Prevention and Preparedness

1. Erosion Control Act (1897)
2. Building Standard Law (1950)
3. Forest Act (1951)
4. Act on Temporary Measures for Disaster Prevention and Development of Special Land Areas (1952)
5. Meteorological Services Act (1952)
6. Seashore Act (1956)
7. Landslide Prevention Act (1958)
8. Act on Special Measures for Disaster Prevention in Typhoon-prone Areas (1958)
9. Act on Special Measures for Heavy Snowfall Areas (1962)
10. River Act (1964)
11. Act on Prevention of Steep Slope Collapse Disaster (1969)
12. Act on Special Measures for Active Volcanoes (1973)
13. Act on Special Financial Measures for Urgent Earthquake Countermeasure Improvement Projects in Areas for Intensified Measures (1980)
14. Act on Special Measures for Earthquake Disaster Countermeasures (1995)
15. Act on Promotion of the Earthquake-proof Retrofit of Buildings (1995)
16. Act on Promotion of Disaster Resilience Improvement in Densely Inhabited Areas (1997)
17. Act on Promotion of Sediment Disaster Countermeasures for Sediment Disaster Prone Areas (2000)
18. Specified Urban River Inundation Countermeasures Act (2003)

Disaster Emergency Response

1. Disaster Relief Act (1947)
2. Fire Services Act (1948)
3. Flood Control Act (1949)

Disaster Recovery and Reconstruction, and Financial Measures

1. Forest National Insurance Act (1937)
2. Agriculture Disaster Compensation Act (1947)
3. Housing Loan Corporation Act (1950)
4. Act on Interim Measures for Subsidizing Recovery Projects for Agriculture, Forestry and Fisheries Facilities Damaged Due to Disasters (1950)

5. Small-Medium Business Credit Insurance Act (1950)
6. Act on National Treasury Share of Expenses for Recovery Projects for Public Civil Engineering Facilities Damaged Due to Disasters (1951)
7. Public Housing Act (1951)
8. Fishing Boat Damage Compensation Act (1952)
9. Agriculture, Forestry and Fisheries Finance Corporation Act (1952)
10. Railway Improvement Act (1953)
11. Act on National Treasury Share of Expenses for Recovery of Public School Facilities Damaged Due to Disasters (1953)
12. Act on Interim Measures for Financing Farmers, Woodsmen and Fishermen Suffering from Natural Disasters (1955)
13. Airport Improvement Act (1956)
14. Small-scale Business Equipment Installation Financial Support Act (1956)
15. Act on Special Financial Support to Deal with Extremely Severe Disasters (1962)
16. Fisheries Disaster Compensation Act (1964)
17. Act on Earthquake Insurance (1966)
18. Act on Special Financial Measures for Group Relocation Promotion Projects for Disaster Mitigation (1972)
19. Act on Payment of Solatia for Disasters (1973)
20. Act on Special Measures for Reconstruction of Disaster-stricken Urban Areas (1995)
21. Act on Special Measures for Reconstruction of Jointly Owned Buildings in Disaster-stricken Areas (1995)
22. Act on Special Measures for Preservation of Rights and Profits of the Victims of Specified Disasters (1996)
23. Act on Support for Livelihood Recovery of Disaster Victims (1998).

List of Acronyms

AMeDAS	Automated Meteorological Data Acquisition System
AWS	Automated Weather Station
CDMRCS	Central Disaster Management Radio Communications System
DIS	Disaster Information System
DMAT	Disaster Medical Assistance Teams
DMO	District Meteorological Observatories
EEW	Earthquake Early Warning
EWS	Early Warning Systems
ICT	Information and Communication Technologies
INDiP	Information Network for Disaster Prevention
ISCDP	Information Service Center for Disaster Prevention
JMA	Japan Meteorological Agency
LMO	Local Meteorological Observatories
MAFF	Ministry of Agriculture, Forestry and Fisheries
MEXT	Ministry of Education, Culture, Sports, Science and Technology
MIC	Ministry of Internal Affairs and Communications
MLIT	Ministry of Land, Infrastructure, Transport and Tourism
NMHS	National Meteorological and Hydrological Service
NWP	Numerical Weather Prediction
PDCA	Plan-Do-Check-Act
PF	Disaster Information Sharing Platform
RAS	Real Damage Analysis System
TIGLD	Technical Investigation Group on Inheritance of Lessons from Disasters

Chapter 9
Synthesis of Seven Good Practices in Multi-Hazard Early Warning Systems

Maryam Golnaraghi

Abstract Earlier chapters of this book have detailed seven good practices in Early Warning Systems (EWS) for hydrometeorological and other hazards that have proven effective in reducing losses from such hazards. The seven cases presented in this book encompass examples from different governance and institutional frameworks, levels of economic development, geographic and climatic regimes, including developed countries such as Germany, France, Japan and the United States of America, the developing country of Bangladesh, the island nation of Cuba and the mega-city of Shanghai-China. Despite the socio-economic, cultural, environmental differences and individualized approaches to the operation of their EWS, all share the common characteristic that their EWS are successful in reducing losses of life and property from hydrometeorological and related hazards within their respective jurisdictions. This chapter synthesizes lessons learned and analyses key principles common to the success of these systems. These Principles are considered to be universally applicable, being not only independent of individual countries' socio-economic, cultural, political and institutional conditions but also allowing for EWS operations that are adapted to local circumstances. They draw attention to fundamental lessons, considerations, systemic issues and practical challenges that need to be addressed when undertaking the task of implementing or improving EWS in any jurisdiction. It is hoped that these principles will prove useful to governments and various agencies who are seeking to establish or strengthen their EWS within their own borders.

9.1 Overview

The underlying premise of this book is that effective Early Warning Systems (EWS) for hydrometeorological and other hazards can significantly reduce losses of lives and property from the impacts of these hazards. The validity of this premise has been clearly demonstrated in earlier chapters where significant reductions in losses have resulted from the successful implementation of EWS in a number of jurisdictions. To cite specific examples, this has been illustrated particularly dramatically in Japan and Bangladesh where the progressive enhancement of EWS has undoubtedly saved many thousands of lives during cyclones, typhoons and earthquakes. The essential political support for the development and improvement of such systems has, in general, been driven by unfortunate experiences with previous, often recurring, disasters in affected nations. Many countries, however, face significant challenges when seeking to develop or strengthen their EWS as an integral part of disaster risk reduction strategies within a multi-hazard framework. The overall objective of this chapter is to provide decision-makers within countries, disaster authorities and other key disaster risk reduction stakeholders with guidelines to assist them in planning, implementing, or further strengthening, EWS for hydrometeorological and related hazards. To that end, it presents, in the form of guiding principles, a synthesis of lessons learned from examples of good practices that have been documented in earlier chapters. It also discusses these Principles, expanding on their general applicability through references to the case studies that demonstrate their validity in countries with widely different socio-economic circumstances and differing approaches to implementation of EWS. The following paragraphs provide a brief overview of the process by which these Principles were developed, refined and, in the end, finalized.

The United Nations International Strategy for Disaster Reduction (UNISDR) definition of an Early Warning System is "The set of capacities needed to generate and disseminate timely and meaningful warning information to enable individuals, communities and organizations threatened by a hazard to prepare and to act appropriately and in sufficient time to re-

---

Maryam Golnaraghi, Chief of Disaster Risk Reduction Programme, World Meteorological Organization

Note: *With contributions from Mr. James Douris and Dr. Desmond O'Neill*

M. Golnaraghi ed., *Institutional Partnerships in Multi-Hazard Early Warning Systems*,
DOI 10.1007/978-3-642-25373-7_9,

duce the possibility of harm or loss". Consideration of the information in the case studies has enabled us to expand on the preceding definition by further clarifying the concept of a Multi-Hazard Early Warning System (MH-EWS). From these cases, we have learned that a well-designed MH-EWS is one that not only targets a number of hazards but in doing so seeks to take optimum advantage of the collective capacities of all stakeholders involved in different components of the EWS. It leverages resources, monitoring and prediction networks, risk analysis capacities, communications and dissemination networks and preparedness and response systems through well-defined coordination structures and protocols to ensure the greatest possible efficiency, effectiveness, inter-operability and sustainability for the system as a whole.

It is worth reiterating here that the preceding case studies illustrate successful EWS in Cuba, Bangladesh, France, Germany, Japan and the United States and the densely populated urban region of Shanghai, China. These examples of EWS span a broad spectrum of geographic and climatic conditions in both developed and developing countries and address a variety of hydrometeorological and other hazards. Their organization reflects varying political structures and differing assignments of roles and responsibilities in relation to disaster risk reduction. Two of the EWS, those in Cuba and Bangladesh, are heavily oriented towards the dominant threat presented to these nations by hurricanes and tropical cyclones. The EWS in France, Germany and the United States are more broadly targeted at hydrometeorological hazards in general, with the US system also encompassing the tsunami hazard. Reflecting its exposure, Japan's EWS not only addresses hydrometeorological hazards but also places major emphasis on earthquakes, tsunamis and volcanic activity. The mega-city of Shanghai, China, with its huge population confined into a relatively small area, poses special challenges from an early warning perspective. In response, the municipal and national governments have collaborated in implementation of a highly organized and effective EWS that operates down to the scale of a Shanghai city block, building on a dense hydrometeorological observation network and a comprehensive, city wide communications/dissemination platform. The ten guiding Principles for successful Early Warning Systems discussed in subsequent sections incorporate common characteristics and universal good practices evident in the seven cases presented here.

9.2 Ten Common Principles for Successful Early Warning Systems

The development and sustainability of successful EWS requires strong political commitment reflected in clear direction to governmental authorities and disaster risk management agencies and in targeted and sustained allocation of appropriate resources. Consequently, EWS should be an integral part of governments' disaster risk management plans and budgeting processes. They must be underpinned by plans at national to local levels and supported by legislation that explicitly defines the roles and responsibilities of the various stakeholders. In addition, EWS establishment requires a clear concept of operations that enables effective coordination to take place among agencies across all components of the system. Systematic evaluation and feedback on various processes and mechanisms are also required to provide a solid basis for improvement of the EWS over time. Moreover, the various stakeholders must come together in collective engagement to optimize their efforts to produce and deliver timely and effective early warnings of hazards to populations at risk and ensure that appropriate preparedness measures and response actions are taken. The following sections expand on these themes, synthesizing lessons learned from successful EWS into a set of guiding principles.

9.2.1 Political Support is Vital for the Success of an Early Warning System

> ***Principle No. 1*** – *There is a strong political recognition of the benefits of early warning systems reflected in harmonized national to local disaster risk management policies, planning, legislation and budgeting.*

The cases cited in earlier chapters demonstrate that the establishment and operation of successful early warning systems is based on an underlying recognition by the highest levels of governments that such systems can contribute substantially to reduction of losses from disasters caused by natural hazards. In all seven of the cases, a history of significant, recurring disasters resulted in political awareness of the need for a coordinated structure and processes to ensure effectiveness and efficiency in EWS planning, preparedness and response at all levels. This has been accompanied by an appreciation of institutional capacities and recognition of the need to align these capacities and resources across national to local levels to enable development of effective EWS and emergency response mechanisms. This high level political awareness has, moreover, been given practical effect through legislation and directives to finance and planning, disaster management, and other relevant authorities and technical agencies.

Box 9.1 *Principle No. 1 – There is a strong political recognition of the benefits of EWS reflected in harmonized national to local disaster risk management policies, planning, legislation and government budgeting.*

The following examples from the countries of good practices, illustrate the strong political recognition and the different approaches toward the establishment and development of their EWS.

Bangladesh

In response to the 1970 super cyclone that killed over 300,000 people, the Government of Bangladesh (GoB) and the Bangladesh Red Crescent Society (BDRCS) created a partnership that led to the creation of the Cyclone Preparedness Programme (CPP). The CPP is administered through a joint GoB/BDRCS policy Committee and Implementation board with the operations being under the BDRCS. The Policy Committee is headed by the Minister of the Food and Disaster Management Ministry.

Since 1970, the Bangladesh Government has implemented several legal instruments that regulate the countries disaster risk reduction framework, including:

- *Disaster Management Act provides the legal basis under which disaster risk and emergency management actions are identified, undertaken and managed. The Act also establishes the roles and responsibilities of Ministries, committees and appointments.*
- *National Plan for Disaster Management, which aims to develop a culture of prevention through introducing disaster management in school curricula, increasing hazard awareness at all levels, and enhancing capacities of disaster managers.*
- *National Disaster Management Policy, which defines the strategic framework on disaster risk reduction in the country.*
- *Standing Orders on Disasters (SOD), which details the roles and responsibilities of Ministries, committees, and other organization involved in disaster risk reduction.*
- *Guidelines for Government at all Levels, based on best practice models, used to assist Ministries, non-government organizations, civil society, etc in implementing disaster risk management.*

France

Following two major storms in December 1999 which cost the lives of nearly 100 people and caused significant damage to the country, Government of France, after a detailed disaster review, accepted the proposal of Météo-France to draw up a Vigilance map that consisted of colour coding various levels of risk. This proposal was enhanced through an institutional dialogue among the main ministries and bodies involved with respect to the definition of follow-up procedures and safety guidelines for the public and began operationally in October 2001. Since 2001, new legislation has been enacted that strengthened the Vigilance system, including:

- *The "natural risks" component of Law No. 2003-699 (2003) relating to the Prevention of Technological and Natural Risks and Repairing the Damage Caused was designed to remedy shortcomings identified during floods in the south of France in September 2002.*
- *Law No. 2004-811: Addressed the organizational structure of the nationwide civil defence systems. Article 1 of this law stipulates that the State guarantees the coherence of the civil defence nationwide but that the citizens are the key players.*

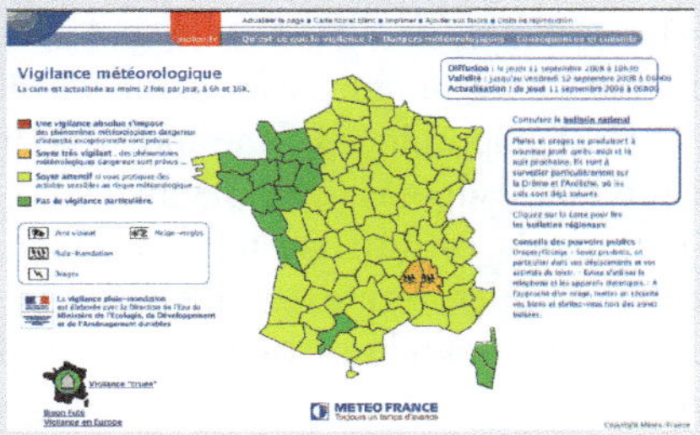

United States

In the United States emergency planning and preparedness is established by the President as a national security priority, which must be conducted in an atmosphere of trust and mutual understanding, supported by relevant Executive Orders and legislation assigning specific EWS roles and responsibilities to designated agencies and organizations including the National Weather Service. Emergency planning and operations are, moreover, undertaken within the well-defined U.S. National Response Framework *that provides the guiding principles that enable all response partners to prepare for and deliver a unified national response to disasters and emergencies.*

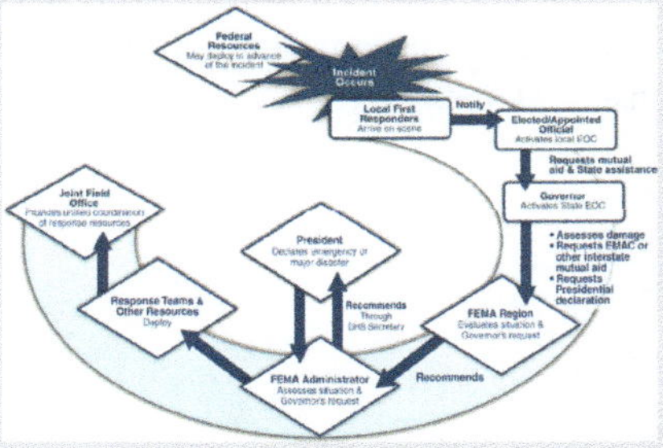

Cuba

Through observing the devastation in neighbouring countries caused by hurricanes in 1995, the Cuban Government identified the need to provide early alerts for tropical cyclones to increase awareness and preparation time. In this context several legal instruments were enacted to establish the Cuban Early Warning System, such as:

- *Law No. 75/94 (1994) of National Defence which establishes the main missions and measures of the Civil Defence System and the principals for it territorial and institutional organization.*
- *Guideline No. 01/05 (1995) which establishes the regulations for the disaster reduction process and the organization of response and recovery at all levels.*
- *Law 75, Decree-law No. 170/97 (1997) which regulates the role and position of state agencies and organization, economic entities and social institutions in the disaster reduction process.*
- *Resolution 106/99 (1999) established the Institute of Meteorology as the authorised agency to give meteorological and climatic information.*

Resolution No 43/06 (2006) establishes the Environment Agency with the mandate to organize, lead and conduct hazard, vulnerability and disaster risk studies.

9.2.2 Operational Capacities to Detect and Predict Hazards Must Be Matched by Community Preparedness

Principle No. 2 – *Effective early warning systems are built upon four components: (i) hazard detection, monitoring and forecasting; (ii) analyzing risks and incorporation of risk information in emergency planning and warnings; (iii) disseminating timely and "authoritative" warnings; and, (iv) community planning and preparedness and the ability to activate emergency plans to prepare and respond, with coordination across agencies, at national to local levels.*

The four components in Principle No. 2 represent the essential building blocks for a coherent, integrated and successful EWS. Weaknesses or deficiencies in any of these four elements can undercut the integrity of the EWS as a whole, leading to diminished overall performance and effectiveness. A failure to detect a hazard and predict its future evolution would, obviously, make it impossible to issue a timely and accurate early warning message. Inadequate consideration and application of risk information can expose people to hazards that would otherwise be avoidable. Equally, any breakdown of the communications system can adversely affect the warning dissemination process with the result that EWS stakeholders and people at risk may not be warned of imminent hazards in sufficient time to take avoidance or mitigating actions. Poor emergency planning, ineffective emergency preparedness and training programmes can also reduce the contribution of the EWS to saving lives and reducing property losses.

The case of Japan provides a good illustration of the presence of the preceding four elements in an effective EWS. The Japan Meteorological Agency (JMA) has the mandate for monitoring, forecasting and development of warnings for various high risk natural hazards in Japan. These include meteorological hazards such as heavy rain, heavy snow and storm surge, as well as earthquakes and tsunamis. Responsibility for preparation and issuance of flood warnings and sediment alerts is, in addition, shared between JMA and other agencies. In carrying out its mandate, JMA draws on an extensive observational network, for detection and monitoring of hazards, a state of the art forecasting capability, to generate forecasts and early warnings on a 24/7 basis, and a modern and reliable telecommunications system, with built-in redundancy, for dissemination of warning messages and for domestic and international relay of data and derived products. An indication of the level of importance of the threat is, moreover, included in JMA's hazard warning messages. These warnings and related products support well-planned and coordinated disaster preparedness and response measures, within the framework of Japan's Basic Disaster Management Plan (Chap. 7; Sect. 7.2.1, Sect. 7.2.3, Sect. 7.5). Japanese agencies have, furthermore, undertaken extensive risk analyses for tsunamis, tidal waves, flooding, volcanic eruptions, earthquakes and other hazards to provide a solid basis for planning hazard avoidance, disaster mitigation and preparedness and response measures.

The system in France provides a further illustration of the existence of the four foundational elements of a successful EWS, but in a very different geographic and organizational context. Within the framework of France's emergency planning and response system, the NMHS, Météo-France, is established as an authoritative provider of meteorological and climate information products and services to the government authorities responsible to ensure safety of persons and property in case of disasters caused by caused by natural hazards. In carrying out this responsibility, Météo-France draws on its well-developed observational and forecasting capacities to operate the "Vigilance" Early Warning System (EWS) for meteorological and hydrological risks. A nation-wide Vigilance map that is disseminated twice daily to warn citizens (Chap. 5; Sect. 5.2.3) incorporates risk information. The agency's Vigilance warnings and related products support comprehensive emergency planning and response systems at departmental and commune levels (Chap. 4; Sect. 4.2.1). Moreover, France's *Plan de Prévention des Risques Naturels* (PPR) takes a close look at vulnerability and exposure to the various risks from natural hazards, in particular flooding, at the communal level.

Other case studies outlined in earlier chapters (Cuba (Chap. 2; Sect. 2.2.3), USA (Chap. 6; Sect. 6.2.3, Sect. 6.2.4), Shanghai (Chap. 7; Sect. 7.1, Sect. 7.2.1), Bangladesh (Chap. 3; Sect. 3.6) also display well-developed observational and forecasting capacities, reliable communications/dissemination systems, some level of consideration of risk information, comprehensive emergency planning and preparedness and effective response systems.

Box 9.2 *Principle No. 2 – Effective early warning systems are built upon four components: (i) hazard detection, monitoring and forecasting; (ii) analyzing risks and incorporation of risk information in emergency planning and warnings; (iii) disseminating timely and "authoritative" warnings; and, (iv) community planning and preparedness and the ability to activate emergency plans to prepare and respond, with coordination across agencies, at national to local levels.*

The following examples from the countries of good practices, illustrate the four components of their EWS.

Japan

JMA's operational services include all four components of EWS.

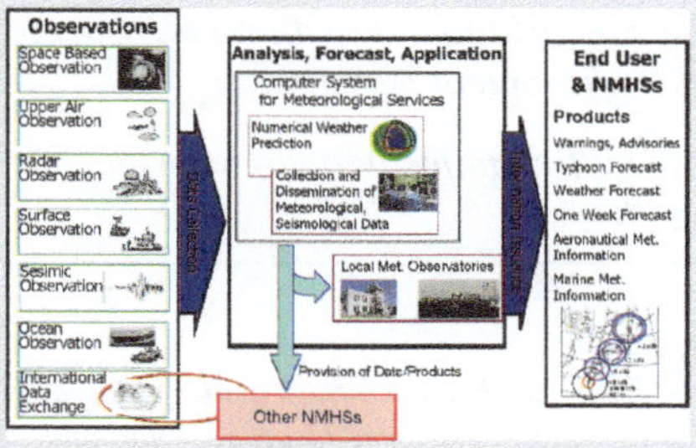

France

The French EWS is built upon all four components of an effective EWS.

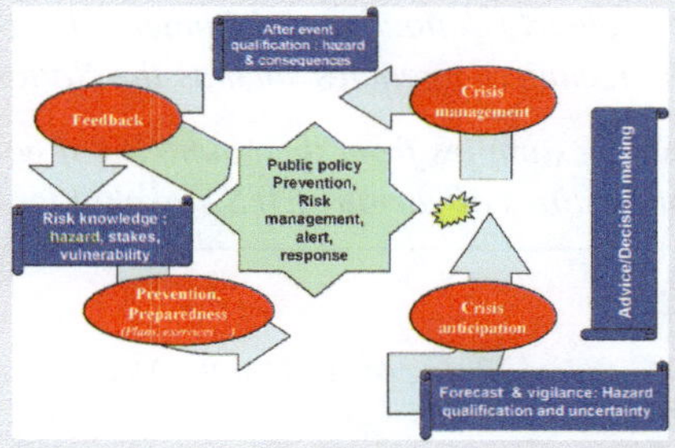

Shanghai, China

The Shanghai MHEWS focuses on the domino effects of high impact weather such as typhoon, severe convective weather, heat wave, etc through a multi-agency management level and multi-hazard detection and monitoring, forecasting and dissemination system at the technical level.

9.2.3 EWS Roles and Responsibilities Must Be Clear

Principle No. 3 *– EWS stakeholders are identified and their roles and responsibilities and coordination mechanisms clearly defined and documented within national to local plans, legislation, directives, MOUs, etc, including those of the technical agencies such as the National Meteorological and Hydrological Services.*

Successful EWS are underpinned by legislation and legal frameworks that enable the enforcement of the legislation. These provide the governing structure under which emergency planning and preparedness is carried out and response to issued warnings is implemented. Successful EWS also depend upon national, provincial and local governments creating an integrated and comprehensive framework, which clarifies roles, responsibilities and relationships for improved decisions. This can be accomplished through integrated emergency planning that involves all governmental, non governmental, private sector and community-based disaster risk stakeholders to develop, implement, validate and improve plans and procedures. Such planning involves prioritization and coordination based on a concept of operations reflecting different EWS stakeholders' responsibilities for the implementation, monitoring and improvement of the plans.

In the good practices, operations and actions are supported by agreements that formalize commitments from public authorities to implement emergency preparedness and response policies, procedures and actions on-the-ground, in response to increasing levels of threat. Each case has been established within the framework of its respective country's disaster risk management planning, legislation, and institutional framework, building upon a scalable implementation plan or governing document which enables decision processes across all levels of government to address the impacts of natural hazards. In addition, the operations of each EWS are detailed within disaster risk management-related plans and legislation. In all cases NMHS role and responsibilities are reflected clearly in the policies, plans, legislation or orders/directives.

Box 9.3 *Principle No. 3 – EWS stakeholders are identified and their roles and responsibilities and coordination mechanisms clearly defined and documented within national to local plans, legislation, directives, MOUs, etc, including those of the technical agencies such as the National Meteorological and Hydrological Services.*

The following examples from the countries of good practices, illustrate the different methodologies with which the roles and responsibilities of various EWS stakeholders are defined and managed.

France
Law No. 2004-811 of 13 August 2004 on the Modernization of Civil Defence gave a new territorial dimension to the ORSEC (Organisation de la Réponse de Sécurité Civile) disaster contingency plans aimed at organizing the civil defence response through coordinated mobilization of all actors. At the local level, this response is structured by means of communal safeguard plans which determine immediate protection and security measures and stipulate modalities for the dissemination of warnings. The public is mobilized via the creation of communal civil defence reserves that help build awareness, inform the public and intervene in post-crisis phases.

United States
The National Response Framework provides the guiding principles that enable all response partners to prepare for and deliver a unified national response to disasters and emergencies. It is built upon scalable, flexible, and adaptable coordinating structures to align key roles and responsibilities across the Nation, linking all levels of government, non-governmental organizations, and the private sector. It outlines the key response principles, participants, roles, and provides structures for implementing national-level policy and operational coordination for domestic incident response.

Bangladesh
Standing Orders on Disaster (SOD) guide, and give the ability to monitor, disaster management activities in Bangladesh. The Standing Orders have been prepared with the objective of making concerned persons understand their duties and responsibilities regarding disaster management at all levels and how to accomplish them. All Ministries, Divisions/Departments and Agencies prepare their own Action Plans with respect to their responsibilities under the Standing Orders, for efficient implementation.

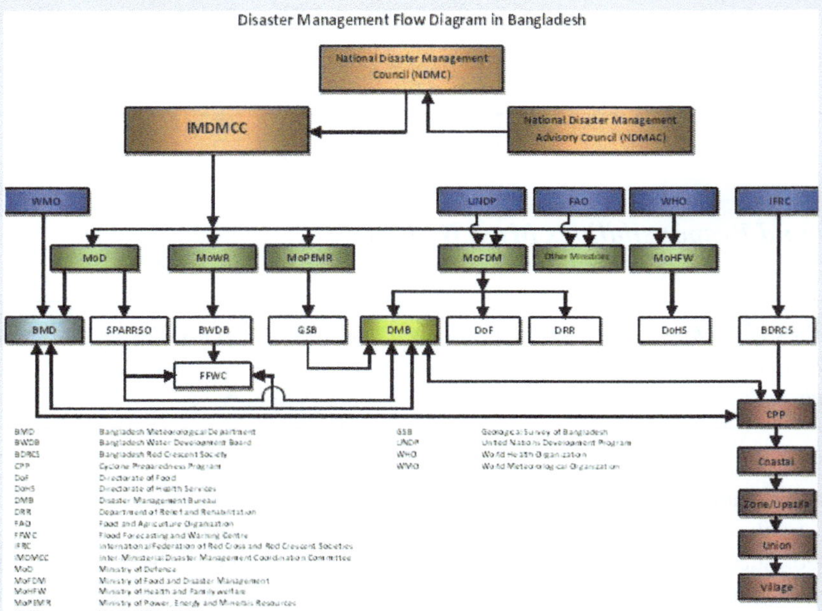

While several approaches to collaboration and operational coordination are evident in the cases presented in this book, well defined institutional structures and clear assignments of roles and responsibilities exist in all seven cases, at all levels. Furthermore, documented procedures and standard operating procedures govern the interactions between the various EWS stakeholders.

The case of Bangladesh provides a good illustration of clarity in the definitions of the roles of EWS stakeholders and specific assignments of responsibilities to individual agencies. In Bangladesh, the Disaster Management Bureau is responsible for implementing the nation's emergency plans, which are very comprehensive in detailing the roles and responsibilities of all EWS stakeholders. The Standing Orders on Disaster (SOD) document all stakeholders, including the Cyclone Preparedness Programme, clarifying their duties and responsibilities regarding disaster management at all levels. As a particular example relevant to technical agencies, the assigned role of the Bangladesh Meteorology Department is to support emergency planning and response through provision of meteorological information, advice and services (Chap. 3; Sect. 3.2.4).

Operating within a different political and administrative system, and in an island environment, the Cuban EWS reflects a similar focus on clarity regarding roles and responsibilities of EWS stakeholders. In Cuba, Ordinance Law No. 279/07 (March 19, 2007) of the Council of Ministers lays down that the "Hydrometeorological System for Exceptional Situations" is the group of hydrological and meteorological entities in the country that has as its main mission to obtain, analyze, evaluate, process and issue the necessary hydrological and meteorological information for the execution of protective measures to mitigation of the effects of disasters caused by natural hazards. As a further refinement, Resolution 106/99 (December 6, 1999) of the Ministry of Science, Technology and Environment specifically designates the Institute of Meteorology as the only authorized institution with nation-wide responsibilities to issue through the media, warnings and forecasts of meteorological hazards and related information for the protection of human life, material goods, and the economy and for the development of the country (Chap. 2; Sect. 2.2.1). Other case studies Japan (Chap. 8; Sect. 8.2.1), France (Chap. 4; Sect. 4.2.1, Sect. 4.2.2, Sect. 4.2.3), Shanghai (Chap. 7; Sect. 7.2.1) provide corresponding illustrations of the emphasis on ensuring clarity regarding stakeholders' roles and responsibilities that is a feature of all successful EWS.

9.2.4 Adequate Resources Must Be Provided

Principle No. 4 *– EWS capacities are supported by adequate resources (e.g., human, financial, equipment, etc.) across national to local levels and the system is designed and implemented accounting for long-term sustainability factors.*

The provision of adequate human, financial, equipment and communications resources is absolutely essential for the operation and sustainability of successful EWS. As pointed out earlier, the development and sustainability of effective EWS requires that a political commitment to such programs is accompanied by the clear assignment of high budgetary priority to them, reflected in continuing allocation of appropriate resources for their implementation and operation. In practice, therefore, funding of EWS should be an integral part of governments' disaster risk management planning and budgeting processes.

Virtually all of the case studies cited in this publication reflect solidly-established governmental funding arrangements that sustain the operations of their respective EWS. As a particular example, funding for the all-hazards early warning system in the United States is driven by the annual Congressional Appropriations process that provides federal agencies with their operating budgets. Funding is then transferred to state and local government entities through various grant programs. Moreover, private sector entities that support development or maintenance of the US EWS are reimbursed through competitively-awarded contracts (Chap. 6; Sect. 6.2.6). Similarly, disaster risk reduction in Japan is funded in the budgets of national and local governments with collaboration between agencies being promoted to optimize resources for improvement of EWS. Following the actual occurrence of a disaster in Japan, an additional supplementary budget is usually approved for disaster restoration and disaster prevention and the JMA has improved its systems by utilizing these supplementary budgetary allocations (Chap. 8; Sect. 8.2.6). As further examples of governmental funding mechanisms, the Early Warning Systems in Shanghai (Chap. 7; Sect. 7.2.5) and Cuba (Chap. 2; Sect. 2.2.5) can also be cited.

Box 9.4 *Principle No. 4 – EWS capacities are supported by adequate resources (e.g., human, financial, equipment, etc.) across national to local levels and the system is designed and implemented accounting for long-term sustainability factors.*

The following examples from the countries of good practices, describe the different methodologies with which their national EWS is supported. These include direct funding, cost recovery and government/non-governmental organization co-financing.

France
Within the framework of the Organic Law relating to the Finance Laws (LOLF in French), various programmes or missions include a reference to risk prevention responsibilities. The LOLF includes a "meteorology" programme under the "Transport" mission. Thanks to contracts signed with the government (MEEDDAT and the ministry of budget), the French national meteorological service receives a global governmental subsidy covering some 55 per cent of its expenditures. The balance of its budgetary requirement must be financed by cost recovery of services to air navigation and by revenues generated by contracts for specific products and services that Météo-France is free to sign with private or public customers. With regard to the financing of some infrastructure (e.g. weather radar facilities), specific arrangements make it possible to raise funds from local authorities, the State and European Union on a case-by-case basis.

Japan
Disaster risk reduction is covered in the budget of national and local governments. The budget is requested by each organization independently. The national budget for disaster management is Approximately 4.5 trillion yen (average annual budget From 1995 to 2004), accounting for approximately 5% of the total amount of the budget for general accounts.
After a severe disaster, a certain amount of budget for disaster restoration and disaster prevention is approved. JMA improved its systems by utilizing these budgets. For example, after the earthquake and tsunami in north-west off of Hokkaido area in 1993, supplement budget for installing more than 150 remote seismic observation stations was allocated. Another example is a supplement budget for installing automatic seismic intensity meters for more than 600 points, which was allotted after the Great Hanshin-Awaji Earthquake in 1995.

Bangladesh
The national budget allocates funds to government institutions to support their roles in disaster management. These institutions include:

- *Ministry of Food and Disaster Management*
- *Disaster management Bureau-*
- *Directorate of Relief and Rehabilitation*
- *Directorate of Food*

The Ministry of Food and Disaster Management has established special funds for disaster response and recovery, risk reduction, sectoral planning and disaster management The salaries of officers and staff of BMD and other EWS, along with the costs of routine operation and maintenance of the establishments, are met directly from the regular budget of the Government of Bangladesh. In addition, the costs of upgrading and improvement of the existing MHEWS and of overseas training are mostly covered by WMO, UNDP, JICA, KOICA, CMA etc. as donations through VCP, grant aid and counterpart training. The Cyclone Preparedness Programme is co-financed between the government of Bangladesh and the Bangladesh Red Crescent Society. Furthermore, the system builds on the force of communities and network of community volunteers who are actively engaged in the implementation of the system.

United States
Funding mechanisms for the all-hazards early warning system in the United States are driven by the annual Congressional Appropriations process that provides federal agencies with their operating budget (http://www. senate.gov/reference/resources/pdf/97-684.pdf). This is how the agencies responsible for the Emergency Alert System (EAS) and Integrated Public Alert Warning System (IPAWS) development (FEMA, FCC and NWS) derive their funding for the development, maintenance, sustainability and improvement of the early warning system and emergency preparedness activities. Funding is then transferred to state and local government entities through various grant programs. Private sector entities that are supporting development or maintenance of the U.S. early warning system are reimbursed through the individual contracts awarded through a competitive bid process.

9.2.5 Risk Information Should Be Used to Enhance Emergency Planning and Warnings

Principle No. 5 *– Hazard, exposure and vulnerability information are used to carry-out risk assessments at different levels, as critical input into emergency planning and development of warning messages.*

Consideration of risk information, including hazards, exposures and vulnerabilities, is a critical step in emergency planning and in the development of warning messages to ensure that authorities and the public have a better understanding of the potential risks associated with a hazard events. Such information provides a solid foundation for the identification and prioritization of vulnerable communities and for the development of emergency preparedness and response plans. Hazard, exposure and vulnerability information, along with risk analysis tools, are required to produce risk information, estimating the likely impacts of hazards on people, infrastructure and economic activities. Consequently, different types of hazards, vulnerability and asset information need to be collected, quality assured, and compiled into a system (based on standards) that is readily accessible by agencies responsible for carrying out risk analysis. Furthermore, effective review and feedback processes should be established to support the progressive development and improvement of risk information.

Calculating the probability of occurrence of hazard events in terms of their magnitude, duration, location, and frequency requires availability of high quality historical meteorological, hydrological and other hazard data. Once the elements exposed to specific hazards have been identified, detailed exploration can be undertaken of their vulnerabilities to these hazards.

In most of the seven documented case studies, institutional roles and responsibilities for the development of risk information are clearly established. In general, the NMHS is the primary source for collection, quality assurance, and archiving of the hydrometeorological hazard data and information. Moreover, there is "usually" a system in place for collection and archiving of vulnerability and disaster impact information by various agencies (e.g., health, agriculture, disaster risk management).

The EWS in Japan, France and Shanghai illustrate effective applications of risk information in emergency planning and warnings. Japanese agencies at national and municipal levels have carried out extensive risk analyses for tsunamis, tidal waves, flooding, volcanic eruptions, earthquakes and other hazards and hazard maps are available to stakeholders and the general public. In addition, the JMA includes an indication of the level of importance of the threat in its hazard warning messages. As a related refinement, JMA has also sub-divided its warning areas to match local government boundaries in order to strengthen the connection between weather warnings and the evacuation orders issued by these governments (Chap. 8; Sect. 8.3).

For its part, France's *Plan de Prévention des Risques Naturels* (PPR) takes a close look at vulnerability and exposure to the various risks from natural hazards, in particular flooding, at the communal level. In addition, Météo-France incor-

Box 9.5 *Principle No. 5 – Hazard, exposure and vulnerability information are used to carry-out risk assessments at different levels, as critical input into emergency planning and development of warning messages.*

The following examples from the countries of good practices show the different aspects of hazard, exposure and vulnerability data are developed, utilized in development of warnings and public awareness.

Japan

The Disaster Countermeasures Basic Act has enabled Japanese organizations to conduct synthesized and systematic disaster countermeasures. A variety of organizations, such as those of the national government and research institutes, disseminate data and information on hazards and vulnerability to the public, including the international community, through web sites and by publications. In particular, JMA provides timely and precise meteorological data as well as historical data and statistics.

The relevant Ministries encourage and support municipalities in producing vulnerability maps through provision of related guidelines and information. Municipalities produce vulnerability maps taking into account not only the ministries' advice and guidelines but also their own local characteristics. In some instances, residents are advised to assist in producing these maps in order to raise their awareness of disaster prevention. The JMA provides timely and precise information on meteorology to municipalities that contributes to municipalities' vulnerability assessments.

Shanghai, China

A comprehensive examination of "hidden risks" in Shanghai is in progress. The Safety Administration has the responsibility for the distribution of data and information related to dangerous chemical sources. The Real Estate Department maintains data on buildings and houses that are vulnerable to disasters. SMB is carrying out a weather disaster distribution project. Through multi-agency coordination and cooperation, these departments effectively share monitoring data and technical advances. These efforts have increased the number of disaster data categories to 65, addressing meteorological, hydrologic and marine, traffic, food, epidemic diseases and agriculture related hazards. The SMB's transformation from weather monitoring to comprehensive monitoring has benefited in the establishment of a multi-hazard and early monitoring information database available to all agencies participating in the Shanghai MHEWS. The development of the MHEWS database is the basis for all disaster risk assessment operations and research.

Cuba

As regulated by law, the Meteorological Service of Cuba is the only entity in the country that issues meteorological information on dangerous phenomena. Its researchers and technicians participate in hazard, vulnerability and risk studies that are undertaken by multidisciplinary technical groups in the Agency of the Environment, a component of the Ministry of Science, Technology and Environment (CITMA) that has the national mandate for such studies. The participation of the NMS in these studies relates, mainly, to investigations of hazards and to numerical simulation of storm surge and winds performed for vulnerability and risk studies.

porates risk information in its Vigilance map that is disseminated twice daily to warn citizens of hydrometeorological hazards (Chap. 4; Sect. 4.3). Similarly, in China, the Shanghai Municipality Emergency Response Management Plan utilizes disaster risk information in carrying out mitigation or preventive actions, with special attention being given to high-risk areas. Risk information such as the possible impact, intensity, scope of influence and potential losses is also included in early warnings of hazards, along with disaster prevention guidelines (Chap. 8; Sect. 8.3). Applications of risk information are, in addition, a feature of EWS in the United States (Chap. 6; Sect. 6.3) and Cuba (Chap. 2; Sect. 2.3).

9.2.6 Warning Messages Must Be Effective, Authoritative and Actionable

> ***Principle No. 6** – Warning messages are; (i) clear, consistent and include risk information, (ii) designed with consideration for linking threat levels to emergency preparedness and response actions (e.g., using colour, flags, etc) and understood by authorities and the population, (iii) issued from a single (or unified), recognized and "authoritative" source.*

Increasingly as the governments are taking ownership in the development of multi-hazard early warning systems, warning authorities could vary from nation-to-nation and in some countries is shifting from technical agencies such as the National Meteorological and Hydrological Services, to multi-hazard warning authorities that incorporate risk information for development of warnings (e.g., disaster risk management agencies, health authorities, etc). In this emerging framework, for example, NMHS are critical providers of "authoritative" science-based hydro-meteorological information, forecasts, alerts, and warning guidance. The warning message development cycle starts from the time when a hazard is detected. Depending on the hazard in question, the detection can vary from minutes (e.g. tornadoes) to days (e.g. tropical cyclones). Following detection, based on the available technology, technical agencies produce forecasts of location, strength and timing of the hazards as the basis for issuing statements, bulletins and warnings at pre-defined thresholds as the hazard approaches.

Warning thresholds are normally established through coordination and collaboration among agencies responsible for monitoring and detecting the hazards and those designated to address emergency preparedness and response on the ground. The indicated level of threat is generally linked to clear actions at national to local levels.

Warning messages must be developed based on specific needs of target EWS stakeholders (e.g., disaster risk management authorities, economic sectors, and the public). In particular, activation of emergency preparedness and response plans in different sectors and at different levels (national to local) requires different information, must contain specific information, analysis, and advice to the government authorities beyond what may be available in the public warnings.[55] Consequently, for all target stakeholders, warning messages should be relevant and readily understandable – the format and content of the message needs to be aligned with the information needs and decision-processes of the target audience. The warning messages should be authoritative – emanating from a recognized single or unified authority that is known to all stakeholders and the general public. They should also be recognizable – the message should be in a consistent format so as to be easily authenticated by the target audience. Finally, warning messages should be timely – reaching their audience as soon as possible after issue and providing sufficient advance warning of the hazard to permit preparedness measures and response actions. Furthermore, regular monitoring and review should be undertaken of the authorities' and the public's understanding of and response to the warning message content and thresholds to ensure their relevance and utility. The results of these assessments should be used to improve future warning messages.

Technical agencies such as NMHS are responsible for round the clock monitoring of hazards to detect potential threats to lives and livelihoods. The mandates of these technical agencies for monitoring, forecasting and development and issuance of warnings, however, vary from country-to-country and also, as noted earlier, depend on whether the hazard is categorized as Type I, Type II or Type III. Depending on the responsibility of the NMHS, for example, organizational collaboration and coordination in relation to the development of hazard warnings occurs on several levels. It takes place between the technical agencies on the one hand and with the emergency services on the other. Regardless of the mandate, close collaboration and cooperation is essential between the NMHS and other EWS stakeholders who will use the information to activate various preparedness and response actions.

[55] EWC-II (Bonn, 2003) "Early Warning as a Matter of Policy – The Conclusions of the Second International Conference on Early Warning", Area of focus 4: "Development of people-centered warning systems"

Where the mandate is shared between different technical agencies, the roles, responsibilities and coordination mechanisms for each agency need to be clearly defined, documented in procedures and reflected in institutional agreements. In most cases discussed here, EWS stakeholders have developed a specific process to discuss and define the roles and responsibilities of each technical agency in order to avoid overlap. Usually NMHS have a clear mandate for monitoring, forecasting and developing warnings related to meteorological hazards and sometimes for hydrological hazards. However, the organizational split that exists in those countries with a separate National Meteorological Service and National Hydrological Service leads to a particular requirement for well defined arrangements for coordination between them in addressing issues such as flooding.

Box 9.6 *Principle No. 6 – Warning messages are; (i) clear, consistent and include risk information, (ii) designed with consideration for linking threat levels to emergency preparedness and response actions (e.g., using colour, flags, etc) and understood by authorities and the population, (iii) issued from a single (or unified), recognized and "authoritative" source.*

The following examples from the countries of good practices illustrate the different aspects of warning messages.

Shanghai, China

The following table depicts the levels of warning for typhoon and flood control and the corresponding actions to be taken by the emergency response authorities.

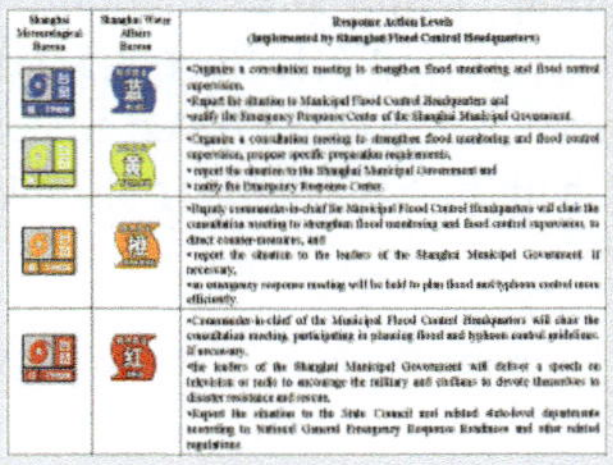

Cuba

In Cuba warnings are issued by a recognized unified authoritative source.

Cuban Meteorological Service has the organizational responsibility for monitoring, forecasting and developing the hazard warning and communicating it to the public from the scientific and operational point of view

National Civil Defense is responsible for the development of the warning in terms of the mobilization of all national and local resources, including all logistics for protective measures and evacuations

France

In the Météo-France Vigilance map is consistent warning and advisory interface that includes risk information with specific safety guidelines as soon as an orange or a red warning level is reached in any of the 95 French metropolitan départments. The map is issued twice a day but can be updated more frequently as dangerous conditions evolve. Follow-up bulletins with even more detailed and updated risk information are also frequently released in this case. A recent survey, undertaken for Météo-France, indicated that the Vigilance map is known and understood by 86 % of the French people.

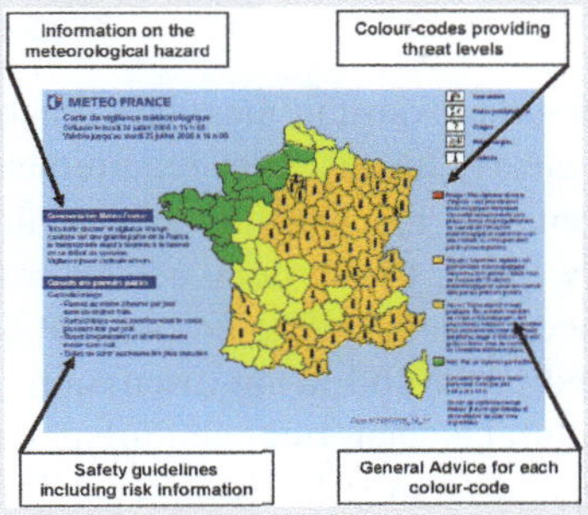

The National Centre for hydro-meteorological Flood Forecasting (SCHAPI) network, in France, provides risk level information on all waterways for which it has measurement networks and hydrological models. Météo-France provides SCHAPI with data on observed and forecast precipitation. SCHAPI produces a national flood monitoring map for the main waterways. The "rainfall-flood" Vigilance warning is produced following prior coordination between meteorological and hydrological experts in the two agencies, at local and national levels (Chap. 5;Sect. 5.4.2).

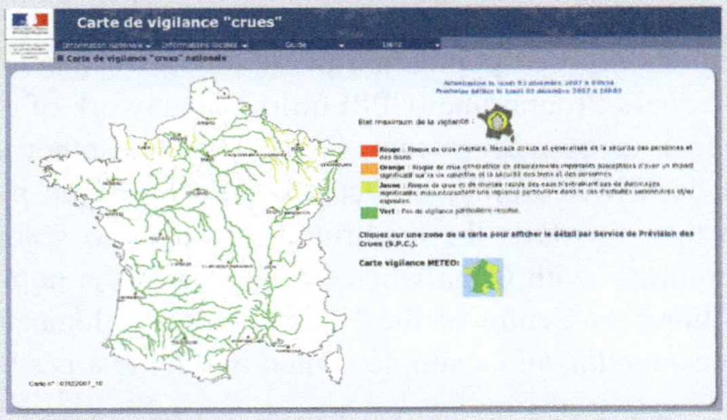

The case studies in earlier chapters illustrate practical applications of the preceding considerations. In all cases, established and documented mandates are in place for the development and issuance of warnings. Warnings are clear, consistent and actionable and generally formulated so as to take into account the requirements of other EWS stakeholders, particularly disaster management authorities, and provide information about potential impacts. A clear and recognizable incremental warning system is in place that facilitates the activation of appropriate emergency response actions, based on the level of threat. Plain language warning messages are often reinforced by coding protocols involving, for example, the use of colours, flags, siren intervals, etc. Moreover, each EWS incorporates a review process aimed at the improvement of warning messages.

The EWS in Shanghai and France provide good illustrations of the previous considerations in practical application. Shanghai's warnings are authoritative, being issued via a communications platform under the aegis of the municipal government. Shanghai uses a four-color coding system for warnings in which each colour is linked to specific response actions. Results of studies regarding how the public interprets early warning messages are used to help avoid confusion and panic during potential disaster events. Message formats are also reviewed following events and corrective actions are taken where necessary. Weak links in the early warning process are identified by means of reviews and assessments and remedial measures are implemented when required (Chap. 7; Sect. 7.5.1, Sect. 7.5.2).

As indicated earlier, weather monitoring in metropolitan France is built around a Vigilance map showing the required warning level for each départment in terms of the meteorological and hydrological dangers for the upcoming 24 hours. Warnings are authoritative, being issued by mandated agencies such as Météo-France, sometimes in collaboration with other agencies such as SCHAPI (The National Centre of France for hydro-meteorological Flood Forecasting). Warning levels are identified by means of four colours that correspond to risk thresholds that meet an operational requirement of the emergency services and are also well known to the general public. The criteria defining the four Vigilance levels and the accompanying status reports are constantly being refined and a follow-up group, with representation from all categories of risk management actors, meets three times a year for this purpose (Chap. 4; Sect. 4.5.1, Sect. 4.5.2). Warning messages in other EWS jurisdictions discussed in the case studies Cuba (Chap. 2; Sect. 2.5.1), the United States (Chap. 6; Sec 6.5.1, Sect. 6.5.2), Bangladesh, (Chap. 3; Sect. 3.5.1, Sect. 3.5.2), Germany (Chap. 5; Sect. 5.5.1)) also wholly or partly reflect the considerations outlined in Principle No. 6.

9.2.7 Warnings Must Reach Stakeholders and People at Risk in Time to Take Action

Principle No. 7 *– Warning dissemination mechanisms are able to reach the authorities, other EWS stakeholders and the population at risk in a timely and reliable fashion.*

Ensuring that warning messages reach the target authorities, EWS stakeholders and the population at risk in a timely fashion is clearly an essential requirement for an effective EWS. Consequently, reliable and redundant communication and dissemination mechanisms are required, which must be sustainable over time based on the available resources of the country. In all of the documented case studies, systems and procedures for the dissemination of alert/warning messages have been established to ensure that both EWS stakeholders and the general public are notified in a timely manner. As illustrated in the following discussion, mechanisms for the dissemination of warnings can and do vary based on the political, cultural, social, rural and urban characteristics as well as financial and technical capacities of the particular nation and the population at risk. However, all of these mechanisms and systems reflect a serious and sustained commitment to ensuring that warning messages reach their intended audiences in a timely fashion for activation of preparedness and response actions.

With an exposed population distributed in vulnerable coastal and low-lying communities in Bangladesh, the Cyclone Preparedness Programme (CPP) utilizes a network of over 42,000 volunteers to communicate warnings directly to the village level, using megaphones, hand sirens and other systems to address the public. Warning messages prepared by the Bangladesh Meteorological Department are relayed by fax and telephone to the Cyclone Preparedness Program (CPP), which then forwards these warning messages to volunteers in the communities by HF/VHF radio (Chap. 3; Sect. 3.2). This contrasts with Cuba where over 90 % of the population has access to TV or radio, and warnings issued by the National Forecast Centre of the National Meteorological Service are largely communicated to the public by means of the broadcast media, albeit supplemented in remote areas by local warning points (Chap. 2; Sect. 2.6).

Box 9.7 *Principle No. 7 – Warning dissemination mechanisms are able to reach the authorities, other EWS stakeholders and the population at risk in a timely and reliable fashion.*

The following examples from the countries of good practices, describe the different methods used for warning dissemination.

Germany

The Deutscher Wetterdienst's (DWD) warnings are disseminated via the Internet, fax-on-demand services, text messages, e-mail and, in specific cases, by telephone. They are, moreover, placed on an FTP server from which they can be downloaded by customers. The specialized FeWIS and SWIS weather information systems for fire brigades and road and winter maintenance services should also mentioned here as examples of dissemination mechanisms for closed online user groups. For information to the general public, the DWD uses the Internet (http://www.dwd.de) and relies on the voluntary support of radio and television companies. Broadcast stations are, however, generally under no obligation to disseminate the DWD's warnings on television and radio since, except in Bavaria, there are no legal regulations applicable to this area. Nevertheless, public service stations, in particular, have increasingly included DWD warnings of critical weather situations in their news bulletins and several broadcasters also display these warnings on their websites.

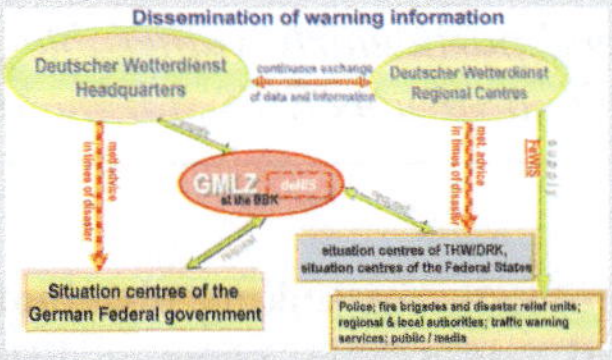

United States

Emergency Alert System is the alert dissemination mechanism for the multi-hazard early warning system in the United States. The EAS is jointly coordinated by the Federal Communications Commission (FCC), Department of Homeland Security, the National Weather Service and the commercial broadcast industry in the United States. The latter two organizations utilize the EAS as a primary means for warning the U.S. public of severe weather (e.g., tornadoes, floods), natural hazards (e.g., tsunamis, earthquakes, wildfires and volcanoes), technological hazards (e.g., chemical releases, oil spills, and nuclear power plant emergencies), and national emergencies (e.g., terrorist attacks). The NWS triggers the Emergency Alert System through its NOAA Weather Radio All Hazards system that broadcasts across 1000 + transmitters throughout the United States. The Department of Homeland Security manages the entire EAS system. The EAS regulations and standards are governed by the Public Safety and Homeland Security Bureau of the FCC. Each State and several territories have their own EAS plan which administers the system in that jurisdiction. Coordination of the State EAS plans typically takes place through a State EAS committee that includes the state Emergency Management Agency, the commercial broadcast industry, and the NWS.

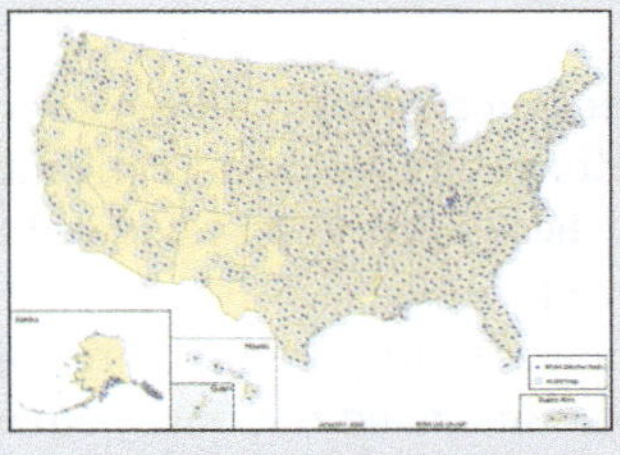

NOAA Weather Radio coverage

Bangladesh

In the event of a tropical cyclone, the Cyclone Preparedness Programme receives cyclone warning signals from the Storm Warning Centre (SWC) of Bangladesh Meteorological Department (BMD) as soon as a depression has formed in the Bay of Bengal. The information is transmitted to the six Zonal offices over HF radio. The Assistant Directors in turn pass it on to Unions through VHF radio. Where VHF radio has not yet been installed, messengers pass on the message. The Union Team Leaders contact the Unit Team Leaders immediately. The Unit Team Leaders, with their volunteers, spread out in the villages and disseminate the cyclone warnings, almost door-to-door, using megaphones, hand sirens and public address systems. The Team Leaders at the same time keep track of the approaching cyclone by listening to national radio broadcasts over transistor radios. The volunteers keep on announcing the special weather bulletins on the characteristics of the approaching cyclone, as per their action plan.

When the situation turns serious, the GoB passes the order for evacuation. The volunteers implement the order and advise and assist people to seek safety in cyclone shelters or other available safe places. The entire process of communicating a warning issued by the BMD through all levels of the CPP communications network takes about 15 minutes. After the cyclone is over, the volunteers rescue injured and marooned people, provide first aid to the injured, send serious cases to the local hospitals and assist in post-cyclone emergency relief operations.

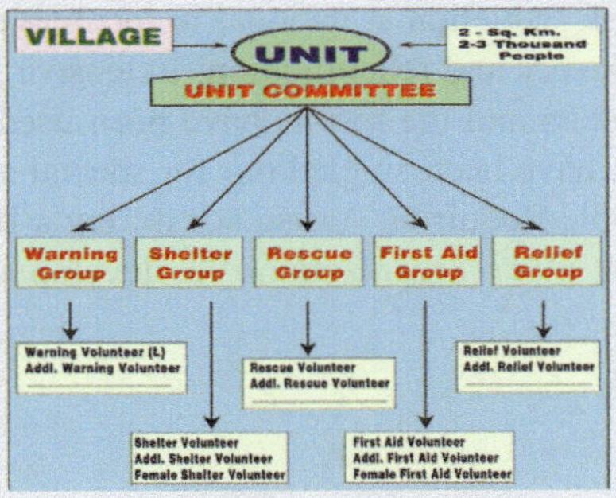

A further variation is evident in the United States of America, with its highly developed infrastructure, where the USA – National Weather Service (USA-NWS) maintains a nationwide NOAA Weather Radio network that continuously broadcasts official warnings, watches, forecasts and other hazard-related information. Like other NMHS, the USA-NWS also utilizes radio, television and cable broadcasts, the Internet, emergency telecommunications networks, wire services, automated telephone systems and sirens (Chap. 6; Sect. 6.6) in a continuing effort to ensure that critical hazard warnings reach all stakeholders as well as the public in a timely manner.

Faced with the unique challenges associated with emergency preparedness and response in a densely populated urban area, mega city of Shanghai, for its part, has implemented a sophisticated hazard warning system in which the Shanghai Meteorological Bureau and other technical agencies relay their hazard warnings to a multi-layer dissemination system which then distributes the warning messages to the authorities and the public via dedicated emergency communication lines, SMSs, messages to hand-held devices as well as TV screens in public locations, broadcast and print media (Chap. 7; Sect. 7.6). As a final note, various combinations of other disseminations systems such as Internet, fax, automatic telephones, and dedicated emergency communications lines and media and broadcast networks are also in use in Japan (Chap. 8; Sect. 8.6), France (Chap. 4; Sect. 4.6) and Germany (Chap. 5; Sect. 5.6).

9.2.8 Disaster Preparedness and Response Planning Must Be Emphasized

> ***Principle No. 8*** *– Emergency response plans are developed with consideration for hazard/risk levels, characteristics of the exposed communities (e.g., urban, rural, ethnic populations, tourists, and particularly vulnerable groups such as women, children, the elderly and the hospitalized), coordination mechanisms and various EWS stakeholders.*

Effective disaster preparedness and response requires clear procedures and protocols for linking different levels of threat, based on pre-defined thresholds, to activation of emergency preparedness and contingency plans at different levels, in a timely fashion. These need to be reviewed regularly through exercises and drills for evaluation of effectiveness and ongoing improvements.

Effective community response requires that the general public is well informed and ready to take action with regard to emergency preparedness and response activities linked to the different levels of threats indicated in the warning messages. To accomplish this, emergency response plans should be targeted to the individual needs of communities, authorities and emergency responders, paying particular attention to especially vulnerable groups such as the aged and infirm, and institutions such as hospitals, schools and campgrounds. Response plans should be widely disseminated to the communities at-risk and, where appropriate, include planning for large and small scale evacuations. Hazard and vulnerability maps should be utilized in developing the community-based emergency preparedness and response plans.

Taken overall, the case studies presented in this book illustrate well-developed emergency response planning that take into account risks, coordination requirements and the needs of all stakeholders, including the most vulnerable. In France, for example, the ORSEC (Organisation de la Réponse de Sécurité Civile) departmental and communal level disaster contingency plans are developed under the lead of the prefects and mayors, respectively, within a highly structured and coordinated national and regional framework. In their definition phase, these plans are based primarily on risk assessments and they are implemented in conjunction with emergency preparedness and civil protection actors, including citizens who can mobilize within the framework of the Communal Civil Defence Reserves. Specific plans are in place for all sites hosting particularly vulnerable population groups, such as hospitals and senior citizens' residences all plans call for regular exercises and post-crisis evaluations (Chap. 4; Sect. 4.7.1). Equally, Japan's Basic Disaster Management Plan provides a comprehensive framework for disaster risk reduction at national to local levels addressing disaster management and coordination systems, disaster mitigation, recovery and rehabilitation. Extensive risk analysis has been undertaken for earthquakes, tsunamis, sediment and other disasters and the results have been used to develop and target mitigation measures and preparedness plans. Vulnerable groups have been singled out for special attention with the Cabinet Office publishing "Guidelines for Evacuation Support of People Requiring Assistance during a Disaster". Disaster reduction drills and exercises are used by local governments and other organizations to develop effective community response capabilities (Chap. 8; Sect. 8.7).

Box 9.8 *Principle No. 8 – Emergency response plans are developed with consideration for hazard/risk levels, characteristics of the exposed communities (e.g., urban, rural, ethnic populations, tourists, and particularly vulnerable groups such as women, children, the elderly and the hospitalized), coordination mechanisms and various EWS stakeholders.*

The following examples from the countries of good practices, describe the various methods and mechanisms for the development, coordination and maintenance of emergency response plans.

United States

Emergency preparedness and response plans are targeted to individual needs of vulnerable communities, authorities and emergency respondents in two phases:

1. *Historical data for vulnerable communities and the response capabilities of emergency respondents is used in planning for hazardous mitigation for vulnerable areas and individuals. For example, historical data available in the hydrological section of the NWS weather forecast internet page provides data showing historical flood levels and current river flow levels. This data allows advance and "current time status" evaluations of potential flooding issues and aids in emergency alerts, advance preparation and response. Historical data also allows a more potent response to areas considered "vulnerable" due to repeated crisis issues for that area and/or individuals.*
2. *Evaluation of need and potential need are considered when developing advance planning for specific areas of vulnerable populations or geographic locations. Emergency Preparedness plans include vulnerable populations (economic, medical, physical limitations, age and handicap, non-English speaking and concentrated areas of need such as assisted living centres) in advance. Priority response preparation includes table-top exercises, review meetings with emergency responders and volunteers. An example of advance preparation for medical needs would include planning for dialysis patients and dialysis clinics if there would be wide spread power outages in the area.*

The role of NWS within disaster response planning includes programmes such as:

- *Citizen Corps: Brings together local leaders from governmental and non-governmental organizations, and the private sector, to prepare for and respond to incidents.*
- *StormReady: Which helps arm America's communities with the communication and safety skills needed to save lives and property before and during weather events. It also helps community leaders and emergency managers strengthen local safety programs. Communities are better prepared to save lives from the onslaught of severe weather through advanced planning, education and awareness.*

Japan

Japan's basic Disaster Management Plan addresses comprehensive and long-term disaster reduction issues such as disaster management related systems, disaster reduction projects, early and appropriate disaster recovery and rehabilitation, and scientific and technical research. The Plan was revised entirely in 1995, based on the experiences of the Great Hanshin-Awaji Earthquake, and further modified as needed to meet the specific needs of vulnerable communities, authorities and emergency respondents. In view of Japan's aging society and the increasing number of the elderly being killed or injured by disasters, measures are being reinforced to provide necessary assistance during disasters to those such as the elderly and physically impaired. The Cabinet Office in cooperation with relevant ministries published "Guidelines for Evacuation Support of People Requiring Assistance during a Disaster" in 2005 (revised in 2006), for implementation at the municipal level. These Guidelines address:

- *Improving the information communications system;*
- *Sharing of information concerning people requiring assistance during a disaster;*
- *Creating a tangible evacuation support plan for those people;*
- *Assistance at evacuation centres; and*
- *Collaboration among related organizations.*

Tangible countermeasures are included such as the issuance of evacuation preparation information that calls for the early evacuation of people requiring assistance and sharing of information on people requiring assistance between disaster management and social welfare-related organizations (exceptional use of social welfare-related personal information to prepare evacuation support systems for the elderly and others).

The Flood Control Act and other relevant Acts require the delivery of hazard maps to facilities used by people requiring assistance. The Ministry of Land, Infrastructure, Transport and Tourism (MLIT) launched a web portal for hazard maps to assist these efforts to enhance preparedness.

In addition, the Act on Sediment Disaster Countermeasures for Sediment Disaster Prone Areas requires prefectural and city governments to investigate the geography, precipitation, land use and other relevant factors in vulnerable areas approximately every five years. Information from these investigations is needed to support appropriate preparedness and countermeasures for sediment disasters.

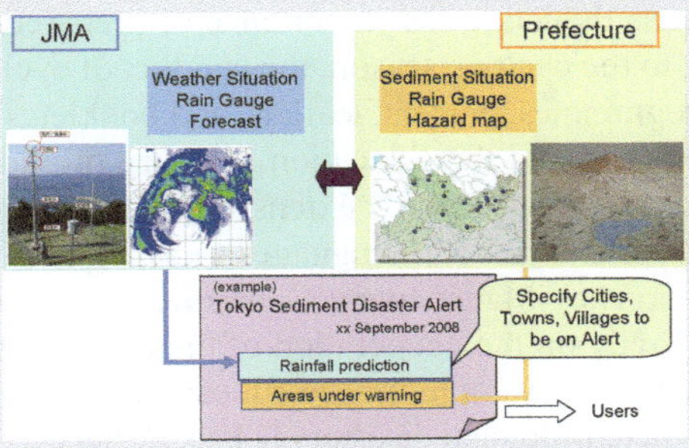

The emergency response plans in the remaining case studies not only incorporate most or all of the major considerations and priorities highlighted in Principle No. 8 but also reflect particular adaptation to their specific circumstances. Bangladesh's Cyclone Preparedness Programme, for example, draws on over 42,000 trained volunteers in drilling and executing its highly coordinated disaster preparedness and response plans that pay special attention to vulnerable groups such as women, children, and the elderly and disabled persons (Chap. 3; Sect. 3.7, Appendix 1). Similarly, Shanghai's Community Safety Strategy addresses schools, rural areas, enterprises, and grass-roots management units as well as the public at large, working within the framework of emergency response plans issued in the form of government regulations (Chap. 7; Sect. 7.7). In Cuba, the disaster reduction plans that are in place at all levels from the people's council to the national scale are updated annually based on risk assessments that take into account changes in vulnerabilities during the preceding year. Moreover, community preparedness is organized by the Civil Defence in each municipality, with the assistance of the local social and mass organizations, and is tested and refined each year during a nationwide exercise (Chap. 2; Sect. 2.7.1, Sect. 2.7.2). Emergency preparedness and response plans in the United States address the individual needs of communities, authorities and emergency responders and the particularly vulnerable, normally include evacuation plans and are also exercised yearly by all emergency response/emergency management agencies. In this case programmes such as StormReady have been critical in providing clear guidance to ensure that communities can build their resilience and obtain certification for their preparedness for certain hazards (Chap. 6; Sect. 6.7). Finally, LÜKEX crisis management exercises involving German Federal and Land (State) emergency task forces and crisis management teams, private operators of critical infrastructure and disaster relief organisations are held roughly every two years to enhance disaster preparedness in that country (Chap. 5; Sect. 5.7.1).

9.2.9 Training is Essential to Operational Readiness

> ***Principle No. 9*** *– Training on risk awareness, hazard recognition and related emergency response actions is integrated in various formal and informal educational programmes and linked to regularly conducted drills and tests across the system to ensure operational readiness at any time.*

Public awareness and understanding of early warnings is, as already mentioned, essential to effective implementation of emergency plans in the event of a potential threat. It is critical that the public is aware of the threats posed by various hazards to their community, knows the potential impacts, and understands the triggering signals for hazards, threat levels reflected in the warning messages and the recommended immediate responses. Members of the public must be clear as to how to interpret warning information and what this means for them at individual, household and community levels. They should also be well-informed regarding the emergency preparedness and response procedures that should be followed at these levels.

The achievement of the preceding level of understanding requires continuing efforts using a variety of training and education strategies and approaches including programmes that are built into school curricula at all levels and informal education initiatives. They should include campaigns directed at specific hazards and campaigns tailored to the needs of particular audiences, taking into account their specific vulnerabilities (e.g. children, elderly, emergency managers, media, etc). In addition, evaluation mechanisms need to be put in place to assess these public education strategies and programmes and to ensure that improvements in awareness and understanding are achieved over time.

Each of the case studies illustrates the presence of strong training and education programmes targeting public awareness and knowledge in the areas of recognition of hazards and risks and emergency preparedness and response actions. These programmes draw on the collaboration of broadcast and print media in efforts to ensure the widest possible audiences. They also reflect specific consideration for the local to national social characteristics, economic circumstances, and institutional and government structures. Public awareness and community-based preparedness is, for example, a vital component of Bangladesh's Cyclone Preparedness Programme and extensive education and training activities are undertaken in the cyclone-prone coastal communities. These involve initiatives such as the use of local volunteers to motivate village people and the staging of specially written dramas depicting the importance of preparedness along with more widely-used techniques such as publicity campaigns prior to the cyclone season, the conduct of cyclone drills and demonstrations, documentary films/videos, radio and television programmes, posters, leaflets, and booklets (Chap. 3; Sect. 3.7.3). In Japan, the Cabinet Office promotes disaster-related education and the Education Ministry provides related educational tools for students and also sponsors "Safe School" forums. As part of Japan's extensive public awareness efforts directed at disaster prevention, some information materials are provided in several languages to assist visitors and other foreign residents (Chap. 8; Sect. 8.7.3). For its part, Shanghai's Emergency Management Committee has established a public awareness program related to disaster prevention that draws heavily on the cooperation of the media and the school system. Disaster prevention topics are included in school curricula with, for example, image education being undertaken in primary schools in order to improve safety and awareness regarding meteorological disasters (Chap. 7; Sect. 7.7.3).

Box 9.9 *Principle No. 9 – Training on risk awareness, hazard recognition and related emergency response actions is integrated in various formal and informal educational programmes and linked to regularly conducted drills and tests across the system to ensure operational readiness at any time.*

The following examples from the countries of good practices, illustrate the different methods used to increase hazard awareness through education, training and regular drills.

Bangladesh

Public awareness is an integral and very important part of CPP for its successful implementation. Keeping this in mind, CPP implements the following public awareness activities, in various ways, in the cyclone-prone costal waters:

- *Public Awareness through Volunteers: Cyclone Drills and Demonstrations;*
- *Film/video Shows: Documentary films/videos on disaster and preparedness;*
- *Publicity Campaigns: Immediately before the cyclone season, a publicity campaign is organized with government officials, NGO's and local public to familiarize and create awareness of cyclone preparedness;*
- *Radio and Television: Posters, Leaflets, and Booklets; and*
- *Staging of Dramas:*

CPP volunteers are given training on the Red Cross and Red Crescent movement, cyclones and their behaviour, warning signals and their dissemination, evacuation, sheltering and rescue, first aid and relief operations, gender issues, humanitarian values and climate change issues, in order to maintain a high level of efficiency. On recruitment, CPP officers give preliminary training to the volunteers at the local level. A three-day basic training course is then given to the volunteers on the subjects mentioned above. The Team Leaders undergo a five-day training course on the above subjects, as well as leadership training at Upazila (District level).

The officers and Upazila Team Leaders who are engaged in the training of volunteers are given disaster management and leadership training in addition to their normal duties. The training programme is designed to broaden their technical knowledge about disasters in general and cyclones in particular. It aims to enhance their familiarity with all aspects of the CPP, to familiarize them with the principles of disaster management and their application in the CPP, to promote their leadership abilities and to create better understanding of their roles and responsibilities and the need for coordinating efforts with other agencies and local communities. The Bangladesh Red Crescent Society First Aid instructors provide first aid training to First Aid volunteers.

Shanghai, China

In Shanghai, training on awareness and hazard recognition and emergency response are accomplished through public training activities such as:

- *Awareness raising multi-hazard brochures and posters distributed through the internet, the media*
- *Training and exercises conducted by representatives of meteorological services and other partners as part of their public outreach program.*
- *education programs are built into school curricula*
- *Distribution of extra curricular reading for children about weather and protective measures which raises the awareness of school children regarding natural hazards and their impacts.*

Contest on meteorological knowledge among local pupils

- *Utilization of National Disaster Prevention and reduction Day, Meteorological Day and Fire Prevention Day, to raise the public's awareness about preparing for and mitigating natural disasters.*

Japan

Disaster education has been incorporated in various school curriculums due to the importance of learning necessary disaster knowledge from childhood. Social education at the community level is also important, including town watching and hazard-mapping programs in which residents participate. The Cabinet Office promotes disaster education including sharing good examples of disaster education programs.

Ministry of Education, Culture, Sports, Science and Technology (MEXT) prepares and provides the educational tools for students which enable students to behave appropriately and safely under hazardous conditions by themselves. MEXT holds safe school promotion fora to enhance safety management in schools.

Government transmits information relevant to disasters and disaster prevention in cooperation with mass media. To share the current information relevant to disasters and their prevention between the disaster risk management agencies, Cabinet Office has published "Disaster Management News" since 2001. New disaster prevention information and publications are advertised extensively on TV, radio, newspapers, etc.

Some materials which describe useful tips for disaster countermeasures are provided in several languages. As one of good examples, Shizuoka Prefecture has distributed brochures named 'Earthquake Disaster Prevention Guidebook' and 'Life Passport' to provide foreigners in the prefecture with some tips to protect themselves from earthquake disasters. These brochures are written in English, Spanish, Portuguese, Chinese, Korean, Tagalong and Japanese.

JMA holds seminars to raise social awareness of weather-related disaster prevention and disseminate knowledge of weather and climate. For the disaster risk management agencies, JMA holds explanatory meeting about dissemination of forecasts, warnings and other relevant information.

On-going public education campaigns are, similarly, built into community education programs and the school system in the United States. Special campaigns have also proven effective in educating people in the United States regarding particular dangers, such as the National Weather Service's "Turn Around, Don't Drown" campaign (Chap. 6; Sect. 6.7.3). Educating the public and building risk awareness are, furthermore, cornerstones of risk management policies in France. Interestingly, communes in France are obliged to report the levels reached during severe flooding and purchasers in real estate transactions must be informed of any risks of flooding (Chap. 4; Sect. 4.7.2). In the same vein, citizen preparedness in Cuba is aimed at making all citizens aware of the risks which they may face and the measures they must take to protect their lives and property and civil preparedness is pursued through exercises, drills, workshops, media campaigns and other initiatives. As in some other jurisdictions, disaster-related topics are taught in the national school system and in institutes of higher education (Chap. 2; Sect. 2.7.3).

9.2.10 Feedback and Improvement Must Be Continuing Priorities

Principle No. 10 *– Effective feedback and improvement mechanisms are in place at all levels of EWS to provide systematic evaluation and ensure system improvement over time.*

As reiterated throughout this book, effective feedback mechanisms provide valuable input to planning, coordination, operational, and technical aspects of EWS and are essential to improvement of the overall system. In the case studies documented in this book, each EWS includes a process for assessing its coordination, operational systems and procedures through a series of reviews and exercises. Somewhat varying approaches are, however, used to obtain feedback and improve the performance of individual EWS. The following section draws attention to specific, sometimes unique, aspects of the feedback and improvement mechanisms that are utilized in the various case studies considered here.

Faced with a long history of recurring disasters, the Japanese government has established a Technical Investigation Group on Inheritance of Lessons from Disasters with the objective of learning lessons from historical disasters that could lead to improvements in the overall operation of the nation's Early Warning System and (Chap. 8; Sect. 8.8). To the same ends, Shanghai's Municipal Emergency Response Management Committee holds quarterly working group meetings of EWS partners to review feedback on system operations and plan improvements (Chap. 7; Sect. 7.8). Similar reviews of operations are routinely carried out in the United States. Following disastrous events, however, the US National Weather Service undertakes more formal service assessments that involve members of the emergency management community and NWS peers from other locations with the goals of identifying best practices and lessons learned and providing a road map for nationwide implementation of specific improvements (Chap. 6; Sect. 6.9).

A variation on the preceding approach is evident in Bangladesh, where Standing Orders for Disaster require each stakeholder to report any deficiencies or ideas for improvement in the Cyclone Preparedness Programme up the chain of command (Chap. 3; Sect. 3.9). The adoption by Météo-France of an ISO 9001 quality management system represents a further refinement that makes it possible to track and report on actions taken, from identifying customers' needs to ensuring operational reliability. Moreover, France's *Conseil Supérieur de la Météorologie* includes a "civil defence" commission that is responsible for evaluating Météo-France services, making recommendations for improvements and proposing ways of responding to new needs (Chap. 4; Sect. 4.9).

Feedback and improvement mechanisms in Germany and Cuba are more conventional, and, in that respect, somewhat similar to each other. Cuba's NMS utilizes direct dialogue with users and key stakeholders to obtain feedback with all suggestions being taken into consideration and used to help to improve the warning service (Chap. 2; Sect. 2.9). For its part, the Deutscher Wetterdienst also relies on feedback from meetings with stakeholders and disaster management organisations and from conferences and training programmes for closed user groups. As a further initiative, however, a contact officer has been designated by the DWD (Deutscher Wetterdienst) to coordinate and improve the information systems that support key stakeholders such as fire brigades and relief units and this has proven to be a useful innovation (Chap. 5; Sect. 5.8).

Box 9.10: *Principle No. 10 – Effective feedback and improvement mechanisms are in place at all levels of EWS to provide systematic evaluation and ensure system improvement over time.*

The following examples from the countries of good practices, describe the different methods and mechanisms for systematic evaluation and eliciting feedback from the public and EWS stakeholders.

France

Météo-France contracts a polling institute to undertake an annual survey on recognition of and feelings towards the Vigilance Map in order to obtain a more accurate reading of public awareness. In 2008, the Vigilance recognition rate was 86 per cent, based on a poll of 2000 people. Even though this awareness rate varies depending on lifestyle and region, it is at least 75 per cent regardless of social group.

Two other processes designed to improve the system as a whole are worthy of note, in addition to the above mechanisms for refining Vigilance products. First, the Conseil Supérieur de la Météorologie, chaired at ministerial level, is a nationwide body for coordination between the top management of Météo-France and its public or private contact partners or users. Within this body, a "civil defence" commission, chaired by the Civil Defense, is responsible for the following:

- *Evaluating the services provided;*
- *Drafting requests or recommendations;*
- *Proposing ways of responding to new needs which have emerged;*
- *Making proposals to secure the corresponding funding;*
- *Following up action taken.*

Second, Météo-France has implemented a quality management system that meets ISO 9001 standards. Two processes have been identified in this connection:

- *"meeting the State's needs with regard to civil defence, prevention of major risks, nuclear safety and ensuring the safety of persons and property, on land and sea, in the meteorological field"; and*
- *Matters pertaining to "warning".*

These two processes make it possible to follow and report on actions taken, from identifying customers' needs to guaranteeing the reliability of operational production.

Germany

The Deutscher Wetterdienst relies on feedback from meetings with stakeholders and disaster management organisations and from conferences and training programmes for closed user groups. As a further initiative, however, a contact officer has been designated by the DWD to coordinate and improve the information systems that support key stakeholders such as fire brigades and relief units and this has proven to be a useful innovation.

Cuba

Cuba's NMS utilizes direct dialogue with users and key stakeholders to obtain feedback with all suggestions being taken into consideration and used to help to improve the warning service. The operations of the Cuban EWS is assessed during the annual "METEOR" disaster preparedness exercise, held prior to the hurricane season, and involving all structures from the national to the local levels.

Shanghai

The Shanghai Municipal Government and the Shanghai Meteorological Bureau schedule large-scale training and exercise sessions several times a year in an effort to validate their EWS operations plan and develop and strengthen collaboration and relationships with EWS stakeholders.

United States

There are a variety of formal and informal mechanisms for improving dissemination of warnings utilized in the United States. The Government Accountability Office audit, evaluation and investigation process mentioned previously in response 2.2.6, is an example of a formal mechanism. Informal mechanisms utilized by the NWS would include assimilating stakeholder feedback or social science research into future service improvements. Stakeholder inputs are acquired in a variety of formats but primarily through meetings and workshops, and web-based feedback. In some instances, the NWS has utilized surveys and focus groups to gauge potential service improvements. The NWS dedicates a position in every one of its 122 Weather Forecast Offices, 13 River Forecast Centers, 6 Regional Headquarters and National Headquarters, to serve as the lead for customer feedback and service enhancements. In NWS Weather Forecast Offices and Headquarters, that position is known as the Warning Coordination Meteorologist (WCM) while in the River Forecast Centers it is called the Service Coordination Hydrologist (SCH).

Warning Coordination Meteorologists (WCMs) and Service Coordination Hydrologists (SCHs) lead the effort to identify and meet customer requirements for weather and hydrologic service and information in a number of ways. They conduct area-wide evaluations of WFO products and services by ensuring they are generated according to established national and regional directives. They lead the preparedness program which involves ensuring all people within the county warning area are aware of the weather hazards which threaten their safety and property and are prepared to take action if and when these threats become reality. Teaching is paramount in this endeavour, which involves conducting education and training courses with media, emergency managers, schools and clubs in the area. As needed, they also lead or serve as members on Flood or Disaster Survey teams. In addition, they assist in planning the execution of drills to test federal, state and local operational plans and procedures.

Another way improvement is obtained is by the use of verification statistics after all events. Three main statistics are generated: probability of detection (POD), false alarm ratio (FAR), and critical success index (CSI). For episodes where these numbers are not within satisfactory values, post-event reviews are preformed at the offices involved. This involves a complete review of the event including staffing, radar interpretation, and product preparation. Areas found to be in need of improvement are identified. Office personnel are made aware of the findings with the hope for improvements during the next event.

Improvements are also identified through post-storm service assessments. The NWS conducts service assessments to evaluate its performance after catastrophic weather events. For these events, assessment teams composed of experts in and outside of the organization study what happened before, during and after the event. Interviews with the survivors, local emergency management and media outlets are conducted with the hope of getting a complete picture of the service provided. The team will complete a report recommending changes in NWS procedures, products and services with the hope of improving overall performance in future event.

9.3 The Roles of National Meteorological and Hydrological Services (NMHS) and Other Technical Agencies in Early Warning Systems

Around the world, the majority of disasters are caused by natural phenomena are associated with hydrometeorological hazards (Fig. 9.1).

It is appropriate, therefore, to devote particular attention to hydrometeorological hazards and to the key roles that National Meteorological and Hydrological Services (NMHS) play in relation to such hazards. NMHS are critical partners within the national institutional structure of EWS, providing data, analyses, forecasts and warnings. NMHS, for example, operate networks where key meteorological and hydrological variables are systematically monitored and recorded and operate communications systems whereby these observations are transmitted nationally and globally in accordance with universal standards established by the World Meteorological Organization (WMO). This observational information provides essential inputs for analysis and prediction activities leading to the production of forecasts and early warnings of hydrometeorological hazards such as floods, droughts, tropical cyclones, storm surges, tornadoes, and heat and cold waves. The analysis of time and space series of observational data, moreover, provides probabilistic information on the geographic distribution, frequency of occurrence and potential magnitude of severe or extreme hydrometeorological phenomena. The results of such analyses are used in hazard mapping, vulnerability assessments and quantification of risks with important applications in planning of disaster mitigation (such as land use policies, construction standards, etc.), emergency preparedness, and other risk reduction measures. In addition, NMHS and similar specialized technical agencies are expected to support emergency planning, response and relief operations through provision of relevant hydrometeorological data, analyses, forecasts and consultation.

It emerged from the two expert's symposiums (MHEWS-I & MHEWS-II) that the operational roles of NMHS in supporting the warning process, like those of other technical agencies with similar responsibilities such as hydrological or geophysical authorities, can be subdivided into three categories based on their mandates for specific hazards. In the case of so-called *Type I hazards*, the NMHS, or another technical agency in the case on non-hydrometeorological hazards, has the sole mandate for the development and issuance of the warning for the hazard. In these circumstances, the agency with the Type I mandate communicates and coordinates directly with emergency managers and the government while other technical agencies provide support as appropriate. NMHS, for example, normally exercise Type I responsibilities for meteorological hazards such as high winds, thunderstorms and tropical cyclones. For *Type II hazards*, the mandate is shared between two or more technical agencies and this requires that the agencies coordinate in the development and issuance of warning message for these hazards, with other agencies providing support where appropriate.

In many countries, for example, the hazard of flooding falls into the Type II category with warning responsibilities being shared between meteorological and hydrological authorities. In the case of *Type III hazards*, the NMHS has no direct mandate but may, however, be required to provide information or communications support to the technical agencies that have the warning mandate. In Table 9.1 the warning mandates for each of the good practice countries are shown. It will be evident from the preceding discussion that NMHS coordination, cooperation and operational procedures with other technical, disaster risk management and civil protection agencies may vary depending on the level of mandate as well as on factors such as disaster management protocols and communications infrastructures.

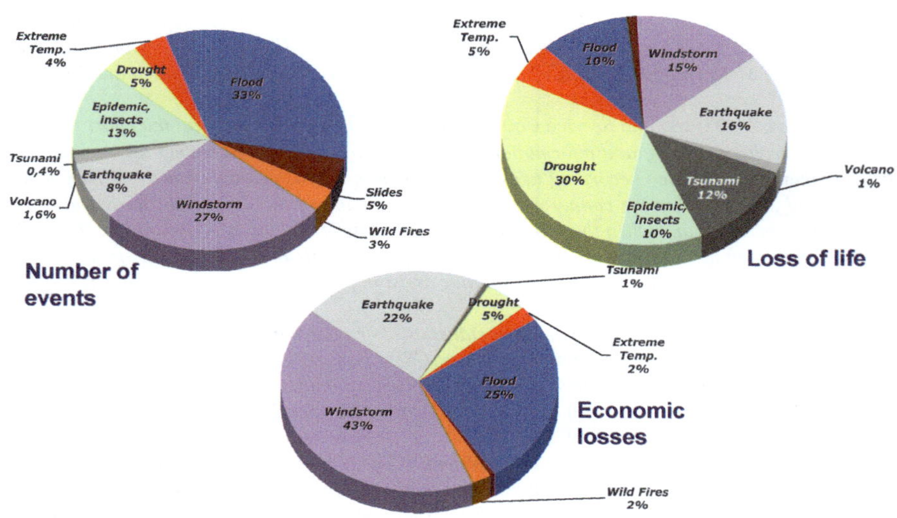

Fig. 9.1 Global Distribution of Disasters Caused by Natural Hazards and their Impacts (1980–2007). 90% of events, 70% of casualties and 75% of economic losses are related to hydro-meteorological hazards and conditions. (Source: EM-DAT: The OFDA/CRED International Disaster Database – http://www.emdat.be)

Table 9.1 Matrix of hazards to mandates for warning development

Classification			Cuba		Bangladesh		France		Germany		United States		Shanghai		Japan	
Primary	Secondary	Hazard	Type	Lead Agency/ supporting agencies	Type	Lead Agency/ supporting agencies	Type	Lead Agency/ supporting agencies	Type	Lead Agency/ supporting agencies	Type	Lead Agency/ supporting agencies	Type	Lead Agency/ supporting agencies	Type	Lead Agency/ supporting agencies
Meteorological		Avalanche					I	Météo-France	III	LfU/ DWD	II	U.S. Land Managment Agencies/ NWS				
Meteorological		Cold wave: period of abnormally low temperatures			I	BMD	I	Météo-France	I	DWD	I	NWS	I	SMB		
Meteorological		Dense fog			I	BMD			I	DWD	I	NWS	I	SMB		
Meteorological		Freezing rain							I	DWD	I	NWS				
Meteorological		Hailstorm			I	BMD	I	Météo-France	I	DWD	I	NWS	I	SMB		
Meteorological		Heat wave: period of abnormally high temperatures	III	EMNDC/ GB, NFC	I	BMD	II	Météo-France, INVS			I	NWS	I	SMB		
Meteorological	Hydrological	Heavy rain					II	SCHAPI - Météo-France	I	DWD			I	SMB	I	JMA
Meteorological		Heavy snow					I	Météo-France	I	DWD	I	NWS	I	SMB	I	JMA
Meteorological		Hydrometeorological hazards to aviation (i.e., turbulence, icing)			I	BMD	II	Météo-France			I	NWS				
Meteorological	Climate	Sandstorm									I	NWS				
Meteorological		Smoke, Dust or Haze			II	DoE					I	NWS	I	SMB		
Meteorological		Strong winds	II	NFC/ EMNDC	I	BMD	I	Météo-France	I	DWD	I	NWS	I	SMB	I	JMA
Meteorological		Thunderstorm or lightning	II	NFC, EMNDC	I	BMD	I	Météo-France	I	DWD	I	NWS	I	SMB		
Meteorological		Tornado (rotational high winds)	II	NFC/ EMNDC	I	BMD	I	Météo-France	I	DWD	I	NWS				
Meteorological		Tropical cyclone	II	NFC/ EMNDC	I	BMD	I	Météo-France			I	NWS	I	SMB	I	JMA
Hydrological	Meteorological	Flash flood	II	NFC/ EMNDC	III	FFWC, BWDB/ BMD	II	SCHAPI - Météo-France	III	Flood Waring Centers of Länder / DWD	I	NWS	II	SMB and Shanghai Flood Prevention Department	I	JMA
Hydrological	Meteorological	River flooding	III	EMNDC/ INRH, NFC	III	FFWC, BWDB/ BMD	II	SCHAPI - Météo-France	III	Flood Waring Centers of Länder / DWD	I	NWS			II	JMA/ MLIT, Prefectural governments
Marine	Hydrological Meteorological	Coastal flooding	II	NFC/ EMNDC	I	BMD	I	SCHAPI - Météo-France	III	BSH*), Authorities of the Länder /DWD	I	NWS				
Marine	Meteorological	Marine hazards (storm, sea ice, icebergs, etc.)					I	Météo-France	I	DWD (Storm)	I	NWS				
Marine	Meteorological	Storm surge	II	NFC/ EMNDC	I	BMD	I	Météo-France	III	BSH*), Authorities of the Länder /DWD	I	NWS	II	SMB/ SHB and Marine	I	JMA
Climate	Meteorological	Desert locust swarm									III	U.S. Land Managment Agencies/ NWS				
Climate	Hydrological Meteorological	Drought	II	NFC/ EMNDC	II	BMD, DAE	III	MEESDSP/ Météo-France, BRGM, DIREN...	III	Authorities of the Länder / DWD	I	NOAA Climate Service / NWS				
Climate	Meteorological	Forest or wild land fire					III	DSD/ Météo-France	III	Forestry admin of the Länder/ DWD	III	U.S. Land Managment Agencies/ NWS				
Geological		Earthquakes	I	EMNDC CNIS	II	BMD (monitoring only)					II	US Geological Survey/ NWS	III	Seismological Bureau	I	JMA
Geological	Meteorological	Landslide or mudslide	II	NFC/ EMNDC	I	BMD	III	MEESDSP/ Météo-France, BRGM, RTM			II	US Geological Survey/ NWS			II	JMA/ Prefectural governments
Geological	Marine	Tsunami			III	PTWC and JMA/ BMD	III	Prefect/ Météo-France, CEA, IOPG	III	US Geological Survey	I	NWS / U.S. Geological Survey	III	Oceanic Administration	I	JMA
Geological	Meteorological	Volcanic events									II	US Geological Survey/ NWS			I	JMA
Environmental (technological)	Meteorological	Airborne hazardous substances (i.e., nuclear, biological, chemical, etc.)							III	BfS/ BMU	III	DHS/ State EMS Agencies, Nuclear Regulatory Commission/ NWS	II	SMB/ Environmental Protection Bureau		
Environmental (technological)	Marine Meteorological	Waterborne hazards (i.e., nuclear, biological, chemical, oil spills, etc.)			II	DPHE/ BMD			III	BfS/ BMU (Radioactivity)	III	DHS, State EMS Agencies, Nuclear Regulatory Commission/ NWS				
Space		Space Weather									I	NWS				

*) BSH = Bundesamt für Seeschifffahrt und Hydrographie (english: Maritime and Hydrographic Agency)

Type I – NMHS has mandate for warning development, Type II – Warning mandate is shared between NMHS and other agency(ies), Type III – the NMHS has no direct mandate but may be required to provide information or communications support to the lead technical agency(ies). This table shows that the NMHS normally has the mandate or co-mandate for warning development for meteorological, hydrological and climate related hazards. However, in some cases the NMHS shares the responsibility with civil defence or other governmental agency. Additionally, some NMHS have the mandate or co-mandate for geological and environmental hazards.

9.4 Concluding Remarks

The preceding set of guiding principles has been derived from the collective experiences in a number of countries where successful early warning systems are in place that have proven effective in reducing losses of life and property in the face of disasters caused by natural hazards. In all these countries, early warning systems with a multi-hazard approach are a critical component of the comprehensive national disaster risk reduction programmes and planning. These Principles draw attention to fundamental lessons, considerations, systemic issues and practical challenges that need to be addressed when undertaking such initiatives. The Principles should, therefore prove helpful to governments, various agencies engaged in the development and implementation of EWS, technical agencies such as the NMHS and other stakeholders in countries that are seeking to develop, implement or strengthen EWS within their own jurisdictions. At the same time, the Principles and examples from this book can only provide a starting point for the planning, design and operation of an effective EWS. Any such system, to be effective and sustainable, must take into account the particular governmental, social, economic and environmental realities of the country or jurisdiction in question, at all levels from national to local. The importance of this national to local dimension is clearly evident in the case studies presented in earlier chapters which illustrate various adaptations of practices and procedures to differing domestic circumstances. It is also noteworthy to highlight that while these cases are treated as good practices, no system is ever perfect. A critical commonality to the cases presented in this book is that there are feedback mechanisms and investments for continuous evaluation and improvement of as various aspects of these systems over time.

Appendix

Bio Editor

Dr. Maryam Golnaraghi, is the Chief of the Disaster Risk Reduction Programme of the World Meteorological Organization. She has over 25 years of experience working with the public and private sectors and the academia in areas ranging from ocean/climate modelling to applications of science and technology in policy and strategic planning, climate and disaster risk management and sectoral decision-making. She has authored numerous articles in science and policy journals and has served as the editor of a number of publications. Among her activities, she is developing knowledge products based on documentation of good practices that are being utilized to strengthen national early warning systems, and hazard/risk modelling capacities for sectoral decision-making. Her particular focus is on institutional partnerships and applications of science and technology to risk management and planning. Dr. Golnaraghi has served as an advisor to senior executives and officials in the private and public sectors in the United States as well as the United Nations.

Short Abstract of the Book

Early Warning Systems (EWS) save lives. Analysis of global trends on impacts of natural hazards over the last five decades has demonstrated that despite increasing economic losses, there has been a significant decrease in loss of life associated with hydro-meteorological hazards. This has been attributed to advancements in monitoring and forecasting of weather-related hazards, more effective communication and emergency preparedness mechanisms at national to local levels in a number of countries with a history of high-impact weather-related hazards. This book is a unique compilation of seven operational national good practices in Multi-Hazard Early Warning Systems, including Bangladesh, megacity of Shanghai in China, Cuba, France, Germany, Japan and the United States, building on the expertise of WMO Members, UN and international partners. The book makes the case for greater integration of early warning systems in planning at all levels of society, building on lessons learned through international efforts. This work reveals that while the specific design and implementation of EWS are different in the seven cases, they share a number of common principles, which have led to their effectiveness over time. One of the critical messages of this book is that advancements in operational forecasting of meteorological and hydrological hazards is leading to more accurate and longer lead time for the issuance of warnings and that incorporation of potential risk information in the warning massages, has empowered authorities and the public to take early action to save lives!

M. Golnaraghi ed., *Institutional Partnerships in Multi-Hazard Early Warning Systems*,
DOI 10.1007/978-3-642-25373-7, © Springer-Verlag Berlin Heidelberg 2012

Index

I

L

M

N

P

Q

R

S

The manufacturer's authorised representative in the EU is Springer Nature Customer Service Centre GmbH, Europaplatz 3, 69115 Heidelberg, Germany. If you have any concerns regarding our products, please contact ProductSafety@springernature.com

Printed and bound by CPI Group (UK) Ltd, Croydon, CR0 4YY
07/01/2026
02030937-0005